T0258044

Mechanical Properties and Performance of Engineering Ceramics and Composites V

Mechanical Properties and Performance of Engineering Ceramics and Composites V

A Collection of Papers Presented at the 34th International Conference on Advanced Ceramics and Composites
January 24–29, 2010
Daytona Beach, Florida

Edited by
Dileep Singh
Jonathan Salem

Volume Editors
Sanjay Mathur
Tatsuki Ohji

The American Ceramic Society

A John Wiley & Sons, Inc., Publication

Copyright © 2010 by The American Ceramic Society. All rights reserved.

Published by John Wiley & Sons, Inc., Hoboken, New Jersey.
Published simultaneously in Canada.

No part of this publication may be reproduced, stored in a retrieval system, or transmitted in any form or by any means, electronic, mechanical, photocopying, recording, scanning, or otherwise, except as permitted under Section 107 or 108 of the 1976 United States Copyright Act, without either the prior written permission of the Publisher, or authorization through payment of the appropriate per-copy fee to the Copyright Clearance Center, Inc., 222 Rosewood Drive, Danvers, MA 01923, (978) 750-8400, fax (978) 750-4470, or on the web at www.copyright.com. Requests to the Publisher for permission should be addressed to the Permissions Department, John Wiley & Sons, Inc., 111 River Street, Hoboken, NJ 07030, (201) 748-6011, fax (201) 748-6008, or online at http://www.wiley.com/go/permission.

Limit of Liability/Disclaimer of Warranty: While the publisher and author have used their best efforts in preparing this book, they make no representations or warranties with respect to the accuracy or completeness of the contents of this book and specifically disclaim any implied warranties of merchantability or fitness for a particular purpose. No warranty may be created or extended by sales representatives or written sales materials. The advice and strategies contained herein may not be suitable for your situation. You should consult with a professional where appropriate. Neither the publisher nor author shall be liable for any loss of profit or any other commercial damages, including but not limited to special, incidental, consequential, or other damages.

For general information on our other products and services or for technical support, please contact our Customer Care Department within the United States at (800) 762-2974, outside the United States at (317) 572-3993 or fax (317) 572-4002.

Wiley also publishes its books in a variety of electronic formats. Some content that appears in print may not be available in electronic format. For information about Wiley products, visit our web site at www.wiley.com.

Library of Congress Cataloging-in-Publication Data is available.

ISBN 978-0-470-59467-4

Printed in the United States of America.

10 9 8 7 6 5 4 3 2 1

Contents

MODELING

Preface

This volume is a compilation of papers presented in the Mechanical Behavior and Performance of Ceramics and Composites symposium during the 34th International Conference and Exposition on Advanced Ceramics and Composites (ICACC) held January 24–29, 2010, in Daytona Beach, Florida.

The Mechanical Behavior and Performance of Ceramics and Composites symposium was one of the largest symposia in terms of the number (>100) of presentations at the ICACC'10. This symposium covered wide ranging and cutting-edge topics on mechanical properties and reliability of ceramics and composites and their correlations to processing, microstructure, and environmental effects. Symposium topics included:

- Ceramics and composites for engine applications
- Design and life prediction methodologies
- Environmental effects on mechanical properties
- Mechanical behavior of porous ceramics
- Ultra high temperature ceramics
- Ternary compounds
- Mechanics and characterization of nanomaterials and devices
- Novel test methods and equipment
- Processing–microstructure–mechanical properties correlations
- Ceramics and composites joining and testing
- NDE of ceramic components

Significant time and effort is required to organize a symposium and publish a proceeding volume. We would like to extend our sincere thanks and appreciation to the symposium organizers, invited speakers, session chairs, presenters, manuscript reviewers, and conference attendees for their enthusiastic participation and contributions. Finally, credit also goes to the dedicated, tireless and courteous staff at The American Ceramic Society for making this symposium a huge success.

DILEEP SINGH
Argonne National Laboratory

JONATHAN SALEM
NASA Glenn Research Center

Introduction

This CESP issue represents papers that were submitted and approved for the proceedings of the 34th International Conference on Advanced Ceramics and Composites (ICACC), held January 24–29, 2010 in Daytona Beach, Florida. ICACC is the most prominent international meeting in the area of advanced structural, functional, and nanoscopic ceramics, composites, and other emerging ceramic materials and technologies. This prestigious conference has been organized by The American Ceramic Society's (ACerS) Engineering Ceramics Division (ECD) since 1977.

The conference was organized into the following symposia and focused sessions:

Symposium 1	Mechanical Behavior and Performance of Ceramics and Composites
Symposium 2	Advanced Ceramic Coatings for Structural, Environmental, and Functional Applications
Symposium 3	7th International Symposium on Solid Oxide Fuel Cells (SOFC): Materials, Science, and Technology
Symposium 4	Armor Ceramics
Symposium 5	Next Generation Bioceramics
Symposium 6	International Symposium on Ceramics for Electric Energy Generation, Storage, and Distribution
Symposium 7	4th International Symposium on Nanostructured Materials and Nanocomposites: Development and Applications
Symposium 8	4th International Symposium on Advanced Processing and Manufacturing Technologies (APMT) for Structural and Multifunctional Materials and Systems
Symposium 9	Porous Ceramics: Novel Developments and Applications
Symposium 10	Thermal Management Materials and Technologies
Symposium 11	Advanced Sensor Technology, Developments and Applications

Focused Session 1 Geopolymers and other Inorganic Polymers
Focused Session 2 Global Mineral Resources for Strategic and Emerging
                   Technologies
Focused Session 3 Computational Design, Modeling, Simulation and
                   Characterization of Ceramics and Composites
Focused Session 4 Nanolaminated Ternary Carbides and Nitrides (MAX Phases)

The conference proceedings are published into 9 issues of the 2010 Ceramic Engineering and Science Proceedings (CESP); Volume 31, Issues 2–10, 2010 as outlined below:

- Mechanical Properties and Performance of Engineering Ceramics and Composites V, CESP Volume 31, Issue 2 (includes papers from Symposium 1)
- Advanced Ceramic Coatings and Interfaces V, Volume 31, Issue 3 (includes papers from Symposium 2)
- Advances in Solid Oxide Fuel Cells VI, CESP Volume 31, Issue 4 (includes papers from Symposium 3)
- Advances in Ceramic Armor VI, CESP Volume 31, Issue 5 (includes papers from Symposium 4)
- Advances in Bioceramics and Porous Ceramics III, CESP Volume 31, Issue 6 (includes papers from Symposia 5 and 9)
- Nanostructured Materials and Nanotechnology IV, CESP Volume 31, Issue 7 (includes papers from Symposium 7)
- Advanced Processing and Manufacturing Technologies for Structural and Multifunctional Materials IV, CESP Volume 31, Issue 8 (includes papers from Symposium 8)
- Advanced Materials for Sustainable Developments, CESP Volume 31, Issue 9 (includes papers from Symposia 6, 10, and 11)
- Strategic Materials and Computational Design, CESP Volume 31, Issue 10 (includes papers from Focused Sessions 1, 3 and 4)

The organization of the Daytona Beach meeting and the publication of these proceedings were possible thanks to the professional staff of ACerS and the tireless dedication of many ECD members. We would especially like to express our sincere thanks to the symposia organizers, session chairs, presenters and conference attendees, for their efforts and enthusiastic participation in the vibrant and cutting-edge conference.

ACerS and the ECD invite you to attend the 35th International Conference on Advanced Ceramics and Composites (http://www.ceramics.org/icacc-11) January 23–28, 2011 in Daytona Beach, Florida.

Sanjay Mathur and Tatsuki Ohji, Volume Editors
July 2010

Processing

THE EFFECTS OF HEATING RATE ON MAGNESIA DOPED ALUMINA PREPARED BY SPS

Apak B.[a], Goller G.[a], Sahin F. C.[a,b], Yucel O.[a,b]

[a]: Istanbul Technical University, Metallurgy and Materials Engineering Department 34469 Maslak, Istanbul, Turkey.

[b]: Istanbul Technical University Prof. Adnan Tekin Applied Research Center of Materials Science & Production Technologies 34469 Maslak, Istanbul, Turkey.

In order to achieve fine grained sintered products, 0.1 wt% MgO was doped into alumina powders. Magnesium nitrate [Mg(NO$_3$)$_2$. 6H$_2$O] with the amount of corresponding the requisite MgO content was added to alumina and the slurry was prepared in ethanol medium. After ball milling and drying, the powders were heated to 750 °C for 1 h in air to decompose magnesium nitrate to magnesia. Doped alumina powders were sintered by using spark plasma sintering under a vacuum with especially high heating rates (120 and 200°C/min). Characterization studies on hardness, density and microstructure were carried out on the sintered specimens. The effects of two heating rates on microstructure, density and hardness of spark plasma sintered alumina ceramics were investigated.

INTRODUCTION

Sintered aluminium oxide (Al$_2$O$_3$) is one of the most significant high technology ceramics with its high strength, hardness, thermal stability, excellent dielectric properties, corrosion and wear resistance at elevated temperatures. Alumina finds lots of application areas such as high voltage insulators, high temperature electrical insulators, grinding media, ballistic armor, wear pads and orthopedic implants with its mentioned properties. Also because of ease in accessibility, alumina is preferred ceramic for researchers. Al$_2$O$_3$ can be defined as a basic ceramic for sintering techniques to understand the sintering mechanism not only for conventional techniques, but also for novel routes such as microwave sintering[1,2], solar sintering[3], spark plasma sintering[4-10] etc. Spark plasma sintering (SPS) is a consolidation technique that can combine high heating and cooling rates with an uniaxial applied pressure resulting in short processing time. The first unique property of the SPS process is the possibility of using very fast heating rates and very short holding times (minutes) to obtain fully dense samples. Also, when compared with the samples sintered with using conventional densification techniques, spark plasma sintered samples are usually reported to have been produced at lower temperatures. In SPS, heating is achieved through a dc pulsed current passed through the die as well as the sample. Normally, as an external heating element, graphite die and punches are used. In addition, part of the pulsed current passing through the sample results in sparking at interparticle contacts and leads to internal heating of the sample.[4]

Wang et al.[7] and Shen et al.[8] concluded that a high heating rate is desirable for a fine grain size. On the other hand, Kim et al.[9] have suggested that low heating rates results in more densified products and smaller grain size. In other words, the actual mechanism which brings the advanced densification and grain growth in SPS is not yet defined certainly.

Yet some researchers investigated the heating rate effect on alumina ceramics[7-10], in this research the effects of heating rate on magnesia doped spark plasma sintered alumina is investigated. Even though the amount of the doped MgO is low, the achieved results are different from mentioned studies[7-9], it is because of that magnesia acts as a grain growth inhibitor in alumina.[4,11] In this study, obtaining highly dense MgO doped alumina products with high heating rates such as 120 °C/min and 200 °C/min in short processing times is aimed. For maximum densification, minimum grain growth and high hardness, the process have been tried to be optimized and efforts have been made to explain the properties achieved in the samples.

EXPERIMENTAL PROCEDURE

Nanocrystalline α-alumina powder (Inframat Advanced Materials, Farmington, CT) with a purity of 99.99%, an average particle size of 100 nm was used in this study. Experiments were carried out with the α-alumina powder containing 0.1 wt% MgO. Batches were prepared by mixing Merck quality magnesium nitrate [$Mg(NO_3)_2 \cdot 6H_2O$] with alumina in ethanol medium by ball milling with 10 mm diameter alumina balls for 24 h. The slurry was then dried and heated at 750 °C for 1 h in air to decompose $Mg(NO_3)_2 \cdot 6H_2O$ to MgO. After screening the dry powder without special treatment was loaded in a graphite die for SPS sintering

A graphite die with 50mm in inner diameter and 3mm in thickness was filled with the mixture, followed by sintering using an SPS apparatus (SPS-7.40MK-VII, SPS Syntex Inc.) at 1200 – 1300 °C with heating rates of 120 °C/min and 200 °C/min and soaking times of 3-7 minutes in a vacuum. A uniaxial pressure of 80 MPa was applied during the entire process. The temperature of the die was measured by an optical pyrometer. Furthermore, shrinkage, displacement, heating current, and voltage were recorded during sintering in every 5 seconds.

After sintering, the final densities of the compacts were determined by the Archimedes method with deionized water and converted to relative density. The micrographs of fractured surfaces of all samples are examined by scanning electron microscopy (SEM; Model JSM 7000F, JEOL, Tokyo, Japan). Before the hardness measurements, the specimens were carefully polished, by standard diamond polishing techniques, down to a diamond particle size of 1 μm. The hardnesses of the samples at room temperature were evaluated by the Vickers indentation technique at a load of 49 N (Struers, Duramin A300).

RESULTS AND DISCUSSION

The density, grain size and hardness values of all spark plasma sintered samples were evaluated. The density was greater than 95.5 % of the theoretical value for all the samples. The variation of density with temperature is shown in Table I and Figure 1. The highest density value is attained in the MgO doped sample which is spark plasma sintered in 1300 °C for 5 minutes with 120 °C /min heating rate.

Hot pressing or pressureless sintering of Al_2O_3 typically is conducted at 1400 - 1700 °C[8]. The data in Figure 1 clearly demonstrates that Al_2O_3 is fully compacted by the SPS process at a markedly lower sintering temperature with short holding times. It is obvious that, for 120 °C/min heated specimens, higher soaking times cause higher density. Also, the obtained results illustrate that the specimens with a heating rate of 120 °C/min heated specimens are denser than more rapid heated specimens; it may be because of the lack of time for densification. In this connection, if the holding time is extended or a higher pressure is applied, obtaining fully dense compacts at temperatures even lower than 1250 °C can be predicted.

The hardness values of the samples were in the range of 15.4 - 19.0 GPa. The hardness of first set of samples heated with 120 °C/min heating rate at different temperatures and second set of samples heated with 200 °C/min are given in Figure 2. For the slower heated samples, rising the temperature from 1200 to 1250 °C does not markedly affect the hardness of the samples, the hardness values of the samples remain as 15.5 GPa. The change in the hardness in mentioned temperatures are higher for rapid heated specimens. Besides, increasing the temperature to 1300 °C resulted in higher hardness values for both heating rates.

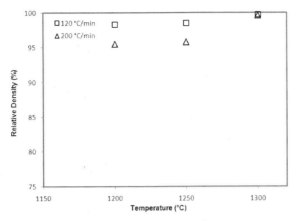

Figure 1. Relative density change versus sintering temperature at spark plasma sintered MgO doped Al₂O₃ with soaking time of 5 minutes under 80 MPa pressure in a vacuum atmosphere.

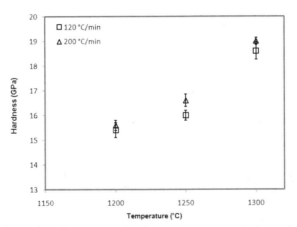

Figure 2. Hardness values change versus sintering temperature at spark plasma sintered MgO doped Al₂O₃ with soaking time of 5 minutes under 80 MPa pressure in a vacuum atmosphere.

The attained results from Figure 1, Figure 2 and the micrographs, it can be predicted that the hardness is influenced from grain size and density. Because the grains can find enough time for higher densification and the pores are removed, the hardness values increase with elevating temperature.

Figure 3. Micrograph fractured surface of spark plasma sintered Al$_2$O$_3$ containing 0.1 % MgO at 1200 °C applying 80 MPa pressure for 5 minutes with (a) 200 °C/min and (b) 120 °C/min heating rates.

Figure 4. Micrograph fractured surface of spark plasma sintered Al$_2$O$_3$ containing 0.1 % MgO at 1250 °C applying 80 MPa pressure for 5 minutes with (a) 200 °C/min and (b) 120 °C/min heating rates.

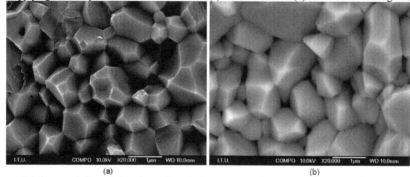

Figure 5. Micrograph fractured surface of spark plasma sintered Al$_2$O$_3$ containing 0.1 % MgO at 1300 °C applying 80 MPa pressure for 5 minutes with (a) 200 °C/min and (b) 120 °C/min heating rates.

Microstructural observations of fractured surfaces were performed by scanning electron microscope. As it is shown in Fig. 3, Fig. 4 and Fig. 5 the grain sizes becomes larger with increasing temperature and depending on the ease in diffusion, the amounts of the pore in the structure are lessened. At the lower temperatures especially at 1200 °C, the grain size difference of the specimens with 120 °C/min and 200 °C/min heating rates is more prominent. In this study, the effect of high heating rate is occurred to be negative for density. The higher porosity values and smaller grains are developed with shorter heating times.

CONCLUSION

It is possible to obtain very dense alumina ceramics (up to 99.7 %) by rapid heated (120 and 200 °C/min) spark plasma sintering of 0.1 wt% MgO doped ultra fine and high purity alumina powders. Lower heating rate resulted in coarser grains and lower porosity for sintering at 1200, 1250 and 1300 °C temperatures under 80 MPa pressure similar with the studies of Wang et al.[7] and Shen et al.[8]. The hardnesses of the samples rised with increasing sintering temperature and increasing heating rate.

REFERENCES
[1]Jiping Cheng, Dinesh Agrawal, Yunjin Zhang, Rustum Roy, Microwave sintering of transparent alumina, *Materials Letters*, **56**, 587– 592 (2002).
[2]Yi Fang, Jiping Cheng, Dinesh K. Agrawal, Effect of powder reactivity on microwave sintering of alumina, *Materials Letters*, **58**, 498– 501 (2004).
[3]R. Roman, I. Canadas, J. Rodriiguez, M.T. Hernandez, M. Gonzalez, Solar sintering of alumina ceramics: Microstructural development, *Solar Energy*, **82**, 893–902 (2008).
[4]Dibyendu Chakravarty, Sandip Bysakh, Kuttanellore Muraleedharan, Tata Narasinga Rao and Ranganathan Sundaresan, Spark Plasma Sintering of Magnesia-Doped Alumina with High Hardness and Fracture Toughness, *J. Am. Ceram. Soc.*, **91 [1]**, 203–208 (2008).
[5]R. S. Mishra and A. K. Mukherjee, Electric Pulse Assisted Rapid Consolidation of Ultrafine Grained Alumina Matrix Composites, *Mater. Sci. Eng. A*, **A287**, 178–82 (2000).
[6]L. Gao, J. S. Hong, H. Miyamoto, and S. D. D. L. Torre, Bending Strength and Microstructure of Alumina Ceramics Densified by Spark Plasma Sintering, *J. Eur. Ceram. Soc.*, **20**, 2149–52 (2000).
[7]S. W. Wang, L. D. Chen, and T. Hirai, Densification of Alumina Powder Using Spark Plasma Sintering, *J. Mater. Res.,* 15 [4] 982–7 (2000).
[8]Z. Shen, M. Johnsson, Z. Zhao, and M. Nygren, Spark Plasma Sintering of Alumina, *J. Am. Ceram. Soc.*, **85 [8]**, 1921–1927 (2002).
[9]Byung-Nam Kim, Keijiro Hiraga, Koji Morita, Hidehiro Yoshida, Effects of heating rate on microstructure and transparency of spark-plasma-sintered alumina, *J. Eur Ceram. Soc.*, **29**, 323–327 (2009).
[10]Zhou, Y., Hirao, K., Yamauchi, Y. and Kanzaki, S., Densification and grain growth in pulse electric current sintering of alumina, *J. Eur. Ceram. Soc.*, **24**, 3465–3470, (2004).
[11]A. H. Heuer, The Role of MgO in the Sintering of Alumina, *J. Am. Ceram. Soc.*, **62** [5,6] 317–8 (1979).

EFFECT OF COKE CALCINATION TEMPERATURE ON THE PROCESSING OF REACTION BONDED SILICON CARBIDE

Rodrigo P. Silva
Universidade Federal do Rio de Janeiro – DMM/POLI/UFRJ
Rio de Janeiro, RJ, BRAZIL

Claudio V. Rocha
Universidade Federal do Rio de Janeiro – POLI/COPPE/UFRJ
Rio de Janeiro, RJ, BRAZIL

Marysilvia F. Costa
Universidade Federal do Rio de Janeiro – POLI/COPPE/UFRJ
Rio de Janeiro, RJ, BRAZIL

Celio A. Costa
Universidade Federal do Rio de Janeiro – POLI/COPPE/UFRJ
Rio de Janeiro, RJ, BRAZIL

ABSTRACT

Reaction bonded silicon carbide (RBSiC) ceramic is a well known material which shows some advantage over its conventional sintered or hot pressed counterpart. On the other hand, its processing can be quite difficult, since microstructure defects, explosion and/or cracks might occur as result of trapped gases, fast reaction between Si+C, low capillarity and low strength preform.

The present work evaluated the effect of trapped gases in the carbon precursor (petroleum coke) in the process RBSiC. The gases are found to be released up to 1600 °C and they will counteract the silicon melt infiltration, which is responsible for the new SiC formation. Petroleum coke can be used to process RBSiC ceramics, but it must have been calcinated to temperatures higher than 1900 °C in order to remove all trapped gases.

INTRODUCTION

The Reaction Bonded Silicon Carbide (RBSiC) is obtained by the infiltration of metallic silicon melt into a porous green body formed by silicon carbide (SiC) and carbon. During the infiltration, the silicon melt must react with the carbon to exothermically form a new SiC. RBSiC is an important advanced ceramic and it has processing and properties that are very promising; for instance, it is processed in near-net complex shapes and requires lower temperature than conventional sintering. Moreover, presents good mechanical and tribological properties [1,2,3].

To process a high quality material is not an easy task since infiltration must occur throughout the microporous green body (preform). During the complex infiltration reaction, the SiC formation is highly volume expansive with respect to the carbon phase; on the other hand, there exists a net volume reduction when the melt Si reacts with the solid carbon to create SiC. Problems related with the process are incomplete infiltration associated to choking off of open porosity and flaws, resulting from thermal stresses and volume expansion stresses [1].

In order to obtain a successful infiltration, it is necessary to control the particles size of the constituents, the green body density, the rate of SiC formation and the preform strength during infiltration. Regarding the constituents, several carbon sources have been studied. In the beginning amorphous carbon or graphite were employed [4], followed by resins [2,3,5] and, lately, petroleum coke (PC) has been reported as a good carbon source [6]. PC has the advantage of being a very cheap carbon source, but it may become a problem during processing due to the very large amount of entrapped gas on it that can be released at different processing stages. Furthermore, the use of SiC with small particle size, to render strength [7], coupled with carbon containing large amount of trapped gases may result in a very difficult material to infiltrate.

This study evaluated the use of petroleum coke (PC) to process RBSiC. Three types of cokes were tested, namely: green coke (no calcinations), calcinated coke at 1400 °C, and calcinated coke at 2100 °C. The results showed that the PC must have been calcinated to the least possible amount of gases in order to produce a homogenously infiltrated material, which for the present study was achieved at about 1900 °C.

EXPERIMENTAL PROCEDURE

Petroleum coke (Petrocoque, Brazil) was used in three different grades: green (no calcination), calcinated by the supplier at 1400 °C (commercial grade) and calcinated at the laboratory at 2100 °C - this last sample was produced by recalcination of the commercial grade in the laboratory furnace.

The three grades were analyzed with respect to ashes (CZ), gases (MV), sulpher (S) and fixed carbon (CF). The ashes and gases were analyzed following procedure described in ASTM D 3174-93 and D 3175-93, respectively, and sulpher as ASTM D 4239. The fixed carbon was calculated as follows: (100-(CZ+MV)). According to the standards, the ashes are the residue of the coke combustion up to 750 °C and the gases are the weight loss of the coke up to 950 °C under inert gas (1 atm). These analysis were performed by *Bureau Veritas do Brasil*. In addition to the above characterization following ASTM standards, the PC was also calcinated in the laboratory furnace to verify the amount gases trapped well above the 950 °C. This analysis measured the variation in the chamber pressure as the temperature was increased up to 2100 °C.

The other silicon carbide grade used was SiC UF-15 (d_{50} = 0.7 µm, BET 15 g/m^2, from the H. C. Starck) and the metallic silicon had purity of 99% (Rima S.A.).

Three different compositions were processed, as shown in Table I. The compositions were individually wet ball-milled in distilled water using PVA+PEG as binder (3+3 wt%) and alumina balls, for 20h. The mixtures were dried in an oven and then sieved. The compositions were uniaxially pressed and the resulting preform had the dimensions of 58.7x 65.2 x 5 mm.

Table I – Identification and chemical compositions used

Preform Identification	Chemical Composition (wt%)
C00	70% SiC + 30% green coke (no calcination)
C14	70% SiC + 30% coke calcinated at 1400 °C
C21	70% SiC + 30% coke calcinated at 2100 °C

The infiltration schedule applied for all compositions is shown in Figure 1. Three plateau temperatures were used, the lower one took 30 minutes and last two took 1 hour each.

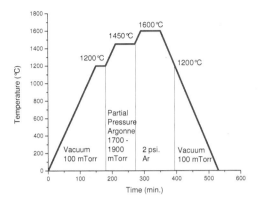

Figure 1 – Sintering schedule used for infiltrations.

RESULTS AND DISCUSSION

The results for the analysis of the three PC grades are shown in Table II, where it can be observed that amount of both ash (CZ) and sulphur (S) is similar for both, in each grade (the CZ was measured up 750 °C). The amount of trapped gases (MV) in green PC is relatively low, but 7.4 and 9.8 times higher than in PC calcinated at 1400 °C and 2100 °C, respectively. The trapped gases were measured up to 950 °C, as ASTM D 3175-93 recommends.

Table II – Analysis of the petroleum coke (PC)

Sample	CZ %	MV %	S %	CF%
Green PC	0.375	0.635	0.824	98.99
PC calcinated at 1400 °C	0.415	0.085	0.893	99.50
PC calcinated at 2100 °C	0.395	0.065	0.768	99.54

The sintering of the composition C00 (green coke) resulted in explosion and/or crack of all plates, as shown in Figure 2. It is also seen that the effect of choking off of porosity, since the infiltration only occurred at the very near border of the plate, none occurred inside it. This kind of behavior has already been reported; for instance, Chiang et al. [1] attributed it to high thermal stress caused by fast exothermic Si+C formation reaction, while Wu et al. [6], working with a high load SiO_2 system, attributed this phenomenon to the high amount of CO generated by the SiO_2 (S) + C (S) reaction and, also, to the gases trapped in the petroleum coke (PC). For the C00 composition, the most probably explanation is the gases trapped in PC, even though the PC analysis (Table II) showed it to be low.

Figure 2 – Example of plate processed with green PC. The trapped gases in the green PC resulted in explosion or cracks of all plates. No infiltration was observed beyond the border.

In order to confirm that explosions and cracks were caused by trapped gas in PC, a series of tests were conducted up to 2100 °C, under continuous vacuum. The first test was carried out in an empty furnace to observe its behavior since the porous heat shields degasses when is heated above 1800 °C, as shown in Figure 3. The degassing of heat shields is due to previous processes which used resins (phenolic, PVA, PEG, etc), and they are adsorbed and trapped. When the green PC was tested, it was observed that a large amount of gases were still present above 950 °C, ie the maximum temperature recommended by ASTM. Gas release by the coke was observed even up to 1600 °C. For the calcinated PC at 1400 °C, there is no gases release until 1100 °C, but in between 1100 and 1450 °C there was an increase in the chamber pressure, showing that gases were being released. After 1450 °C, the curve followed the empty furnace behavior. The PC calcinated at 2100 °C followed the empty furnace behavior during the whole cycle.

It is important to note the any gases released at 1450 °C will counter act the infiltration process, which naturally happens at this temperature and, consequently, shall lead to either an heterogeneous microstructure or a material with elevated porosity or cracks.

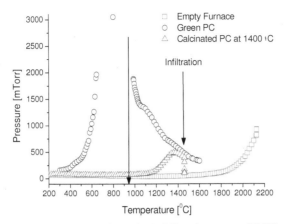

Figure 3 – Effects of gases in the PC used to process RBSiC.

Compositions C14 (calcinated at 1400 °C) and C21 (calcinated at 2100 °C) were sintered with the cycle presented in Figure 1. For both compositions, no explosion or cracks were seen, which confirms that trapped gases were the major drawback in sintering composition C00. The resulting microstructure are shown in Figure 4, where larger pores are seen in the composition C14 (Figure 4 - A) than in C21 (Figure 4 – B). These larger pores may be the result of gases going out of the preform while the melted Si is trying to infiltrate at 1450 °C, since there still gases at this temperature. The microstructures are also different from each other. The amount of residual metallic Si was not yet determined.

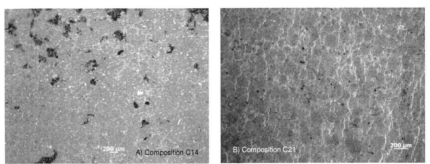

Figure 4 – Microstructure of the composition C14 (left) and the C21 (right)

The presence of gases is very detrimental to the infiltration process, since it might be retained inside the preform and build up the internal pressure, which will lead to explosion or cracks. The Si + C reaction is very exothermic and it can raise the local temperature in 400 °C; in this case, if there are gases trapped, the internal pressure will be raised or more gases will be released. So, the presence of trapped gases in the PC is not desired and calcinations at at least 1900 °C must be carried out before RBSiC processing. This calcination temperature is the combination of infiltration (1450 °C) plus Si+C reaction (400 °C) plus a safety factor (50 °C).

The results obtained so far indicate that the coke must have the least possible amount of gases in order to produce a homogenously infiltrated material.

CONCLUSIONS

Petroleum coke can be used to process reaction bonded silicon carbide ceramics, but it must have been calcinated to temperatures higher than 1900 °C in order to remove all trapped gases.

To measure trapped gases in petroleum coke to process RBSiC using the ASTM D 3175-93 standard shall be avoided, since it evaluates the material up to 950 °C, but trapped gases persist until at least 1600 °C.

ACKNOWLEDGMENT

This work was supported by FINEP (contract 01.05.0966.00), PETROBRAS, ANP-PRH 35 and DEPROCER. Thanks to engineer Elcio Santana and PETROCOQUE for donating the Petroleum Coke.

REFERENCES

1- Y. M. Chiang, R. P. Messner, C. D. Terwillger, Reaction-formed silicon carbide, *Materials Science and Engineering,* A144 (1991) 63-74.
2- Y. Wang, S. Tan, D. Jiang, The effect of porous carbon perform and the infiltration process on the properties of reaction-formed SiC; *Carbon,* V. 42 (2004) 1833-1839.
3- S. Xu, G. Qiao, D. Li, H. Yang, Y. Liu, T. Lu, Reaction forming of silicon carbide ceramic using phenolic resin derived porous carbon perform; *Journal of the European Ceramic Society* 29 (2009) 2395-2402.
4- W. G. Brown, Method for manufacturing silicon carbide bodies; *United States Patent 4,154,878;* May 15, 1979.
5- M. K. Aghajanian, B. N. Morgan, J. R. Singh, R. J. Mears, R. A. Wolffe, A new family of reaction bonded ceramics for armor applications, *In Ceramic Armor Materials by Design, Ceramic Transactions,* **134**, 527-539 (2002).
6- WU Qi-de, Guo Binj-jian, Yan Yong-gao, ZHAO Xiu-jian, HONG Xiao-lin, Effect of SiO_2 on the preparation and properties of pure carbon reaction bonded silicon carbide ceramics; *Journal of Wuhan University of Technology – Mater. Sci. Ed.;* Vol 19, N° 1, (2004), 54-57.
7- M. Wilhelm, M. Kornfeld, W. Wruss, Development of SiC-Si Composites with Fine-Grained SiC Microstructures, *Journal of the European Ceramic Society,* 19 (1999), 2155-2163.

PRESSURELESS SINTERING OF MULLITE-CERIA-DOPED ZIRCONIA-SILICON CARBIDE COMPOSITES

Barry A. Bender, Michael Vick, and Ming-Jen Pan
Naval Research Lab
Washington, DC 20375

ABSTRACT

A small gas turbine is being developed at NRL. The high inlet temperature of 1225°C on a micro engine places stringent conditions on what materials can be used for this design. Mullite is an excellent candidate for the heat exchanger due to its low thermal conductivity and thermal shock resistance. To improve its toughness the mullite is reinforced with 18 vol% ceria-doped ZrO_2. Research has shown that hot-pressing of mullite in inert environments with SiC (particles or whiskers) can further improve its toughness and creep resistance. However, due to the complexity of the design of the heat exchanger, fabrication must be done by pressureless sintering. Pressureless sintering in air of SiC-reinforced zirconia-mullite composites lead to zircon formation. Pressureless sintering in vacuum led to the decomposition of mullite to alumina. Decomposition of the mullite matrix could be inhibited by embedding the green composites in a mixture of alumina and zircon powders followed by sintering at low temperatures in a less reducing atmosphere. SiC-reinforced mullite composites could be sintered in flowing nitrogen at 1375 to 1400°C to high densities with minimal decomposition of the mullite matrix or SiC reinforcements. Addition of the SiC reinforcements led to a 15 to 30% improvement in toughness of the 1400°C-sintered zirconia-mullite composite.

INTRODUCTION

A major thrust of the Navy's unmanned aerial vehicle (UAV) program is to improve the current reconnaissance capabilities of the UAVs by increasing their range and endurance. To achieve this goal researchers at NRL have developed a proprietary design for a 4 hp recuperated micro gas turbine engine that is projected to operate at an efficiency of 25%. To achieve this goal the engine has been designed to operate at a temperature of 1225°C. To operate at such a high temperature requires that the ceramic recuperator material have excellent thermal shock and creep resistance, excellent thermal stability, low thermal conductivity, low thermal expansion, strength of 200 MPa and a toughness of 5 MPa-m½.

Mullite ($3Al_2O_3$-$2SiO_2$) is one ceramic material that satisfies the majority of the above criterion. It has good high temperature properties as it is refractory (melting point as high as 1905°C), and exhibits both good thermal shock and creep resistance.[1-3] It also has excellent intrinsic thermal stability under oxidizing conditions and shows little degradation of strength with temperatures as high as 1500°C.[2,4] Mullite is inexpensive and has a density of 3.17 g/cm³ which is about one third the densities of super alloys.

However, there are two major shortcomings of mullite- its poor sinterability and poor fracture toughness. Commercial undoped stoichiometric mullite powders have to be sintered at high temperatures of 1650°C or higher.[1,2] While its room temperature strength of 200 MPa meets design requirements its fracture toughness of 2.2 MPa-m½ is deficient.[5] Researchers have overcome these difficulties via hot-pressing mullite with zirconia and/or SiC reinforcements. Becher et al.[5] were able to improve the toughness of mullite from 2.0 to 3.2 MPa-m½ by hot-pressing mullite with 15 vol% zirconia. They also incorporated 20 vol% SiC whiskers to make a mullite composite with a fracture toughness of 4.7 MPa-m½. Jin et al.[6] were able to improve the toughness of mullite from 2.8 to 4.2 MPa-m½ via hot pressing mullite with 12 vol% nanoparticles of SiC. Hong et al.[7] used one micron size particles of SiC at 30 vol% to toughen mullite as the fracture toughness of the hot-pressed

15

composites increased from 2.9 to 4.6 MPa-m½. Unfortunately, the complicated design of the recuperator precludes consolidation via hot pressing. Pressureless sintering must be used for fabrication of the intricately-channeled recuperator. Researchers have sintered mullite composites reinforced with SiC whiskers, SiC nanoparticles, and SiC particles but with very limited mechanical data being reported.[8-11] Earlier research[12] on an 18 vol% zirconia- (partially stabilized with 10 mol% ceria) mullite composite (CZM) had shown that a zirconia-reinforced mullite could be pressureless-sintered with a fracture toughness of 4.7 MPa-m½. Seminal research by Becher et al.[5] showed that by adding SiC reinforcements to a zirconia-toughened mullite composite that a synergetic toughening effect occurred resulting in a significant enhancement in fracture toughness. This paper reports on the effort to fabricate a SiC-reinforced (whiskers, nanoparticles, or one micron-size particles) ceria-doped zirconia-mullite composite via pressureless sintering.

EXPERIMENTAL PROCEDURE

The test samples were prepared via conventional solid state reaction processing of mullite and zirconia (18 vol%) powders mixed with SiC whiskers, particles, or nanoparticles. The mullite powder is a fine particle size (0.7 μm) stoichiometric (71.83% alumina) mullite powder (KM101, KCM Corp., Nagoya, Japan). The submicron zirconia powder (CeSZ) was partially stabilized with 10 mol% ceria (Zircar Inc., New York). The SiC nanoparticles were 30 nm in size (95.0%, Alfa Aesar, Ward Hill, MA). The SiC particles were 2 μm in size (99.8%, Alfa Aesar, Ward Hill, MA). The SiC whiskers were typically 1.5 μm x 18 μm in size (95.0%, Alfa Aesar, Ward Hill, MA).

Mullite-CeSZ-SiC nanocomposites (SiCnp) were made with the addition of 5 vol% SiC nanoparticles. This vol% was chosen because Warrier et al.[10] had shown that the higher addition of 10 vol% nanoparticles inhibited densification of the composites via pressureless sintering and that even small amounts of nanoparticles can significantly improve the properties of mullite composites.[6] The composites were first made by treating the SiC nanoparticles in 1% HF followed by rinsing in water and acetone. This was done to remove any trace layer of silica.[10] The nanoparticles, mullite, and CeSZ powders were then attrition-milled for 60 minutes. The mixed slurry was dried at 90°C. A 2% PVA binder solution was mixed with the composite powder and sieved to eliminate any large agglomerates. The dried powder was uniaxially pressed into discs typically 13 mm in diameter and 1 mm in thickness and the binder was then burnout before sintering took place.

Mullite-CeSZ-SiC particle composites (SiCp) were made with the addition of 10 vol% SiC particles. This vol% was chosen because Kamiaka et al.[9] encountered problems sintering mullite-SiC composites with amounts greater this and at the same time 10 vol% is the minimum amount of SiC needed for crack healing in mullite-SiC composites.[13] The composites were made by first attrition-milling the SiC particles by themselves for 30 minutes in order to reduce the typical particle size by 50% to one micron or less to improve their sinterability. Then the SiC particles, CeSZ, and mullite powders were combined together and attrition-milled for 60 minutes.

Mullite-CeSZ-SiC whisker composites (SiCw) were fabricated with the addition of 5 vol% SiC whiskers. The Alfa Aesar SiC whiskers (Fig. 1a) were chosen because they have a low aspect ratio which is needed to facilitate pressureless sintering of whisker composites.[14] The 5 vol% addition was also chosen to avoid the problem of sintering higher loadings of whisker where the whiskers start forming networks which make them hard to breakup without applying additional pressure during fabrication.[15] The whiskers were first cleaned in 1% HF. Next they were de-nested by attrition milling the whiskers at low speeds for a short period of time.[14] The whiskers were further detangled by making a 5% solution of whiskers in distilled water which contained a dispersant. This solution was sonicated for 60 minutes. The whiskers were then fractionated by waiting for 60 minutes for the large whiskers and large unwanted SiC particles to settle and decanting off the solution with the finer whiskers. The whiskers were then added to the CeSZ and mullite powders and attrition-milled for only

30 minutes and at a lower speed to prevent breakage of the whiskers. The resultant powder was oven-dried at 90°C. This powder was then mixed with ethanol and stir-dried on a hot plate in an effort to prevent preferential settling of the larger SiC whiskers. Binder was added and the pellets were pressed, burnt-out and sintered resulting in a well-dispersed SiC whisker-mullite composite (Fig. 1b).

Figure 1. (A) SEM micrograph of the SiC whiskers after HF washing and (B) optical micrograph of the polished surface of the SiCw sample after sintering showing the good distribution of the whiskers.

All of the composite pellets were sintered in vacuum, air, nitrogen, or argon at temperatures ranging from 1375 to 1500°C. The pellets were either sintered in ambient, carbon powder, mullite powder, or a 50/50 mol% mixture of alumina and zircon powders. Material characterization was done on the discs after processing. Fracture toughness was determined using an indentation microcrack technique assuming an elastic modulus of 200 GPa which has been measured for a mullite dispersed with 20 vol% zirconia.[5] Microstructures were observed by scanning electron microscopy (SEM) on fracture surfaces, polished surfaces, and thermally-etched polished surfaces. X-ray diffraction (XRD) scans were made from both as-fired and polished surfaces using monochromated CuKα radiation. The densities of the fired-specimens were measured by the Archimedes method using water.

RESULTS AND DISCUSSION
Sintering in Air
CZM (no SiC reinforcements) and SiCp (CZM with 10% SiC particles) composites were sintered in air at 1475°C for 4 h. The CZM composite densified well (> 98% relative theoretical density (TD)) with a weight loss of 0.9%. XRD detected only mullite and monoclinic and tetragonal zirconia. SiCp composites had a weight gain of 2.9% and only mullite and zircon were detected by XRD. The unexpected weight gain is a result of the oxidation of the SiC particles by air to form silica. Luthra and Park[16] observed oxidation of SiC particles in SiC-reinforced mullite composites in oxidation experiments at temperatures as low as 1375°C. The silica can then diffuse and react with the grains of zirconia to form zircon which was detected by XRD. The above reaction explains the weight gain and the presence of zircon in the composite after sintering. The presence of zircon in oxidation of SiC-ZrO$_2$-mullite composites has also been observed by Lin.[17] Packing the green disks in mullite powder did nothing to inhibit the oxidation of the composite. SiCw (CZM with 5% SiC whiskers) sintered in air at 1475°C lead to a weight gain of only 1%. This is because there is less SiC present to form silica to react with the zirconia to form zircon. This should result in a composite with zircon, zirconia, and mullite present. This was indeed verified by XRD as zircon was detected while the amount of zirconia as compared to CZM was reduced by about 50%.

Sintering in Reducing Atmospheres- Vacuum and Carbon Powder

CZM and SiCw composites were vacuum-sintered for four hours at 1487°C to inhibit oxidation of the silicon carbide. The CZM composite lost only 0.4% weight and had a TD > 99%. On the other hand, the SiCw composite lost 16.9%. XRD detected typical amounts of zirconia, very little mullite, and significant amounts of alumina, which was in contrast to XRD of CZM which showed mullite but no alumina. This data supports the research of Park and McNallan[15] who modeled the decomposition of mullite with and without SiC reinforcements in a reducing atmosphere. They showed that in a certain temperature range that mullite by itself won't decompose in a reducing atmosphere. However, if SiC is present it will act as a reducing agent and react with mullite to form alumina and SiO and CO gas, which is what we observed. Park and Luthra[3] also conjectured that the present of a liquid reaction phase should drive the rate of decomposition of mullite and SiC even faster. This is because gaseous diffusion of the decomposition products is faster through a liquid than a solid which would favor the decomposition of the mullite. This tendency was observed in the experiments in sintering mullite composites packed in carbon powder at 1467°C for 4 h. The standard SiCp composite (green diameter- 10 mm) fabricated with ceria-doped zirconia sintered well to a final diameter of 8.45 mm, while a SiCp composite made with undoped zirconia sintered to a disc diameter of only 9.74 mm. This is because previous research[12] had shown that in CeSZ-mullite composites that a transient liquid phase (TLP) forms which enhances the sintering of the CZM composites. However, this TLP also enhances the decomposition of mullite composites reinforced with SiC because no alumina was detected in the undoped zirconia-SiC-mullite composite while alumina was the predominant phase in the SiCp composite sintered in the reducing atmosphere of carbon powder.

Sintering in Flowing Argon or Nitrogen

In an effort to reduce the amount of decomposition of the composites during sintering, lower sintering temperatures and less reductive atmospheres were tried. CZM and SiCp composites were sintered in flowing Ar at 1425°C for 4 h. Despite the low processing temperature the CZM composite sintered well to a TD > 97% and showed no signs of decomposition. However, the SiCp composite lost 11.7% weight and XRD detected decomposition as no mullite was found- just Al_2O_3 and ZrO_2.

In a further effort to inhibit the decomposition of the mullite matrix the sintering temperature was reduced to 1375°C and samples were embedded in various powders. Excellent densification still occurred at this sintering temperature as the CZM composite had a TD >98%. For the SiCp composite embedded in mullite powder during sintering, decomposition of the mullite was inhibited. Though the sample had a weight loss of 6.7% the resultant alumina peak was reduced considerably and mullite was the major phase detected. In an attempt to further reduce the decomposition of the mullite matrix Le Chatelier's principle was used. A powder bed of 50-50 mole percent alumina and zircon was tested. The results were encouraging as the weight loss was reduced to 2.5%. XRD detected the presence of the SiC particles and showed no signs of any alumina. However the sample had only a TD of 94%. Efforts to increase the densification of the SiCp composite by sintering at 1400°C in flowing Ar led to an increase in weight loss to 4.5% and the formation of a surface layer showing the presence of Al_2O_3.

An atmosphere of flowing nitrogen was tried at 1400°C in an attempt to sinter a SiCp composite that was dense but did not form a surface layer. This is important because the complex design of the recuperator involves the presence of many fine thin channels and the presence of a thin compromised surface layer could lead to degradation of the mechanical properties of the mullite composite. CZM composites sintered in flowing nitrogen for 4h at 1400°C sintered well. The composite lost only 0.7% weight and had a TD > 99%. The SiCp composite sintered in a bed of alumina and zircon powder sintered well (TD > 96%) and had a weight loss of only 2.9%. XRD of the surface detected no Al_2O_3 and detected the presence of the SiC particles. SiCnp and SiCw composites were sintered too at this temperature. However, XRD of the composites did not detect the presence of the SiC reinforcements. Therefore, sintering at 1375°C for 4 h was tried. The average weight loss of the two composites was reduced by 50% to 1.6%. XRD did not detect the presence of alumina. It detected the existence of the SiC nanoparticles and the SiC whiskers whose presence was also confirmed by optical microscopy (Fig. 1b).

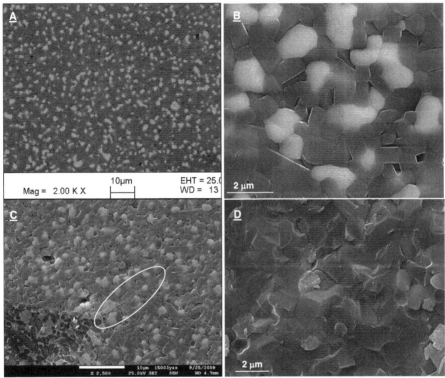

Figure 2. (A) Back-scattered (BSE) SEM image of the polished surface of the 1500 CZM composite. (B&C) are the BSE images of the composite after thermal etching where due to atomic number contrast the mullite grains are darker. Fig. C shows the crack emanating from the test indent. Fig. D is the fracture surface of the 1500 CZM composite.

Characterization of the SiC-ZrO$_2$-Mullite Composites

Figure 2a shows that the 1500°C-sintered CZM composite densified quite well (TD >99%) with little porosity present and excellent dispersion of the zirconia grains in the mullite matrix. Despite the significantly 100 degree lower processing temperature the 1400°C-sintered CZM composite also sintered well (TD >99%) and exhibited little porosity and fine dispersion of zirconia in the mullite (Fig. 3a). However, the grain size of the mullite and zirconia is about 50% smaller in the 1400 CZM which should be expected due to its lowering sintering temperature (Figs. 2b, 3b&c). As a result, the grain size affects the amount of zirconia that transforms from tetragonal to monoclinic due to stress. Table I shows that little transformation toughening is occurring in the 1400 CZM composite as the vol% of monoclinic does not change at all from the vol% detected in the starting CeSZ powder. However, in the 1500 CZM composite the larger zirconia grain size is closer to the critical size needed for phase transformation resulting in a larger vol% of monoclinic being detected on the polished surfaces of the composites. As a result, the 1500 CZM composite is 50% tougher than the 1400 CZM composite (see Table I). Figures 2c and 3d confirm this trend as the crack emanating from the test indent in the 1400 CZM composite is almost twice as long as the crack observed in the 1500 CZM sample.

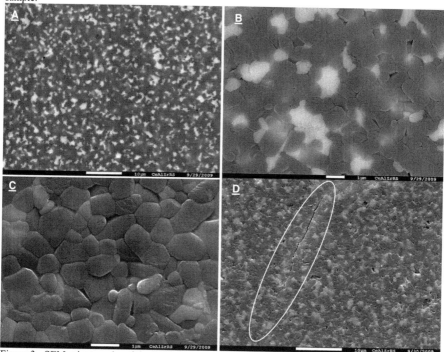

Figure 3. SEM micrographs of the 1400 CZM composite- (A) is a BSE image of the polished surface, (B-D) are micrographs of the polished surface that has been thermally etched where (B) is a BSE image, (C) is a secondary electron image (SE), and (D) shows the indent crack emanating from the edge of the test indent.

Table I. Properties of SiC-Zirconia-Mullite Composites

SiC Reinforcement	Sintering Conditions	Hardness (GPa)	Weight Loss	Diameter (mm)*	Density (g/cm³)	^Tet (%)/ Mono(%)	K_{ic} (MPa-m½)
No SiC	1500°C- air	10.87	0.9%	8.33	3.63	40/100	4.7
No SiC	1400°C- N$_2$	10.73	0.7%	8.30	3.62	100/19	3.1
SiCw	1375°C- N$_2$	9.57	1.5%	8.61	3.46	100/65	3.6
SiCnp	1375°C- N$_2$	10.64	1.75%	8.40	3.54	100/64	4.1
SiCp	1400°C- N$_2$	10.44	2.9%	8.40	3.52	100/94	4.1

*Sintered disc diameter; ^XRD intensity ratio of tetragonal to monoclinic zirconia- polished surface

Figure 4a shows that the SiCw composite did not densify as well as the 1400 CZM composite (Fig. 3a). This is also reflected in the much lower hardness of the composite (see Table I) which is expected due to the increase in the porosity of the composite. This increase in porosity is expected in pressureless-sintered whisker-reinforced composites due to the presence of the whiskers which can interact with each other and form semi-rigid networks that can only be broken via applied pressure during fabrication. However, careful processing did result in good dispersion of the SiC whiskers as shown in Fig. 1b. It appears from Figs. 4b&c that the SiC whiskers bond well to the matrix. This is not unexpected as the presence of a transient liquid silicate phase which enables enhanced sintering[12] also would promote bonding of the SiC whisker to the mullite matrix. This is reflected in fractograph of the SiCw composite (Fig. 4d) as there are no signs of whisker pullout. In fact the fracture surface (Fig. 4d) is similar in appearance to that of the un-reinforced CZM composite (Fig. 2d). Despite the paucity of SiC whisker reinforcements the toughness of the SiCw composite was 3.6 MPa-m½- a 15% improvement over the fracture toughness of the 1400 CZM composite.

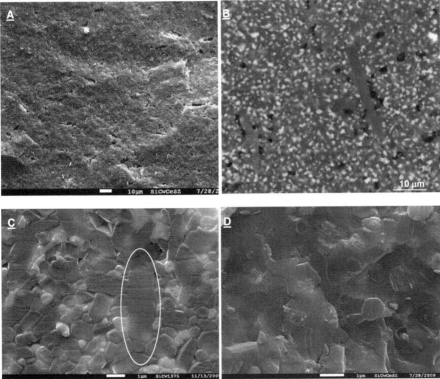

Figure 4. SEM micrographs of the SiCw composite showing (A) the more porous microstructure, (B) the whiskers as the darker elongated phase in the BSE image, (C) a large whisker in a SE image of a thermally-etched surface, and (D) no evidence of whisker pullout on a relatively flat fracture surface.

Figure 5a shows that the SiCp composite sintered better than the SiCw composite (Fig. 4a). Table I indicates that disc shrunk to a diameter of 8.40 mm as compared to 8.61 mm for the SiCw composite. The lower porosity of the composite explains the higher hardness of the composite as compared to the SiCw composite (see Table I). Weight loss for the sample was the greatest of all the fabricated reinforced-composites. This is probably due to the higher vol% of reinforcements and higher sintering temperature which lead to degradation of some of the SiC particles. Attrition-milling of the SiC powder resulted in a starting powder where the typical SiC particle size was one micron or smaller (Fig. 5b). As a result, the SiC particles mixed well with the starting composite powders and led to a composite with finely-dispersed zirconia (Fig. 5c). However, due to the atomic number contrast of SiC and mullite being the same it was hard to differentiate between the two phases in the SEM. Micrographs of the fracture surface (Fig. 5d) appear to be more tortuous than the fracture surfaces of the SiCw composites (Fig. 4d). This implies that the SiC particles are causing some crack deflection which results in a composite with a toughness of 4.1 MPa-m½ which is 30% tougher than the 1400 CZM composite.

Figure 5. SEM micrographs of the SiCp sample showing (A) a relatively dense microstructure, (B) the SiC particles after attrition milling, (C) a BSE image of a polished surface, and (D) the more tortuous fracture surface of the SiCp composite as compared to the fracture surface of the CZM composite.

Figure 6a shows little porosity as the SiCnp composite densified to a similar degree as the SiCp composite but to even a higher density than the SiCw composite (Table I). The SiCnp composite sintered better than the SiCw composite because of the extremely fine size of the starting SiC powder (30 nm- see Fig. 6b). As a result of the fine particle size, the surface area of the 5 vol% SiC particles was much higher than the surface area for 5 vol% whiskers and therefore more reactive which explains the higher weight loss during processing of the SiCnp composite as compared to the SiCw composite. Due to the fine size of the SiC reinforcement it was hard to detect their presence and where the particles were distributed in the microstructure of the composite. However, as seen from the back-scattered and secondary electron SEM image pair (Fig. 6c&d) some of the nanoparticles are located inside of the larger mullite grains. No signs of matrix grain refinement are observed in the SiCnp composite which has been observed in other mullite composites reinforced with SiC nanoparticles.[6] Despite being reinforced with only 5 vol% SiC particles the SiCnp composite was as tough as the SiCp composite and it toughness of 4.1 MPa-m½ was 14% higher than the SiCw composite with a similar reinforcement loading. This is not unexpected because researchers have shown that small amounts of SiC nanoparticles can significantly impact the mechanical properties of mullite composites.[6,10]

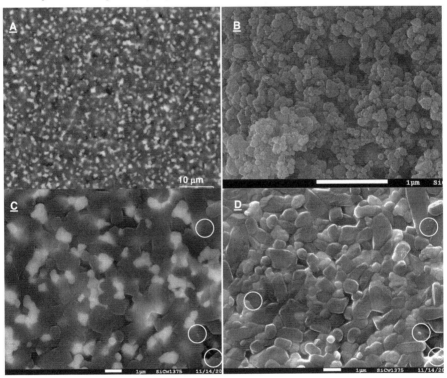

Figure 6. SEM micrographs of the SiCnp sample showing (A) a relatively dense microstructure, (B) the as-received SiC nanoparticles, and (C &D) a BSE-SE micrograph pair of the thermally-etched surface showing that some of the SiC nanoparticles have been entrapped in larger mullite grains.

CONCLUSIONS

Careful processing was needed in order to fabricate ceria-doped zirconia-mullite composites reinforced with either SiC whiskers, or SiC particles, or SiC nanoparticles. Sintering in air at higher temperatures led to oxidation of the SiC reinforcements and the subsequent formation of silica. The silica then reacted with the zirconia to form zircon. In reducing atmospheres (vacuum) the SiC reinforcements acted as a reducing agent leading to the decomposition of the mullite and SiC particles to form alumina and SiO and CO gas. It was also shown that the transient liquid silicate phase that promotes sintering also enhances the decomposition of the mullite. However, by embedding the samples in a protective bed of zircon and alumina powder and sintering the composites at a lower temperature in a less reducing atmosphere such as flowing argon or nitrogen that the amount of decomposition could be significantly reduced. CZM composites could be sintered in flowing nitrogen at 1375 to 1400°C to greater than 98% TD with a toughness of 3.1 MPa-m½. Under similar processing conditions CZM composites reinforced with 5% SiC whiskers could be successfully fabricated and led to a mullite composite with a fracture toughness of 3.6 MPa-m½. CZM composites reinforced with either 5% SiC nanoparticles or 10% SiC particles could also be fabricated at these low temperatures and the presence of the SiC reinforcements resulted in composites that had a fracture toughness of 4.1 MPa-m½ which was a 30% improvement over the toughness of the 1400 CZM composite without any SiC reinforcements. However, the 4.7 MPa-m½ fracture toughness of the CZM composite sintered at 1500 °C was higher than any of the composites fabricated at 1400 °C due to the larger grain size of its zirconia particles which enabled an increase in K_{ic} via transformation toughening.

REFERENCES
[1] S. Prochazka and F.J. Kug, Infrared-Transparent Mullite Ceramic, *J. Am. Ceram. Soc.*, **66**, 874-80 (1983).
[2] I.A. Askay, D.M. Dabbs, and M. Sarikaya, Mullite for Structural, Electronic, and Optical Applications, *J. Am. Ceram. Soc.*, **74**, 2343-58 (1991).
[3] P.A. Lessing, R.S. Gordon, and K.S. Mazdiyasni, Creep of Polycrystalline Mullite, *J. Am. Ceram. Soc.*, **58**, 149 (1975).
[4] G. Orange, G. Fantozzi, F. Cambier, C. Leblud, M.R. Anseau, and A. Leriche, High Temperature Mechanical Properties of Reaction-Sintered Mullite/Zirconia and Mullite/Alumina/Zirconia Composites, *J. Mat Sci.*, **20**, 2533-40 (1985).
[5] P.F. Becher and T.N. Tiegs, Toughening Behavior Involving Multiple Mechanisms: Whisker Reinforcement and Zirconia Toughening, *J. Am. Ceram. Soc.*, **70**, 651-54 (1987).
[6] X.H. Jin, L. Gao, L.H. Gui, and J.K. Guo, Microstructure and Mechanical Properties of SiC/Zirconia-Toughened Mullite Nanocomposites Prepared from Mixtures of Mullite Gel, 2Y-TZP, and SiC Nanopowders, *J. Mater. Res.*, **17**, 1024-29 (2002).
[7] J.S. Hong, X.X. Huang, J.K. Guo, Toughening Mechanisms and Properties of Mullite Matrix Composites Reinforced by the Addition of SiC Particles and Y-TZP, *J. Mat. Sci.*, **31**, 4847-52 (1996).
[8] T. Tiegs, P. Becher, and P. Angelini, Microstructures and Properties of SiC Whisker-Reinforced Mullite Composites, *Ceram. Trans.*, **6**, 463-72 (1990).
[9] H. Kamiaka, C. Yamagishi, and J. Asaumi, Mechanical Properties and Microstructure of Mullite-SiC-ZrO_2 Particulate Composites, *Ceram. Trans.*, **6**, 509-17 (1990).
[10] K.G.K. Warrier, G.M.A. Kumar, and S. Ananthakumar, Densification and Mechanical Properties of Mullite-SiC Nanocomposites Synthesized Through Sol-Gel Coated Precursors, *Bull. Mater. Sci.*, **24**, 191-95 (2001).
[11] M.D. Sacks, N. Bozkurt, G.W. Scheiffele, Fabrication of Mullite and Mullite-Matrix Composites by Transient Viscous Sintering of Composite Powders, *J. Am. Ceram. Soc.*, **74**, 248-37 (1991).
[12] B.A. Bender and M.-J. Pan, Selection of a Toughened Mullite for a Miniature Gas Turbine Engine, *Ceram. Eng. Sci. Proc.*, **30**, 167-176 (2009).

[13]T. Osada, W. Kakao, K. Takahashi, and K. Andl, Healing Behavior of Machining Cracks in Oxide-Based Composite Containing SiC Particles, *Ceram. Eng. Sci. Proc.*, **30**, 45-55 (2009).

[14] T. Tiegs, "SiC Whisker Reinforced Alumina," p. 307-323 in Handbook of Composites, Edited by N.P. Bansal, Kluwer Academic Publishers, Boston, MA 2005.

[15] Y.S. Park and M.J. McNallan, Chemical Reactions in Mullite Matrix SiC Whisker Reinforced Composites in RF Plasma, *J. Mater. Sci.*, **32**, 523-28 (1997).

[16] K. Luthra and H.D. Park, Oxidation of Silicon Carbide-Reinforced Oxide-Matrix Composites at 1375 to 1575°C, *J. Am. Ceram. Soc.*, **73**, 1014-23 (1990).

[17] C.-C. Lin, A. Zangvil, and R. Ruh, Phase Evolution in SiC-Whisker-Reinforced Mullite/ZrO_2 Composite during Long-Term Oxidation at 1000 to 1350°C, *J. Am. Ceram. Soc.*, **83**, 1797-803 (2000).

THE ROLE OF CARBON IN PROCESSING HOT PRESSED ALUMINIUM NITRIDE DOPED SILICON CARBIDE

N. Ur-rehman*, P. Brown[°,*], L.J. Vandeperre*

* UK Centre for Structural Ceramics & Department of Materials, Imperial College London, London SW7 2AZ, UK
° Defence Science and Technology Laboratory, Porton Down, Salisbury, SP4 0JQ, UK

ABSTRACT

In this study the effect of carbon, added as a phenolic resin, on the processing, microstructure and properties on silicon carbide after hot pressing was investigated. Mixtures of an α-silicon carbide (SiC) powder, an aluminium nitride (AlN) powder and phenolic resin were prepared by ball milling and drying in a rotary evaporator. Variations in the green density of cold pressed powders with uni-axial pressure were analysed to determine the effect of the resin on the pressing characteristics of the powder. The microstructures after vacuum hot pressing were studied by optical and scanning electron microscopy. Variations in hardness with applied load were determined using nano-, micro- and macro-hardness measurement equipment.

It is shown that the phenolic resin acts as a binder which enhances the green strength of cold pressed powders. It also lowers their green density at low applied pressures by binding powders together, hindering effective packing. During vacuum hot pressing, the evolution of vacuum strength is consistent with the reduction of silica on the surface of SiC powders by carbon formed during resin pyrolysis. It is found that carbon also stops exaggerated coarsening of the microstructure despite more complete transformation from the 6H to the 4H SiC polytype. The hardness of samples processed without additional carbon is strongly dependent on applied load. This is not the case for samples with additional carbon.

INTRODUCTION

Producing dense SiC by sintering had long proved difficult until Prochazka's[1] discovery that simultaneous addition of B and C to SiC leads to a high sintered density (>97%). Later work showed that adding Al and C also promotes good densification behaviour. This resulted in the development of ABC SiC to which Al, B and C are added[2]. Other developments include adding oxide mixtures to promote the formation of a liquid phase[3-6].

The use of AlN as a sintering additive to SiC appears to have originated in developments to reproduce the versatility of the Si-Al-O-N system[7-9], by making equivalent SiCAlON compositions[2]. The motivation for the work was the realisation that hexagonal SiC of the 2H polytype can incorporate substantial amounts of AlN and Al₂OC. This was further helped by the discovery that mixtures of SiC and Al₂OC could be sintered without the aid of pressure owing to a liquid phase forming between Al₂O₃ and Al₂OC.

Some authors have used AlN as an alternative to Al₂O₃ in additive systems yielding liquid phase sintering[10,11,12], but other workers have focused on producing a solid solution of SiC and AlN. For example, Rafaniello et al.[13] produced SiC-AlN solid solution powders and investigated their densification by hot pressing, and Zangvil and Ruh[14] confirmed that AlN dissolves as AlN into SiC and created a tentative phase diagram for the binary SiC-AlN system. In a recent extensive study on the influence of different additives by Ray and Cutler[15], SiC was hot pressed to high density using only AlN as an additive.

Surprisingly the combination of AlN and carbon without further additions was not investigated. Therefore, in this paper, the effect of adding carbon as well as AlN to SiC to promote sintering is investigated.

EXPERIMENTAL

Powder mixtures of α-SiC (H.C. Starck, Grade UF-15), AlN (H.C. Starck, Grade C) and a phenolic resin (CR-96 Novolak, Crios Resinas, Brazil) were prepared by ball milling for 24 h in methyl ethyl ketone using Si_3N_4 milling media (Union Process Inc., USA) followed by drying in a rotary evaporator (Rotavapor R-210/215, Büchi, Germany). The characteristics of the powders are shown in Table I. The phenolic resin was added to the mixtures as a source of carbon and the level of addition is therefore expressed as the weight percentage carbon yield, where the carbon yield was taken as 61 wt.%. To ascertain whether hard agglomerates form during drying, the cold densification behaviour of the mixtures was determined by uni-axial pressing in a cylindrical die (Ø 13 mm) to different pressures. Prior to hot pressing, powder containing resin was heated under protection of Ar gas to 600 °C to pyrolise the resin.

Table I. Producer's data for oxygen content, specific surface area, pressed green density and powder size distribution.

	α-SiC	AlN
Grade	H.C. Starck UF-15	H.C.Starck C
Oxygen content (wt.%)	1.5	0.1
Specific surface area ($m^2 g^{-1}$)	14-16	4-8
Green density at 100 MPa (g cm^{-3})	1.55-1.75	
Particle size (μm)		
D 90%	1.20	2.30-4.50
D 50%	0.55	0.80-1.80
D 10%	0.20	0.20-0.35

Hot pressing was carried out in a KCE hot-press equipped with a pyrometer (Impac 140) and a Mitutoyo micrometer sensor to follow the ram displacement in order to estimate the density evolution during processing. Approximately 5 g of the powder mixture was placed in a graphitic die (Ø 30 mm). A minimum load of 6 kN was applied before starting the hot pressing programme reported in Table II, which was based on the work of Ezis et al.[16]. To enable comparison with data by Ray et al.[15], an additional hot pressing experiment was carried out using continuous heating at approximately 50 °C min^{-1} to the sintering temperature of 2045 °C, i.e. without dwelling at intermediate temperatures. These samples were held at the sintering temperature for 1 h.

Ground and polished cross-sections were prepared for observation of microstructures using optical microscopy. The polytype content of the SiC was determined from x-ray diffraction measurements using the method developed by Ruska et al.[17]. The x-ray diffraction was carried out using Cu Kα-radiation on a powder diffractometer (PW 1729, Phillips, The Netherlands).

Hardness measurements were carried out using loads ranging from 50 mN to 50 N. The low load indentations (50 mN-500 mN) were made using a Berkovich pyramidal indenter attached to a depth sensing nano-indenter (Nanotest, Micromaterials, UK). Intermediate load indentations (5 N-20 N) were made using a Berkovich indenter attached to the depth-sensing micro-indenter on the same indentation platform. The data was corrected for thermal drift and analysed using the method developed by Oliver and Pharr[18]. Indents, which were found to have been centred on a pore or on the edge of a pore were discarded. Macroscopic indentations (~50N) were made with a Vickers indenter mounted on a standard macro-indentation instrument (Indentec, Zwick). For the latter, the hardness was determined by measuring the size of the indentations using an optical microscope. Vickers hardness is defined as force over total contact area, whereas in nano-indentation instruments hardness is defined as force over projected area. To remove this inconsistency, the hardness values obtained from Vickers indentation were recalculated to force over projected area by dividing through 0.9272[19].

Table II. Heating rates, dwell times and applied pressures for most of the hot pressing experiments.

Step	Heating rate ($°C \, min^{-1}$)	Dwell (min)	Pressure (MPa)
25°C - 388°C	n/a		8.5
388°C -1630°C	10	30	8.5
1630°C - 1820°C	10	30	14.2
1820°C - 1895°C	5	30	21.2
1895°C - 2045 °C	5	30	21.2
2045°C - 25°C	-25	n/a	21.2

RESULTS AND DISCUSSION

The variation of the green density of uni-axially cold pressed SiC powders, with different resin contents, versus applied pressure is shown in Figure 1. At low applied pressures the density decreases with increasing resin content, which is an indication that as more resin is added some agglomeration occurs. However, as the pressure is increased, the agglomerates can be broken up and the density of samples with resin is now higher because the resin lubricates the particle rearrangements and fills the spaces between the particles. Moreover, without the resin present end-capping was observed from 200 MPa onwards, whereas no damage was observed for pressures in excess of 700MPa when resin was present confirming that the resin acts as a lubricant and binder.

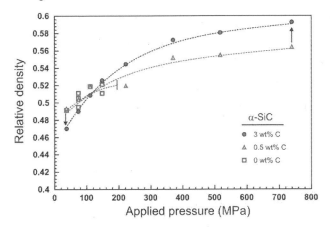

Figure 1. Relative green density of uni-axially cold pressed SiC powders, with different resin contents, versus applied pressure. The arrows indicate the trends from 0 wt.% to 3 wt.% carbon. At lower pressures, higher resin contents lead to lower densities. At higher pressures this trend is reversed.

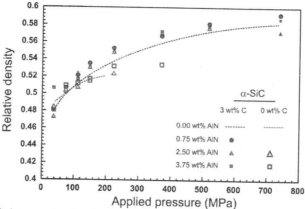

Figure 2. Relative green density of uni-axially cold pressed SiC powders, with different AlN and carbon contents, versus applied pressure.

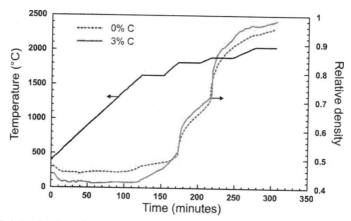

Figure 3. Relative density and temperature versus time during hot pressing of samples with 3.75 wt.% AlN and 0 wt.% or 3 wt.% carbon.

Figure 2 illustrates that the small amounts of AlN added to the mixtures do not alter the pressing characteristics of the SiC powder. Hence, adding carbon as a phenolic resin enhances the processability of the mixtures.

The evolution of the relative density with temperature during hot pressing is illustrated with data for SiC with 3.75 wt.% AlN with and without 3 wt.% C in Figure 3. In both cases a high density (>98%) was achieved and overall the densification traces are very similar: every time the pressure is increased a spur of densification ensues followed by slower densification during the intermediate and final temperature dwells.

The evolution of the pressure inside the hot pressing vacuum chamber, however, differs substantially, Figure 4. Whereas for samples without resin the pressure in the vacuum remains constant up to the first temperature hold at 1630 °C, the pressure in the vacuum chamber increases twice for samples containing resin. The first instance occurs at temperatures close to 600 °C, Figure 4 – peak (a), and can be attributed to further pyrolysis of the resin. The vacuum deteriorates again near 1500 °C, Figure 4 - peak (b), and this can be attributed to CO/CO_2 evolution from the reduction of the surface silica on SiC by the carbon of the resin[20]. From then onwards, the pressure signal for samples with or without carbon is very similar. The decreases in pressure are the result of further evacuation by the vacuum pump during isothermal dwells, whereas the increases in pressure are a result of gas expansion and continued degassing owing to the increase in temperature. At the final sintering temperature, some volatilisation of SiC might also be contributing to the increase in pressure.

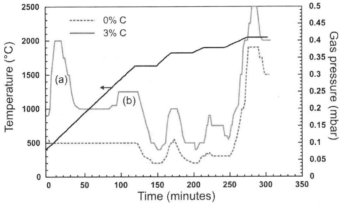

Figure 4. Evolution of the vacuum and temperature versus time during hot pressing of α-SiC with 3.75 wt.% AlN and 0 wt.% or 3 wt.% carbon.

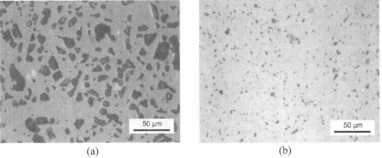

(a)   (b)

Figure 5. Optical micrographs of the microstructures obtained after hot pressing α-SiC with 3.75% AlN containing (a) no carbon and (b) 3 wt% carbon.

Optical microscopy reveals that in the absence of carbon a coarse microstructure forms, which consists of elongated SiC platelets and large pores, Figure 5a. In samples with 3 wt.% carbon the residual porosity is much smaller, Figure 5b. The coarsening of SiC microstructures during solid state sintering without carbon is in agreement with observations elsewhere that grains remain finer when more carbon is added[1, 21, 22]. Clegg[20] has shown that one role of the carbon is to reduce the silica on the surface of the SiC before self-reduction by the underlying SiC can occur. This is crucial because self-reduction leads to the formation of large pores. A second role of carbon, when added in slight excess as is the case here, is that carbon limits grain growth because carbon inclusions pin the grain boundaries[21]. Hence, much finer microstructures can be produced by adding carbon than without it.

Fitting of the x-ray diffraction patterns obtained after sintering, Figure 6, showed that the α-SiC powder, which consisted almost entirely of 6H-SiC (96%) before sintering, had extensively transformed into 4H-SiC: the samples with carbon consisted of 4H-SiC alone whereas the sample containing no carbon was 86% 4H-SiC and 14% 6H-SiC.

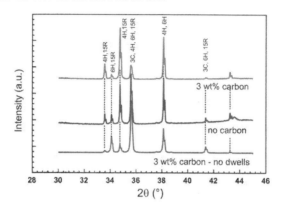

Figure 6. X-ray diffraction patterns for SiC with 3.75 wt.% AlN and either 3 wt.% carbon or no added carbon. To make the differences between the patterns clearer, the data is normalised against the highest count observed.

The extensive transformation towards 4H in the presence of AlN is consistent with the tentative phase diagram by Zangvil and Ruh[14] who for 3.75 wt.% AlN predict a 4H-polytype alone. It should be noted that the extent of the transformation observed here is much higher than in the work of Ray et al.[15], who hot pressed α-SiC with 2.5 wt.% and 5 wt.% AlN and found that about 80% of the 6H remained after sintering. However, in their work no intermediate temperature dwells were used, i.e. the SiC was heated directly to the sintering temperature. Hence, it is possible that the much larger degree of transformation observed here occurs during the dwells at 1820°C and 1895 °C, where 4H-SiC is reported to be the stable polytype for pure SiC[14, 23]. To confirm this interpretation, a sample with 3 wt.% carbon was hot pressed by heating to 2045°C without dwells at intermediate temperatures. The x-ray diffraction pattern for this sample is also shown in Figure 6, and fitting showed that the residual 6H content in this sample was 78%, which is in agreement with the level of transformation observed by Ray et al.[15] for continuous heating to the sintering temperature. This confirms that the more extensive transformation observed here can be attributed to the intermediate holds. This finding supports the proposal by Jepps that a combination of a solid state transformation mechanism with transport along the surface aids faster polytype transformation[23].

The results of hardness measurements at different applied load are presented in Figure 7. For low load indentations, the hardness of SiC without additional carbon is higher than the hardness of samples with carbon additions. Secondly, the hardness decreases with applied load for all types of material tested. However, it is clear that the decrease in hardness is more pronounced for the SiC where no carbon was added than for the two types to which 3 wt% carbon was added.

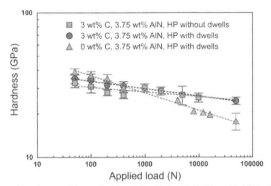

Figure 7. Variation of hardness with applied load for SiC with 3.75 wt.% AlN with or without 3 wt.% carbon processed using the hot pressing schedule shown in Table 2 (with dwells), and for SiC with 3.75 wt.% AlN and 3 wt.% carbon processed by heating continuously to the hot pressing temperature (without dwells).

The high hardness (~35 GPa) at low loads is consistent with estimates for the hardness of dense SiC (36 GPa)[24]. The higher hardness for the SiC without additional carbon for low applied loads and hence small indentations can therefore be explained by the presence of large, dense, grains of SiC. When small indents are made in the central regions of such grains, the porosity does not influence the hardness obtained. In contrast, when carbon is added, the refinement of the microstructure causes much finer, more evenly distributed porosity. Therefore the porosity influences the hardness even for the smaller indentations made.

As the load is increased, the size of the indents increase and even in the coarse microstructure, the effect of porosity on hardness becomes noticeable. At the same time the increase in load also causes the formation of cracks and crushing of the material underneath the indentation; two mechanisms which contribute to a further decrease in hardness. The coarse microstructure of the SiC with no added carbon is clearly less resistant to such damage mechanisms, which explains the lower hardness compared to the SiC samples to which carbon was added for larger applied loads. Hence, while the refinement of the microstructure obtained by adding carbon in the form of a resin causes the porosity to influence the measured hardness for even small indentations, this type of microstructure can retain its hardness better as the scale of the contact is increased and hence will resist contact damage better in a wider range of applications.

CONCLUSIONS
Addition of a phenolic resin to SiC powder mixtures containing up to 3.75 wt.% AlN improves their cold pressing behaviour by acting as a binder and plasticiser. When pyrolysed at 600°C it also chars to give a carbon yield of up to 61 wt.% which facilitates reduction of silica on the surface of SiC powders during vacuum hot pressing. This suppresses the formation of large lath-like SiC grains, promoting a refined, equiaxed microstructure with a hardness that is not strongly load dependant.

ACKNOWLEDGEMENTS
Naeem Ur-rehman and Luc Vandeperre thank the Defence Science and Technology Laboratory (Dstl) for providing the financial support for this work under contract No Dstlx-1000015398.

© Crown copyright 2010. Published with the permission of the Defence Science and Technology Laboratory on behalf of the Controller of HMSO

REFERENCES
1. S. Prochazka and R.J. Charles, Strength of Boron-doped hot-pressed SiC. *Ceramic Bulletin*, **52**, 885-891 (1973).
2. T.B. Jackson, A.C. Hudford, S.L. Bruner, and R.A. Cutler, SiC based ceramics with improved strength, in *Silicon carbide '87*, J.D. Cawley and C.E. Semler, Editors. 1989, American Ceramic Society: Westerville. p. 227-240.
3. F.K.v. Dijen and E. Mayer, Liquid phase sintering of SiC. *Journal of the European Ceramic Society*, **16**, 413-420 (1996).
4. R.A. Cutler and T.B. Jackson, Liquid phase sintered SiC. *Ceramic Materials and Components for Engines*, **42**, 309-318 (1994).
5. L.S. Sigl and H.J. Kleebe, Core/Rim Structure of Liquid-Phase sintered SiC. *Journal of the American Ceramic Society*, **76**, 773-776 (1993).
6. A. Kerber and S.V. Velken, Liquid phase sintered SiC : production, properties and possible applications. *Fourth Euro Ceramics*, **2**, 177-184 (1995).
7. S. Hampshire, R.A.L. Drew, and K.H. Jack, Viscosities, Glass Transition Temperatures, and Microhardness of Y-Si-Al-O-N Glasses. *Journal of the American Ceramic Society*, **67**, C46-C47 (1984).
8. S. Hampshire, R. Flynn, V. Morrisey, M. Pomeroy, T. Rouxel, J.L. Besson, and P. Goursat, Effects of temperature on viscosities and elastic modulus of oxynitride glasses. *Key Engineering Materials*, **89-91**, 351-356 (1994).
9. J. Vleugels, L. Vandeperre, and O. Van der Biest, Influence of alloying elements on the chemical reactivity between Si-Al-O-N ceramics and iron-based alloys. *Journal of Materials Research*, **11**, 1265-1275 (1996).
10. J. Lee, H. Tanaka, and H. Kim, Formation of solid solutions between SiC and AlN during liquid-phase sintering. *Materials letters*, **29**, 1-6 (1996).
11. R. Huang, H. Gu, Z.M. Chen, T.H. Shouhong, and D.L. Jiang, The sintering mechanism and microstructure evolution in SiC-AlN ceramics studied by EFTEM. *International Journal of Materials Research*, **97**, 614-620 (2006).
12. M. Flinders, D. Ray, A. Anderson, and R.A. Cutler, High-Toughness Silicon Carbide as Armour. *Journal of the American Ceramic Society*, **88**, 2217-2226 (2005).
13. W. Rafaniello, K. Cho, and A.V. Virkar, Fabrication and characterization of SiC-AlN alloys. *Journal of Materials Science*, **16**, 3479-3488 (1981).
14. A. Zangvil and R. Ruh, Phase relationships in the silicon carbide - aluminium nitride system. *Journal of the American Ceramic Society*, **71**, 884-90 (1988).
15. D.A. Ray, S. Kaur, R.A. Cutler, and D.K. Shetty, Effects of additives on the pressure-assisted densification and properties of silicon carbide. *Journal of the American Ceramic Society*, **91**, 2163-2169 (2008).
16. A. Ezis, Monolithic, fully dense silicon carbide material, method of manufacturing and end uses. 1994, Cercom Inc: USA.
17. J. Ruska, L.J. Gauckler, J. Lorenz, and H.U. Rexer, The quantitative calculation of SiC polytypes from measurement of X-ray diffraction peak intensities. *Journal of Materials Science*, **14**, 2013-2017 (1979).
18. W.C. Oliver and G.M. Pharr, An Improved Technique for Determining Hardness and Elastic Modulus using Load and Displacement Sensing Indentation Experiments. *J. Mater. Res.*, **7**, 1564-1583 (1992).
19. D. Tabor, The Physical Meaning of Indentation and Scratch Tests. *Br. J. Appl. Phys.*, **7**, 159-166 (1956).
20. W. Clegg, Role of carbon in the sintering of boron-doped silicon carbide. *Journal of the American Ceramic Society*, **83**, 1039-1043 (2000).
21. E. Ermer, P. Wieslaw, and S. Ludoslaw, Influence of sintering activators on structure of silicon carbide. *Solid state Ionics*, **141-142**, 523-528 (2001).
22. W. Böcker and H. Hausner, The influence of boron and carbon additions on the microstructure of sintered a-SiC. *Powder Metallurgy International*, **10**, 87-89 (1978).
23. T.F. Page, Silicon Carbide : structure and polytypic transformations. *The physics and chemistry of carbides, nitrides & borides*, **185**, 197-214 (1990).
24. A. Slutsker, A.B. Sinani, V.I. Betekhtin, A.G. Kadomtsev, and S.S. Ordan'yan, Hardness of microporous SiC ceramics. *Technical physics*, **53**, 1591-1596 (2008).

Characterization

MICROSTRUCTURE AND HIGH-TEMPERATURE PROPERTIES OF SI-B-C-N MA-POWDERS AND CERAMIC

Yu Zhou [1,*], Zhi-Hua Yang[1], De-Chang Jia[1], Xiao-Ming Duan[2,1]
1. Institute for Advanced Ceramics, Harbin Institute of Technology, Harbin 150001, China
2. National Key Laboratory for Precision Hot Processing of Metals, Harbin Institute of Technology, Harbin 150001, China

ABSTRACT:
Nano-sized Si-B-C-N powders nearly amorphous was prepared by mechanical alloying (MA) technique using cubic silicon (c-Si), hexagonal boron nitride (h-BN) and graphite (C) as starting materials, then consolidated at 1900 °C under a pressure of 40 MPa for 30 min in N_2 by hot-pressing. The MA Si-B-C-N powders and ceramics show good oxidation resistance. The weight gain of the Si-B-C-N powder was only 12.3 wt% till 1400 °C in air for TG analysis. And it was found that the existence of B-C-N bond in the MA Si-B-C-N powder and ceramics could improve their oxidation resistance. For the Si-B-C-N ceramics, dense and coherent oxide layer could be formed after being oxidized at high temperature and very good oxidation resistance was shown. When oxidized at 1200 °C for 85 h in dry atmosphere for the ceramic, a oxidation layer of 9.6 μm in thickness formed. When oxidized at 1050 °C for 85 h in humid atmosphere, however, the oxidation layer reached 20.2 μm in thickness.

1. INTRODUCTION

During the last two decades, non-oxide ceramics, mainly in the system of Si-B-C-N, derived from organo-elemented precursors has aroused ever increasing attention. These kinds of materials are prepared by thermolysis of pre-ceramic polymers [1-3], and the resulting ceramics are free of sintering additives and exhibit outstanding high temperature stability [4-6] and superior high-temperature compression creep property [7-9]. But the overwhelming majority of synthesis and pyrolysis process for the precursors need to be done in inert gases and the corresponding processes are very complicated. In addition, the starting materials are very expensive. So it is necessary to find another route of synthesis for Si-B-C-N powders and ceramics.

Powder processing by mechanical alloying (MA) has attracted wide practical interest as it offers a simple but powerful way to synthesize non-equilibrium phases and microstructures, from nanograin materials to extended solid solutions, amorphous phases, chemically disordered compounds, and nanocomposites [10]. Earlier work of us established that nano-sized Si-B-C-N powders can be produced by two routes of MA using c-Si, h-BN and graphite powders as starting materials [11]. The two routes are one-step milling and two-step milling, respectively. The one-step milling process is raw materials being simultaneously milling for 20 hours, and two-step milling process is Si and graphite powders (with a mole ratio of 1:1) are firstly milled for 15 hours, and then milled with BN and the residual graphite for 5 hours. The Si-B-C-N ceramics have been consolidated to be a high density by hot pressing (HP) [12] or spark plasma sintering (SPS) [13]. Some mechanical and thermal properties of Si-B-C-N ceramics, such as flexural strength, fractural toughness, thermal diffusivity and thermal expansion coefficients, were reported for the first time [12,13]. For the $SiB_{0.5}C_{1.5}N_{0.5}$ fabricated by one-step milling and HP, the flexural strength was 312.8 MPa. And for the $SiB_{0.5}C_{1.5}N_{0.5}$ fabricated by two-step milling and HP, the flexural strength was 423.4 MPa. At high temperature, the strengths of the former reduced 8.2 percent (287.2MPa) and 28.0 percent (225.3MPa) at 1000 °C and 1400 °C, respectively. The reduction rates of strengths for the latter were 11.3 percent (375.4 MPa) and 38.5 percent (260.4 MPa), respectively. For the Si_3BC_4N ceramics fabricated by two-step milling and SPS, the flexural strength, fracture toughness, elastic modulus and Vickers' hardness values was up to 501.5 MPa, 5.64 MPa·m$^{1/2}$, 157 GPa and 5.92GPa, respectively.

Si-B-C-N ceramics had the lowest reported oxidation rate of any non-oxide material known to date. But the oxidation resistance of Si-B-C-N powders and ceramics may be influenced by the composition and crystalline state. The preliminary results of the oxidation research for the Si-B-C-N powders have been reported [11]. And the results indicated that the oxidation resistance of Si-B-C-N powders may be influenced by the content of B-C-N bonds. In this paper, the further research of the oxidations of the Si-B-C-N powders, fabricated by one-step milling will be reported. And oxidation behavior of hot-pressed Si-B-C-N ceramics in different temperatures, times and atmospheres will be presented, too.

2. EXPERIMENTAL PROCEDURE

2.1 Preparation of Si-B-C-N powder and ceramics

As for the Si-B-C-N powders, cubic silicon (c-Si, 99.5% in purity, 45 μm), hexagonal boron nitride (h-BN, 98% in purity, 0.6 μm) and graphite (C, 99.5% in purity, 8.7 μm) were blended in mole ratio of Si:B:C:N being 1:0.5:1.5:0.5 and then subjected to high-energy ball milling in a shaker mill (model ABQ-045-I, Changjiang Anwei Technology Development CO.LTD, Changchun, China) with a ball-to-powder ratio of 20:1. The milling time is 20 h. In order to avoid the increase of temperature in vessel, the milling stopped for ten minutes at intervals of 1 hour for milling. All transfers of powders to and from the vessel were done in a glovebag. Si-B-C-N powders mainly contained amorphous Si-B-C-N, nano-crystalline (4~5 nm) 3C-SiC and 6H-SiC. In addition, about 4.7 percent Si remained in the powders [11].

Sintering was performed by HP at 1900 °C under a pressure of 40 MPa for 30 min in N_2 (1.0×10^5 Pa). Nano-sized 6H-SiC, 3C-SiC and BCN were the main phases for Si-B-C-N ceramic. A little amorphous structure was remained in the ceramic, and a lot of distorted structure lies in the SiC and BCN phase. In addition, a little of ZrO_2, which came from the contamination of ZrO_2 balls and inner surface of vessel, was contained in Si-B-C-N ceramic.

2.2 Characterization methods

Thermogravimetric analysis (TGA) of the milled powders was performed in Al_2O_3 crucible (STA-449C, Netzsch) by heating up to 1400°C at a heating rate of 4, 6, 10, and 12°C/min in air, respectively. Identification of the gaseous species produced during the oxidation was achieved through a continuous process using thermo-gravimetric-mass spectrometry analysis (TG-MS, Model TGA92, Setaram, Caluire, France) coupled to a quadrupole mass spectrometer in a flowing gas atmosphere containing 19.8% oxygen and 80.2% argon; this way the detection of the mass fragment with m/z = 14 can unambiguously assigned to N^+ which released by the oxidation of powders. The maximum temperature was 1000°C for the TG-MS analysis, and the heating rate was 10 °C/min.

Samples of Si-B-C-N ceramic for oxidation were 5×4×3mm^3. Samples were ultrasonically cleaned successively in de-ionized water, acetone and alcohol prior to exposure. Initial sample weights (to an accuracy of 0.0001g) were recorded. Samples were loaded on a Fe-Cr alloy net fixed into a silica boat to prevent the conglutination between the oxidation produce and boat. And then silica boat with samples was set into a silica tube and heated up to 900 °C, 1050 °C, and 1200°C for 85 h. Gas flow was switched from Ar to air when the temperature attained the desired value. Two oxidation environments were investigated: dry and humid air. Dry air was passed into the tube with the flow rate of 51.81cm^3/min after removal of moisture by sequent passage through CaCl and P$_2$O$_5$ gas drying towers. For the humid environment, the atmosphere consisted of air that was saturated with water by bubbling air at 51.81cm^3/min through boiling deionized water.

Scanning electron microscopy (SEM) was performed using a Hitachi model S4800. The surface of ceramics before and after oxidized was analyzed on X-ray diffractometer (XRD, Rigaku D/Max-rB, Tokyo, Japan) with Cu-Kα radiation. X-ray photoelectron spectroscopy (XPS) was introduced to analyze the chemical status of BN and C in the matrix on X-ray photoelectron spectrometer (ESCALABMK II, VG, UK) with an Al-Kα energy source. The signal curves were fitted by peak addition using Gaussian–Lorentzian peak approximations and Shirley background reduction.

3. RESULTS AND DISCUSSION

3.1 Oxidation behavior of the Si-B-C-N powder
(1) Weight change behavior

One of the most important properties of the Si-B-C-N ceramics is they had the lowest reported oxidation rate of any non-oxide material known to date [14]. Generally, low resistance to oxidation is a noticeable disadvantage of non-oxide ceramics. But the Si-B-C-N powders [15], fibers [16,17] and ceramics [18], all of which fabricated by precursors, had very low oxidation rates. So, TGA experiment in air atmosphere has been done for the Si-B-C-N powder. For comparison, oxidation resistance of commercial pure β-SiC and α-Si$_3$N$_4$ was investigated under the same experimental condition. Fig. 1 shows the oxidation resistance of SiC, Si$_3$N$_4$ and as-milled samples. It can be seen that the weight gain of SiC (0.5 μm) and Si$_3$N$_4$ (0.6 μm) powders reached 30.7 % and 20.3 % at 1400 °C, respectively. But for the nano-sized (about 50 nm) Si-B-C-N powder, the weight gain is 12.3 %, though a 1.97 % of weight loss happened at about 680 °C.

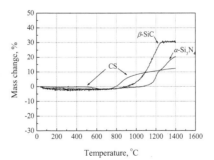

Fig.1 TG curves for the oxidation of pure SiC, Si$_3$N$_4$, and milled Si-B-C-N powders in flowing air

In the Si-B-C-N powders, Si-C, C-C, C-N, B-N, and B-C-N bonds simultaneously existed [11].

Therefore, the following oxide reaction may be occurred at high temperature:

$$C(s) + \tfrac{1}{2}O_2(g) \rightarrow CO(g) \tag{1}$$

$$C(s) + O_2(g) \rightarrow CO_2(g) \tag{2}$$

$$BN(s) + \tfrac{3}{2}O_2(g) \rightarrow \tfrac{1}{2}B_2O_3(s, \ l) + \tfrac{1}{2}N_2(g) \tag{3}$$

$$BCN(s) + \tfrac{7}{4}O_2(g) \rightarrow \tfrac{1}{2}B_2O_3(s, \ l) + CO_2(g) + \tfrac{1}{2}N_2(g) \tag{4}$$

$$SiC(s) + O_2(g) \rightarrow SiO_2(s,l) + C(s) \tag{5}$$

$$\tfrac{2}{3}SiC(s) + O_2(g) \rightarrow \tfrac{2}{3}SiO_2(s,l) + \tfrac{2}{3}CO(g) \tag{6}$$

$$\tfrac{1}{2}SiC(s) + O_2(g) \rightarrow \tfrac{1}{2}SiO_2(s,l) + \tfrac{1}{2}CO_2 \ (g) \tag{7}$$

$$SiC(s) + O_2(g) \rightarrow Si(s,l) + CO_2(g) \tag{8}$$

$$\tfrac{2}{3}SiC(s) + O_2(g) \rightarrow \tfrac{2}{3}SiO(g) + \tfrac{2}{3}CO_2(g) \tag{9}$$

$$SiC(s) + O_2(g) \rightarrow SiO(g) + CO(g) \tag{10}$$

$$2SiC(s) + O_2(g) \rightarrow 2Si(s,l) + 2CO(g) \tag{11}$$

$$2SiC(s) + O_2(g) \rightarrow 2SiO(g) + 2C(s) \tag{12}$$

The reactions (1), (2) and (4) led to weight loss, and the reaction (3) resulted in weight gain. Reactions from (5) to (12) were the possible oxidation reaction of SiC phase. The oxidation behavior of SiC can be divided into two categories: passive and active. Passive oxidation forms a coherent and dense SiO_2 layer on the surface, which drastically reduces oxidation rate. Active oxidation forms gaseous SiO, which dissipates away and the oxidation becomes severe [19]. Active oxidation occurs at high temperatures, but the temperature at which the oxidation changes from passive to active, decreases with the decrease of oxygen partial pressure. For most elevated-temperature application in normal atmosphere, SiC are oxidized to form SiO_2. Furthermore, the reaction (5) had the lowest free energy from room temperature to 1600K [20], so SiC phase was been oxidized to form SiO_2 and C for the Si-B-C-N powders. So we could conclude that the weigh loss came from the oxidation of the C and BCN below 680 °C. And the oxidation of the BN and SiC resulted in the weight gain above 680 °C.

(2) Oxidation activation energy

The heating rate is one of the important parameters in the DSC study. Higher heating rates will result in a shift of the exothermic/endothermic peaks to higher temperatures. Here we applied the peak temperatures at different heating rates to the Kissinger equation [21] as an example for illustrating the activation energies (E_a) changes in Si-B-C-N powders oxidation study. Fig. 2 shows four DSC curves obtained at different heating rates of 4, 6, 10, and 12 °C/min for Si-B-C-N powders. The results clearly showed that the oxidation was shifted to higher temperatures as expected. Two endothermic peaks at about 650 °C and 750 °C. We selected the legible peaks at higher temperature to calculate activation energy. Fig. 3 shows the Kissinger plots for these reactions. The calculated oxidation activation energy was 211.2 KJ/mol for Si-B-C-N powders.

(3) Determinant factor of the excellent oxidation resistance

Si-B-C-N ceramics have excellent high temperature phase stability should be contributed the present of the BCN phase [22,23]. So we think that the good oxidation resistance of powders should connect with B-C-N bonds formed during the high-energy ball milling. To invest the influence of the BCN phase on the oxidation resistance of Si-B-C-N powder, the oxidation resistance of high-energy ball milling BCN powders was tested by TG (as shown in Fig. 4). The mixture of the BN and C was milled for 5 h and 15 h, which were denoted as BCN5 and BCN15, respectively. Weight loss of the BCN15 decreased 6.2% than that of BCN5 at about 800 °C in air. And the high temperature stability of BCN15 was also higher than that of BCN5 from 800 °C to 1400 °C.

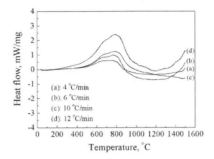

Fig. 2 DSC curves of at different heating rate of Si-B-C-N powders

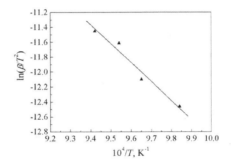

Fig.3 The Kissinger plots for the oxidation reactions

To investigate bonding state of MA BCN powders, XPS analysis was carried out. Fig. 4 gives the deconvolution of the corresponding chemical binding state of each element in the BCN5 and BCN15. Quantitative analysis on each element at different chemical state was carried out by following elemental sensitivity factors: $S_B = 0.13$, $S_C = 0.25$, and $S_N = 0.42$. The results are listed in Table 1. As expected, the deconvolution result revealed the coexistence of chemical bonding formed by reactions of element of B, C, and N.

B1s spectrum of BCN5 and BCN15 were deconvoluted into two distinguishable peaks, respectively (see Fig. 4 (a)). We ascribe the component centered at 190.0 ~ 190.8 eV to B-N bonds of h-BN, which accord with the results of carbon-boron nitrogen fabricated by high-energy ball milling [24]. The components with higher binding energy of 191.7 eV for sample BCN5 and 191.1 eV for sample BCN15 are attributed to B-C-N bonds [24]. Fig. 4 (b) shows the C1s spectra of as-milled samples. A distinct change occurred for C1s spectrums after ball milling treatment. C1s spectra could be deconvoluted into three peaks. The peaks at the about 284.5 eV and 287 eV correspond to the C-C and C-N bonds [25], respectively. The peaks at the highest binding energy correspond to C-O bonds [25]. The XPS analysis of the N1s spectrum of as-milled powders BCN5 and BCN15 is shown in Fig. 4 (c). N1s spectrum is deconvoluted into two distinguishable peaks. Except the B-N bonds, the peaks at about 399.3 eV are similar to the experimental results of Ulrich [26] and they think that the peaks were C-B-N bonds. After the high-energy ball milling, C-B-N and C-N bonds presented and their contents increased with the increase of the milling time (as shown in Table 1).

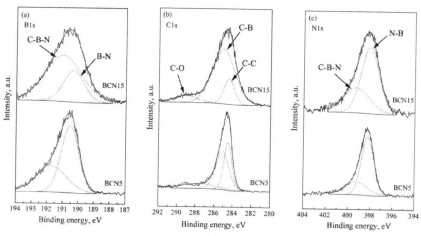

Fig.4 XPS spectra of BCN powders fabricated by different times

The MS curves of BCN5 and BCN15 was also shown in Fig. 5. It can be seen that CO, N_2, CO_2, NO, and NO_2 were detected at the weight loss stage. And the ion current for C and CO_2 decreased, and N, NO, and NO_2 increased at the weight gain stage. In addition, the weight gain of BCN powders was less then their origin weight, so the main weight gain of Si-B-C-N powders should be attributed to the oxidation of SiC phase. The analysis of TG-MS and XPS of BCN samples clearly demonstrated that the content of B-C-N bond had an important effect on the oxidation resistance of Si-B-C-N powders.

3.2 Oxidation behavior of the Si-B-C-N ceramic
(1) Oxidation behavior of the Si-B-C-N ceramic in dry air
Fig. 6 shows weight change of the Si-B-C-N ceramics at elevated temperatures for a 85-h oxidation period in different environments. The weight change of the Si-B-C-N ceramic was small in dry air. For all samples, a little weight loss was observed at the start of the oxidation period, and then the weight loss increased with the increase of the oxidation temperatures. After oxidized for less than 2 hours, the weight gain happened. And the range of weight gain decreased with the increase of the oxidation time.

Table 1 Results of deconvolution and relative concentration of different binding state

Element line	Bonding type	E_B, eV		Relative intensity, %	
		BCN5	BCN15	BCN5	BCN15
B1s	B-N	190.5	190.3	10.9	5.0
	C-B-N	191.7	191.1	7.4	12.8
C1s	C-C	284.5	284.5	23.4	10.6
	C-N	285.1	285.1	30.5	47.9
	C-O	287.7	289.0	9.8	6.0
N1s	N-B/N-C	398.2	398.1	12.8	11.5
	C-B-N	399.3	399.3	5.2	6.3

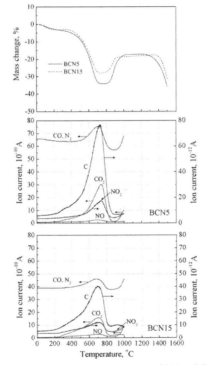

Fig. 5 TG and MS curves for the oxidation of the mixture of BN and C milled for 5 h and 15 h

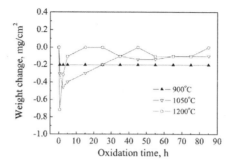

Fig. 6 Weight changes of the Si-B-C-N ceramic as a function of oxidation time in dry air

According the foregoing of the oxidation reaction, the weight loss of the Si-B-C-N ceramics should be attributed to the reactions (1), (2) and (4), and the weight gains were due to the formation of the SiO_2 and B_2O_3 (namely, the results of the reactions of (3) and (5)). The solid oxide products were SiO_2, B_2O_3, and C. SiO_2 and B_2O_3 can react to form a stable borosilicate glass above 372 °C [27], which can form a protective layer which reduces oxidation rate. But B_2O_3 was easy to vapor at high temperature, especially above 900 °C. From X-ray diffraction profiles obtained from the surface of the Si-B-C-N ceramic after oxidation in different temperature, BCN peaks disappeared, and C, SiO_2, SiC peaks were observed (as shown in Fig. 7). And the intensities of C and SiO_2 increased with the increase of the oxidation temperature and time.

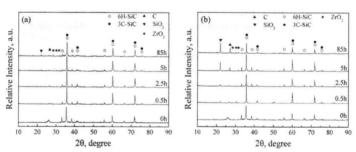

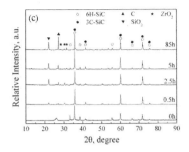

Fig. 7 XRD patterns of the Si-B-C-N ceramics oxidized at (a) 900 °C, (b) 1050 °C, (c) 1200 °C in dry air

The oxidation resistance of the BCN phase is higher than the BN phase, but they were easier to oxidize than SiC phase. So, it is can be seen that the pores formed by the oxidation of BCN phase at the beginning of the oxidation (as shown in Fig. 8). At this time, the amount of SiO_2 is not sufficient to form a dense and coherent layer on the surface of Si-B-C-N ceramics. The amount of pores increased with the increase of the oxidation temperature. Fig. 9 shows SEM images of the oxide scale formed on the Si-B-C-N ceramic after oxidized for 85 h at different temperature. It can be noticed that the oxide layer of the Si-B-C-N oxidized in dry air is dense and coherent. SEM images of fracture sections of the ceramic are shown in Fig. 10. After oxidized at 900 °C for 85 h, the thickness of the oxidized layer was not obvious to observe. And then the thickness of dense oxidized layer increased up to 9.6 μm for oxidized for 85 h at 1200 °C in dry air.

(2) Oxidation behavior of the Si-B-C-N ceramic in humid air

At high temperature, H_2O may be the dominant oxidant of the Si-based materials [28]. Not only are the SiC and BCN oxidation reactions as following accelerated in the presence of water vapor, but reactant diffusivities through the subsequently formed borosilicate glass are high, thereby further increasing the oxidation rate of the ceramic.

$$SiC(s) + 3H_2O(g) = SiO_2(s) + CO\ (g) + 3H_2(g) \tag{13}$$
$$2BCN(s) + 2H_2O(g) = B_2O_3(l) + 2CO(g) + N_2(g) + 2H_2(g) \tag{14}$$
$$B_2O_3(l) + H_2O(g) = 2HBO_2(g) \tag{15}$$
$$B_2O_3(l) + 3H_2O(g) = 2H_3BO_3(g) \tag{16}$$
$$SiO_2(s) + H_2O(g) = H_2SiO_3(l,g) \tag{17}$$
$$C(s) + H_2O(g) = CO(g) + H_2(g) \tag{18}$$

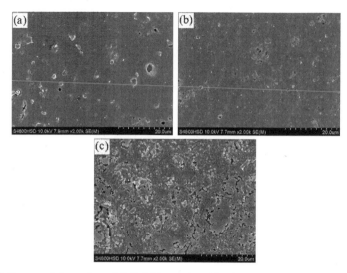

Fig. 8 SEM images of the surface of Si-B-C-N ceramic oxidized at (a) 900 °C, (b) 1050 °C, (c) 1200 °C for 2.5 h in dry air

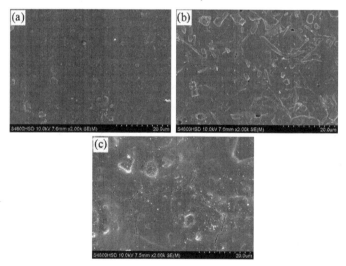

Fig. 9 SEM images of the surface of Si-B-C-N ceramic oxidized at (a) 900 °C, (b) 1050 °C, (c) 1200 °C for 85 h in dry air

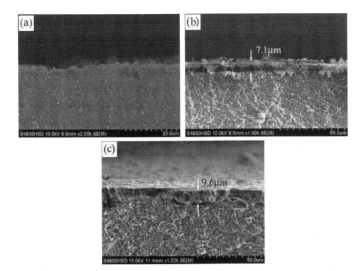

Fig.10 SEM images of the oxidation cross section of the Si-B-C-N at (a) 900 °C, (b) 1050 °C, (c) 1200 °C for 85 h in dry air

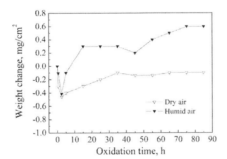

Fig. 11 Weight changes of the Si-B-C-N ceramic as a function of oxidation time at 1050°C in different atmospheres

Besides the accelerated oxidation, excessive amounts of gaseous reaction products resulting in highly defective, nonprotective, surface products. So it can be seen that weight gains of ceramic oxidized in humid atmosphere was higher than those oxidized in dry air (as shown in Fig 11).

Fig. 12 shows SEM images of the oxidation surface and cross section of the Si-B-C-N ceramics oxidized in different atmosphere for 85 h. The oxidation surface of Si-B-C-N ceramics was not as smooth as these formed in dry air, and a few rough areas presented on the surface of Si-B-C-N ceramic after oxidized for 85 h (as shown in Fig. 12 (b)). The thickness of oxidized layer formed in humid air was about three times as much as that formed in dry air, and the oxidized layer became loose and porous under influence of H_2O (as shown in Fig. 12 (c)).

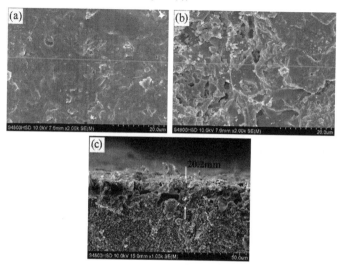

Fig.12 SEM images of the oxidation surface and cross section of the Si-B-C-N ceramics at 1050 °C for 85 h in humid air

(a) surface cross section, (b) surface cross section, and (c) cross section

CONCLUSIONS:

(1) The oxidation resistance of Si-B-C-N powders was higher than SiC and Si_3N_4 powders, and its resistance depended on the content of B-C-N bond. The higher content, the better oxidation resistance was.

(2) SiO_2 and B_2O_3 formed dense and coherent oxide layer formed on the surface of Si-B-C-N ceramics at high temperature in air, which could effectively protect the ceramic to further oxidation.

(3) The thickness of the oxidation layer increased with the increase of the oxidation time and temperature. In dry atmosphere, the largest thickness was 9.6 μm after the Si-B-C-N ceramics were oxidized at 1200 °C for 85 h. But in humid air, H_2O aggravate the oxidation rate of the phases in ceramics, and the oxide layer become loose and porous. When the ceramic was oxidized at 1050 °C for 85 h, the thickness of the oxide layer was up to 20.2 μm, which was about three times as much as that formed in dry air.

ACKNOWLEDGEMENTS:

This work is supported by the Project Supported by Development Program for Outstanding Young Teachers in Harbin Institute of Technology (HITQNJS.2009.064).

REFERENCES:
1. Yun Tang, Jun Wang, Xiaodong Li, Wenhua Li, Hao Wang, Xiaozhou Wang. Thermal stability of polymer derived SiBNC ceramics. Ceram. Int. 2009, 35: 2871-2876.
2. R. Riedel, A. Kienzle, W. Dressler, L. Ruwisch, J. Bill, F. Aldinger. A silicoboron carbonitride ceramic stable to 2000°C. Nature, 1996, 382 (29): 796-798.
3. Y. Cai, S. Prinz, A. Zimmermann, A. Zern, W. sigle, M. Ruhle, F. Aldinger. Electron diffraction study of the local atomic arrangement of as-pyrolysed Si-B-C-N ceramics. Scripta Mater. 2002, 47: 7-11.
4. T. Jäschke, M. Jansen. Improved durability of Si/B/N/C random inorganic networks. J. Eur. Ceram. Soc. 2005, 25: 211-220.
5. Anita Muller, Achim Zern, Peter Gerstel, Joachim Bill, Fritz Aldinger. Boron-modified poly (propenysilazane)-derived Si-B-C-N ceramics: preparation and high temperature properties. J. Eur. Ceram. Soc. 2002, 22: 1631-1643.
6. P. Gerstel, A. Muller, J. Bill, F. Aldinger. Synthesis and High-Temperature Behavior of Si/B/C/N Precursor-Derived Ceramics without "Free Carbon". Chem. Mater. 2003, 15(26): 4980-4986.
7. Ravi Kumar, F. Phillipp, F. Aldinger. Oxidation induced effects on the creep properties of nano-crystalline porous Si–B–C–N ceramics. Mater. Sci. Eng. A. 2007, 445-446: 251-258.
8. B. Baufeld, H. Gu, J. Bill, F. Wakai, F. Aldinger. High temperature deformation of precursor-derived amorphous Si-B-C-N ceramics. J. Eur. Ceram. Soc. 1999, 19: 2797-2814.
9. A. Zimmermann, A. Bauer, M. Christ, Y. Cai, F.Aldinger. High-temperature deformation of amorphous Si-C-N and Si-B-C-N ceramics derived from polymers. Acta Mater. 2002, 50: 1187-1196.
10. D.L. Zhang. Processing of advanced materials using high-energy mechanical milling. Prog. Mater. Sci., 2004, 49: 537-560.
11. Zhi-Hua Yang, De-Chang Jia, Xiao-ming Duan, Yu Zhou. Microstructure and thermal stabilities in various atmospheres of $SiB_{0.5}C_{1.5}N_{0.5}$ nano-sized powders fabricated by mechanical alloying technique. J. Non-cryst. Solids. 2010, 356: 326-333.
12. Zhi-Hua Yang, Yu Zhou, De-Chang Jia, Qing-Chang Meng, Microstructures and properties of $SiB_{0.5}C_{1.5}N_{0.5}$ ceramics consolidated by mechanical alloying and hot pressing. Mater. Sci. Eng. A. 2008, 489(1-2): 187-192.
13. Zhi-Hua Yang, De-Chang Jia, Yu Zhou, Jiu-Xing Zhang. Processing and characterization of $SiB_{0.5}C_{1.5}N_{0.5}$ produced by mechanical alloying and subsequent spark plasma sintering. Mater. Sci. Eng. A. 2008, 488(1-2): 241-246.
14. N.S. Jacobson, E.J. Opila, K.N. Lee. Oxidation and corrosion of ceramics and ceramic matrix composites. Curr. Opin. Solid State Mater. Sci., 2001, 5: 301-309.
15. Weinmann, M.; Schuhmacher, J.; Kummer, H.; Prinz, S.; Peng, J.; Seifert, H. J.; Christ, M.; Muller, K.; Bill, J.; Aldinger, F.; Synthesis and Thermal Behavior of Novel Si-B-C-N Ceramic Precursors. Chem. Mater. 2000, 12(3): 623-632.
16. Michael K. Cinibulk, Triplicane A. Parthasarathy. Characterization of Oxidized Polymer-Derived SiBCN Fibers. J. Am. Ceram. Soc. 2001, 84 (10): 2197-202.
17. Martin Jansen, Hardy Jüngermann. A new class of promising ceramics based on amorphous inorganic networks. Curr. Opin. Solid State Mater. Sci. 1997, 2: 150-157.
18. N.V. Ravi Kumar, R. Mager, Y. Cai, A. Zimmermann, Fritz Aldinger. High temperature deformation

behaviour of crystallized Si-B-C-N ceramics obtained from a boron modified poly(vinyl)silazane polymeric precursor. Scripta Mater. 2004, 51: 65-69.

19. Yung-Jen Lin, Lee-Jen Chen, Oxidation of SiC powders in SiC/alumina/zirconia compacts, Ceram. Int. 2000, 26: 593-598.

20. Zhi-Hua Yang, De-Chang Jia, Yu Zhou, Peng-Yuan Shi, Cheng-Bin Song, Ling Lin. Oxidation resistance of hot-pressed SiC-BN composites. Ceram. Int. 2008, 34: 317-321.

21. H.E. Kissinger. Reaction kinetics in different thermal analysis. Anal. Chem. 1957, 29: 1881-1886.

22. Joachim Bill,Thomas W. Kamphouwe, Anita Müller, Thomas Wichmann, Achim Zern, Artur Jalowieki, Joachim Mayer, Markus Wernamann, Jörg Schuhmacherk, Klaus Müller, Jianqiang Peng, Hans Jürgen Seifert, Fritz Aldinger. Precursor-derived Si-(B-)C-N ceramics: thermolysis, amorphous, state and crystallization. Appl. Organomental. Chem. 2001, 15: 777-793.

23. Anita Müller, Achim Zern, Peter Gerstel, Joachim Bill, Fritz Aldinger. Boron-modified poly (propenysilazane)-derived Si-B-C-N ceramics: preparation and high temperature properties. J. Eur. Ceram. Soc. 2002, 22: 1631-1643.

24. Y.H. Xiong, C.S. Xiong, S.Q. Wei, H.W. Yang, Y.T. Mai, W. Xu, S. Yang, G.H. Dai, S.J. Song, J. Xiong, Z.M. Ren, J. Zhang, H.L. Pi, Z.C. Xia, S.L. Yuan. Study on the bonding state for carbon–boron nitrogen with different ball milling time. Appl. Surf. Sci. 2006, 253: 2515-2521.

25. Y.H. Xiong, S. Yang, C.S. Xiong, H.L. Pi, J. Zhang, Z.M. Ren, Y.T. Mai, W. Xu, G.H. Dai, S.J. Song, L. Zhang, Z.C. Xia, S.L. Yuan. Preparation and characterization of CBN ternary compounds with nano-structure. Physica B. 2006, 382: 151-155.

26. S. Ulrich, A. Kratzsch, H. Leiste, M. Stüber, P. Schloßmacher, H. Holleck, J. Binder, D. Schild, S. Westermeyer, P. Becker, H. Oechsner. Variation of carbon concentration, ion energy, and ion current density of magnetron-sputtered boron carbonitride layers. Surf. Coat. Tech. 1999, 116-119: 742-750.

27. N. S. Jacobson, G. N. Morscher, D. R. Bryant, R. E. Tressler, High-temperature oxidation of boron nitride: II, boron nitride layers in composites, J. Am. Ceram. Soc. 1999, 82 (6): 1473-1482.

28. K. Kobayashi, K. Maeda, H. Sano, Y. Uchiyama. Formation and oxidation resistance of the coating formed on carbon material composed of B_4C-SiC powders. Carbon, 1995, 33 (4): 397-403.

INFLUENCE OF WATER QUALITY ON CORROSION OF MULTI-OXIDE ENGINEERING CERAMICS

Mannila, Marju; Häkkinen, Antti
Department of Chemical Technology, Lappeenranta University of Technology
Lappeenranta, Finland

ABSTRACT

The objective of this study was to test a developed systematic method for comparing the corrosion resistance of various types of technical ceramics in hydrofluoric acid containing aqueous media. This testing was done with acidic solutions made in waters of differing quality with four technical ceramic samples. The samples were shaken in containers for up to two weeks, after which their weight losses and remaining hardness were measured. Daily samples withdrawn from the liquid phase were analyzed for dissolved components, redox potential and fluoride content. Recorded weight losses were 5 to 30 %. For any given material, there is no significant difference in mass loss observed for exposures in different corrosive media. The materials underwent large drops in hardness with remaining hardness values recorded to be under 5 HV3 in the worst case and around 90 HV3 for the best performing sample. The reductions in hardness varied between different media for the materials. For two samples reverse osmosis treated water caused more softening while one was more deformed in tap water based solution. The amount of material dissolved in any given corrosive media was not significantly different for the four ceramics, however, the largest dissolution by far occurred for all materials in the solution containing added salts, which contrary to expectations seemed to accelerate dissolution instead of acting as inhibitors.

INTRODUCTION

As opposed to the corrosion of metals being an electrochemical phenomenon, ceramic corrosion has to do with the solubility of the material. Most often, corrosion results from several disadvantageous factors acting together on the material or from separate process stages that are interacting unfavourably. The structure of the ceramic, the method by which it has been manufactured and the type of corrosive media together determine the final lifetime. The most usual corrodent encountered in the processing industry is water as its presence is a requirement for many types of corrosion. [1] Thus, water quality issues are paramount in the processing industries, especially in developing countries where water supplies, and technology to treat the raw water, are limited.

Silica, hard glass and aluminium are not attacked by pure fluorine[2]. However, in acidic conditions, fluoride ions have a high tendency to form hydrofluoric acid. This acid attacks silica, by protons attacking the bridging oxygens, forming silanol groups. At the same time, the fluoride ion attacks the silicate and replaces hydroxyl groups on the silica surface. As the reaction produces more hydrofluoric acid as it proceeds, the attack continues autocatalytically.

Westendorf experimented with the dissociation and hydrolysis of magnesium hexafluorosilicate. The reaction was buffered with a diethylbarbiturate solution containing sodium and calcium ions. Reduced relative fluoride was observed, an effect that the author concluded to result from the capacity of both Mg^{2+} and Ca^{2+} to form complexes with F^-. [3]Another study has been reviewed where fluoride was found to bind onto trivalent and divalent metal cations such as Fe(III), Al(III), Ca(II) and Mg(II). In a second study, solutions containing silicon, calcium and fluoride remain ionised in aqueous conditions but upon dilution due to hydrolysis, the equilibrium species found were silicic acid, hydrofluoric acid and a precipitate of calcium fluoride. In another study it was discovered that lead, zinc, magnesium and copper form hydrated tetra-aquo and hexa-aquo ions in aqueous solutions containing fluoride and silicon. However, it was concluded that an accelerative effect on fluoride binding by metal cations is not fully proved yet.[4]

MATERIALS AND METHODS

Four different multi-oxide technical ceramic materials were selected for the experiment. Three were porous structures out of which one, ceramic A, had additionally a thin membrane coating, and one, sample D, was a glass-structured non-porous ceramic. The test pieces were cut to squares approximately 2 cm x 2 cm in size from a larger piece of ceramic material. Before and after exposure to corrosion medium each piece was washed with ion-exchanged water and dried for 24 hours at 90 °C.

All the three solutions had pH 1 and contained 10 ppm of HF. The first solution was made in tap water, the second in reverse osmosis treated water while the third solution was also RO-water based but additionally contained 500 ppm sodium, calcium and magnesium as their chloride salts each. The 40 % hydrofluoric acid was supplied by Merck. Acidic pH-levels were adjusted with Merck 65 % nitric acid, and in the third solution with a 50% HNO_3-50 % H_2SO_4 (Merck 95–97 %) mixture. Chloride salts of calcium, sodium and magnesium were from Merck as well. All chemicals were of analytical grade.

The compositions of the materials were determined with scanning electron microscopy combined with energy dispersive spectrometry using the JEOL JSM–5800 instrument. In order to facilitate this, small chips were detached from the surfaces and the substrate layers with a chisel. The chips selected for analysis were consequently sputtered with gold in order to make the surface conductive. The working distance was 10 mm from the filament and the acceleration voltage was 20 kV. Full vacuum and highest sensitivity were used. The results of the SEM-EDS analyses are displayed in Table I, below.

Table I. Material compositions determined by SEM-EDS.

	A	B	C	D
Al_2O_3	75	85	90	24.0
SiO_2	22	12	9.0	65.0
K_2O	0.1	1.0	1.0	2.5
CaO	0.1	2.0	-	-
Na_2O	1.5	-	-	8.0
Fe_2O_3	0.1	-	-	-
TiO_2	0.1	-	-	0.5
MgO	0.1	-	-	-
BaO	0.1	-	-	-

Far-from-equilibrium conditions were created by using 0.8 dm^3 of solution per container and changing them every 24 hours for duration of 14 days. In each container there were initially three pieces of ceramic. After five days, one piece was removed and after 10 days another piece was taken out. Consequently, the pieces were exposed to the corrosive media for 5, 10 or 14 days, respectively. All experiments were conducted at room temperature.

The set-up consisted of a tray with slots for 1 dm^3 jars attached to a shaker. Shaking frequency was kept constant at 60 min^{-1}. Polyethylene jars were selected, as they are fluoride-resistant. Inside each jar was a divider that kept the different pieces separate and distinguishable from each other.

Thermo Scientific Orion 4-Star Benchtop pH/ISE meter accompanied with Orion ionplus 9609BNWP fluoride electrode was used for the determination of remaining fluoride. Before analysis samples were diluted 1:1 with TISAB (Total Ionic Strength Adjustment Buffer) solution which removes the interferences due to the presence of polyvalent cations.

Redox potential was measured from each sample with WTW Microprocessor pH meter pH 537 which on top of pH measurements can quantify redox and temperature. The electrode connected to the meter was WTW SenTix ORP.

The hardness of the samples was measured with a Zwick 3202 instrument using a 3 kg weight. The sample surfaces were marked with a purple highlighter or alternatively with a red permanent marker to facilitate the location of indentations.

The dissolution of ceramic components was recorded by inductively coupled plasma optical emission spectrometry (ICP–OES). A specific method including all the components found in the ceramic materials was created for the Thermo Electron IRIS Intrepid II XDL instrument. The instrument covers wavelengths in the region of 165–1,050 nm. Shorter wavelengths are measured axially and the longer ones radially.

In addition, the weights of the test pieces were recorded before and after exposure to corrosive media and weight losses calculated.

RESULTS AND DISCUSSION

Weight loss

The calculated weight losses are listed in Table II. Material C has lost the least weight, remaining well under 10 % lost in the end of the experiment. Furthermore, there seems to be no significant difference between the types of corrosive media. Losing 50–100 % more weight than C, material B fared the second best whereas D and especially material A lost significantly larger amount of their initial mass. Materials B and C are very similar, and the difference in weight loss might be down to the exclusion of calcium species from the structure of C as calcium ions were detected in the liquid phase for Material B. Material D lost approximately 5 % more mass in both the RO-water solutions than in the tap water solution while ceramic A lost almost 10 % more weight in the solution with chloride salts than in the other two experimental points.

Table II. Recorded weight losses

Material	weight loss, %		
	Tap water		
	5 days	10 days	14 days
A	8,02	15,24	24,51
B	6,10	9,16	11,38
C	3,75	4,64	5,64
D	5,09	11,55	12,62
	Reverse osmosis water		
	5 days	10 days	14 days
A	5,41	16,88	22,51
B	5,68	8,10	9,52
C	3,25	6,08	6,13
D	6,06	11,36	17,27
	RO-water + salts		
	5 days	10 days	14 days
A	6,55	19,49	32,37
B	6,41	8,42	10,26
C	3,13	5,27	6,12
D	5,76	13,29	18,46

Hardness

The material hardness before and after exposures is displayed in Table III. Material A fared much better in tap water than in reverse osmosis water solutions. Especially dissolved salts destroyed the hardness of the membrane layer.. The final hardness after 14 days is still up to 60 HV3, and can be considered fairly satisfactory if compared to e.g. ceramic B. As opposed to the behavior of material A in tap water vs. RO water, ceramic B has initially a better resistance to the RO water–based solution than tap water. As a common factor both A and B are not able to withstand salt solutions without suffering significant damage on their surface layers.

The hardness of ceramic C seems to drop initially and then increase again, according to Table III. This is indicative of a heterogeneous structure. The capability of C to withstand corrodents is quite close to acceptable levels. Material D withstands tap water better than RO-water solutions where the final values indicate a very soft surface.

Table III. The remaining material hardness after exposure to corrosive media

Sample	Initial HV3	Final HV3, 5 days	Final HV3, 10 days	Final HV3, 14 days
		Tap water		
A	170	78	103	61
B	163	74	13	12
C	130	42	57	45
D	122	40	79	58
		RO-water		
A	170	25	3	3
B	163	127	12	14
C	130	123	68	89
D	122	77	36	30
		RO-water + salts		
A	170	89	3	1
B	163	21	15	6
C	130	48	63	71
D	122	47	34	13

Redox potential

Redox potentials of the liquid samples were recorded, and an example of the obtained results is plotted in Figure 1. No large deviations are seen and all the solutions in contact with all the tested materials have potentials ranging from approximately +450 to 550 mV, which is in the oxidizer region. The levels seem to constantly decrease from their initial level maximums.

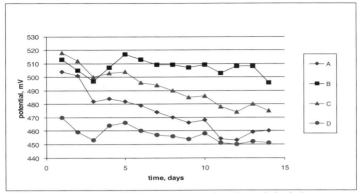

Figure 1. Redox potentials of RO-water based sampled solutions.

Fluoride consumption

The Figure 2 depicting cumulative fluoride consumption from the RO-water based corrodent is a representative example for the behavior recorded during the exposure time. For all three types of solutions experimented with have the largest consumption after having soaked a sample of material D. Their total sums add up to approximately 80 mg. Ceramics A, B and C all fall in a region with a total cumulative consumption of approximately 30 to 50 mg.

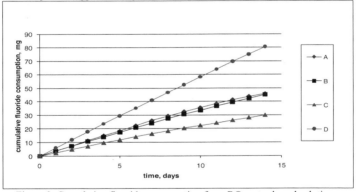

Figure 2. Cumulative fluoride consumption from RO-water based solutions.

Dissolved components

In Figures 3–4, the total cumulative dissolution data for the experiments are displayed. The graphs make no distinction between the soluble species. The order of the curves differs between ceramics somewhat, but the largest amounts of leached ions in liquid phase are detected in the added salts experiments for every type of ceramic.

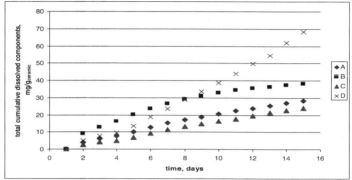

Figure 3. Total cumulative leached components from samples in RO-water based solutions.

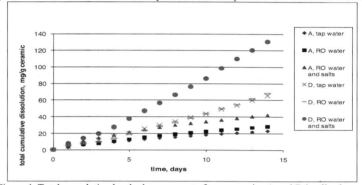

Figure 4. Total cumulative leached components from ceramics A and D in all solvents.

CONCLUSIONS

The ceramic D dissolved in much larger quantities than the other samples, and the dissolution was the strongest in the experiments where chloride salts had been added to the solution for all types of ceramic. This was surprising as the salts were expected to act as inhibitors, reacting with fluoride so that the fluoride would not be free to attack the ceramic structures.

The total dissolved amounts stayed at intermediate levels, yet the structures were badly misshapen. This could indicate the solubility being highest at the grain boundaries where glassy phases are present. After the binder would go to liquid phase after reacting with a component from the solvent, the whole structure could collapse. This is supported by the finding of fine, undissolved, particulate matter dispersed in the liquid phase.

The results were deemed to be of acceptable accuracy based on previous experimentation. The RO-water without salts –experiment had been to a larger set earlier, and those results correlate well with the ones obtained in the current study.

REFERENCES
[1] McCauley R.A., *Corrosion of ceramics*, 2nd ed., Marcel Dekker, 2004, 405 p.

[2] Fontana M.G., Staehle R.W., *Advances in corrosion science and technology,* Vol. 5, Filbert A.M., Hair M.L., ed., Plenum Press, 1976, p.1–51.

[3] Westendorf J., Kinetics of Acetylcholinesterase Inhibition and the Influence of Fluoride and Fluoride Complexes on the Permeability of Erythrocyte Membranes. PhD thesis, University of Hamburg, Germany, 1975.

[4] Urbansky E.T., Fate of fluorosilicate drinking water additives. *Chemical Reviews,* **102**, 8, 2837–2854 (2002).

THE EFFECT OF LOAD AND TEMPERATURE ON HARDNESS OF ZrB$_2$ COMPOSITES

J. Wang*, F. Giuliani*, L.J. Vandeperre*
* UK Centre for Advanced Structural Ceramics and Department of Materials, Imperial College London, South Kensington Campus, London SW7 2AZ, UK

ABSTRACT

Zirconium diboride (ZrB$_2$) and its composites are being investigated as candidate materials for ultra-high temperature applications (>2000 °C). At such high temperatures, it is conceivable that dislocation slip would become an active deformation mechanism. Hence, to predict the mechanical response of these materials both the resistance to plastic flow as well as the resistance to cracking must be estimated. For ceramics, the former is most conveniently estimated from the hardness, but literature values for the hardness of ZrB$_2$ vary widely (14-24 GPa). To address this, a systematic study of the room temperature hardness of ZrB$_2$–based materials was carried out. ZrB$_2$ and ZrB$_2$ with 20 vol% silicon carbide were pressureless sintered at 2000 °C to near-full density with 4 wt% boron carbide and 1.5 wt% carbon as sintering additives. It is shown that the hardness of the constituent phases can be determined from small scale indentation measurements and hence that the hardness of dense ZrB$_2$ is 20-21 GPa. As the load is increased, the hardness decreases and it is shown that the variability in hardness values reported in the literature is consistent with the effect of load on hardness. The relative contributions of cracking and porosity on the measured values will be discussed and initial results at higher temperatures will be presented.

INTRODUCTION

Ultra-high temperature ceramics (UHTC) refers to a class of carbides, nitrides and borides with very high melting temperatures (>3000 °C). UHTCs have good mechanical properties (Young's modulus > 500 GPa, Vickers' hardness >20GPa, strength > 500 MPa), high thermal and electrical conductivities, chemical stability and good thermal shock resistance[1, 2]. Therefore UHTCs are considered as candidate materials for leading edges of the next generation of hypersonic aerospace vehicles and other high temperature applications such as cutting tools and molten metal containers[2,3].

The single-phase borides are limited by their poor oxidation and ablation resistance at elevated temperatures. Therefore composites have been investigated in order to improve the densification, mechanical and physical properties, as well as the oxidation and ablation resistance. Pure fine-grained ZrB$_2$ and HfB$_2$ have strengths of a few hundred MPa, which can be increased to over 1 GPa with the addition of SiC as a reinforcement[4]. Also the addition of silica scale formers such as SiC or MoSi$_2$ improves the oxidation behaviour above 1100 °C, where B$_2$O$_3$ starts to evaporate rapidly and causes linear, dramatic oxidation. Currently the best composition is ZrB$_2$-20vol% SiC which has the best oxidation resistance by forming a protective silica layer at high temperature, and excellent mechanical properties[2, 5].

ZrB$_2$ is difficult to sinter to full density because of the covalent nature of its bonding, which causes low volume and grain boundary diffusion rates. Therefore, historically, hot pressing (HP) has been the main method to sinter ZrB$_2$ ceramics. Recently, it has been shown that by adding sintering additives such as boron carbide (B$_4$C) and carbon, near-full density ZrB$_2$ and its composites can be obtained by pressureless sintering[6]. A key advantage of pressureless sintering over hot pressing is that using pressureless sintering more complex shaped ceramic components can be produced, which reduces the cost of final machining steps.

The mechanical properties of ZrB$_2$-based materials have to date mainly been measured at room temperature and there is only very limited information about their mechanical and thermal properties at the intended service temperatures. As seen in Figure 1, historical data for the high temperature hardness of carbides and diborides[7, 8] suggest that the hardness of these materials drops to half the room temperature value for temperatures as low as 600 °C. Such sharp decrease in hardness of these materials indicates that the resistance to dislocation flow reduces strongly with temperature and hence could be important at even more elevated temperatures. On the other hand, the room temperature hardness of ZrB$_2$ as reported in the literature varies widely (14-24 GPa)[6, 9-12], which renders making estimates for the resistance to dislocation flow difficult. To clarify the origin of these discrepancies, this paper presents the results of a systematic study of the effect of the applied load on the hardness of ZrB$_2$-based materials. In addition, preliminary results of experiments aiming at determining the variation of hardness with temperature for ZrB$_2$-20vol% SiC are reported.

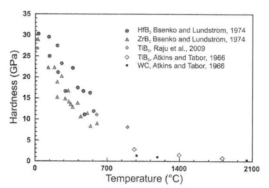

Figure 1. Literature data for the hardness variation with temperature of a range of borides[7, 8, 13] and tungsten carbide[7].

EXPERIMENTAL

Commercially available ZrB$_2$, SiC and B$_4$C powders were used to prepare the materials. The properties of the powders are listed in Table I. Powder blends of typically 50 g were prepared by dispersing the powders in 75 g methyl ethyl ketone (MEK) using a magnetic stirrer. Carbon was added in the form of a liquid phenolic resin (Novolak, CR-96, carbon yield 61 wt%). The latter was slowly added to the mixtures while stirring was continued. The slurry was then ball-milled for 24 h using poly-ethylene bottles and tungsten carbide milling balls to mix the raw materials homogenously and to break agglomerates and large particles. After milling, the slurry was dried under vacuum in a rotary evaporator (BUCHI Rotavapor R-210/215) at 90 °C. The dried powders were ground with a mortar and pestle and then sieved to eliminate coarse agglomerates. To examine the as-received powder and the effect of ball milling, the particle size of raw and ball-milled powders was analysed with a laser light diffraction technique (Mastersizer, Malvern Instruments Ltd, UK). Powders were dispersed in distilled water, and to ensure good dispersion, the slurries were put in an ultra-sonic bath for 5 minutes.

Table I. Characteristics of the raw materials

	ZrB$_2$	α-SiC	B$_4$C
Supplier	H.C. Starck	H.C. Starck	H.C. Starck
Grade	B	UF-25	HS
Density (g cm^{-3})	6.085	3.21	2.52
Mean particle size (μm)	2.7	0.45	0.25
Specific surface area (m^2 g^{-1})	1	25	15.8
Oxygen content (wt%)	0.9	2.0	1.3

Sample shaping was achieved through uni-axially pressing to 30 MPa with a hold time of 30 s at maximum pressure using 13 mm or 40 mm diameter cylindrical dies. Further densification prior to sintering was achieved using cold iso-static pressing (CIP) to 300 MPa for 120 s. Compacts containing resin were charred in a flowing Ar atmosphere initially at 700 °C, but guided by thermogravimetric analysis (Jupiter, Netzsch, Germany) this was lowered to 400 °C for subsequent samples. Finally, the samples were sintered for up to 2 h in a graphite furnace at temperatures from 1800 °C to 2000 °C in flowing argon. A heating rate of 10 °C min^{-1} was employed throughout and additionally two 1 h isothermal holdings at 1450 °C and 1600 °C were used as this is claimed to greatly improve the sintering through B$_2$O$_3$ evaporation[6]. The furnace was then cooled to room temperature with a 20 °C min^{-1} cooling rate.

The densities of sintered specimens were measured by Archimedes's method and the relative density was calculated by dividing the density by the theoretical density calculated by rule of mixtures. Cross-sections of specimens were cut by diamond saw and polished to submicron finish using a colloidal silica solution (OPS, Struers, UK). Vickers hardness measurements were carried out on polished specimens using a micro and a macro indenter (Indentec 6030LKV, Zwick, Germany) at different loads (500 g to 5 kg), dwell time 10 s. Reported values were obtained from an average of at least 10 indents. Nano-indentation was carried out at different loads from 25 mN to 400 mN using a depth-sensing nano-indenter (NANOTEST, Micromaterials Ltd., UK). Typically 25 indentations were taken at each load. The indentations were made 10 μm apart for 100 mN indentations to reduce the effects of one indent on another and 5 μm apart for 25 mN and 50 mN to obtain a hardness map of different phases. The loading time was typically set to 20 s to reduce the influence of thermal drift. The indentations were imaged using a SEM to determine in which phase the indents were made. High temperature nano-indentation was performed using a hot stage and heated diamond Berkovitch tip attached to the nano-indenter after 8 h of thermal stabilisation.

RESULTS AND DISCUSSION

Material preparation

Measurements of the average particle sizes, see Figure 2, confirmed that the as-received ZrB$_2$ indeed has an average particle size of 2.7 μm. This reduces to ~1 μm after one day of ball milling. Since further milling did not give any further reductions in size, 24 h of ball milling was used for all further work.

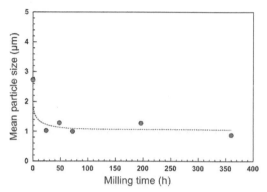

Figure 2. Mean particle size of the ZrB$_2$ powder as a function of ball milling time.

Initially pyrolisis of the resin prior to sintering was carried out by heating in flowing argon to 700 °C. After sintering up to 2000 °C, the highest relative density for ZrB$_2$ without SiC was only 92%. The latter was achieved with addition of 2 wt% B$_4$C and 1.5 wt% carbon and sintering at 2000 °C for 2 hours. For the same conditions, the ZrB$_2$-20 vol% SiC composites achieved 95% relative density with 4 wt% B$_4$C and 1.5 wt% carbon as sintering additives. For comparison, additive-free ZrB$_2$ reached only a relative density of 62% at the same sintering temperature. This supports the idea that the surface oxide impurities in the form of B$_2$O$_3$ and ZrO$_2$ inhibit the densification by promoting coarsening through evaporation-condensation at intermediate temperatures, and B$_4$C and carbon can help remove the surface oxygen contaminations. The processes through which these sintering additives reduce the oxide impurities have been proposed to be[6]:

$$7ZrO_2 + 5B_4C \rightarrow 7ZrB_2 + 5CO(g) + 3B_2O_3(l)$$
$$2ZrO_2 + B_4C + 3C \rightarrow 2ZrB_2 + 4CO(g)$$
$$ZrO_2 + B_2O_3(l) + 5C \rightarrow ZrB_2 + 5CO(g)$$

However, the densities achieved here are lower than what has been reported for the same ZrB$_2$ powder: according to Fahrenholtz, Hilmas, Zhang and Zhu this powder should sinter to near full density (>99%) with 2 wt% B$_4$C and 1 wt% carbon at 1900 °C[6].

Since the microstructures contained what appeared to be small crystallites embedded in glassy phases underneath particles broken out during polishing, see Figure 3, TGA was carried out to determine whether oxidation occurs during the charring of the resin. Although the TGA experiment was carried out in flowing Ar gas environment, it is clear from the increase in mass shown in Figure 4 that ZrB$_2$ starts to oxidise from approximately 550 °C and B$_4$C oxidises from as low as 400 °C onwards. Only SiC does not gain weight below 700 °C. This suggested that for the quality of the argon gas used and for the charring furnace used, the charring temperature of the mixed powders needed to be limited to be 400 °C to avoid extensive oxidation of the powders.

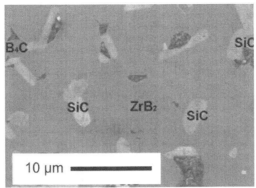

Figure 3. Secondary electron micrograph of ZrB$_2$ with 20 vol% SiC , 4 wt% B$_4$C and 1.5 wt% carbon after charring at 700 °C and sintering for 2 h at 2000 °C.

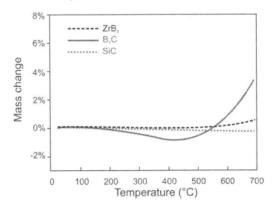

Figure 4. Thermogravimetric analysis of ZrB$_2$, B$_4$C and SiC powders.

Indeed, lowering the pyrolysis temperature to 400 °C to reduce oxidation, led to near-full density (>99% RD) for ZrB$_2$ - 20 vol% SiC with addition of 4 wt% B$_4$C and 1 wt% carbon and sintering at 2000°C for 2 hrs. As shown in Figure 5, the microstructure consists of three distinct phases, which are well distributed throughout the sample. The ZrB$_2$ grains are the largest (4 to 6 μm), while the B$_4$C grains range from submicron to >10 μm. The SiC particles have become elongated with an aspect ratio close to 3 to 1 (approximately 1-3 μm wide, 5-15 μm long). This agrees with the work of Zhang, Hilmas and Fahrenholtz[11], who observed that the morphology of SiC grains in ZrB$_2$-SiC ceramics varies from equi-axed to whisker-like depending on the initial size of the SiC particles. The aspect ratios of SiC grains in their work were 1.05, 1.75, and 3.05 for starting particle sizes of 1.45, 1.05, and 0.45 μm respectively and this variation was attributed to smaller particles being able to form an interconnected network between the ZrB$_2$ grains more easily, whereas large SiC particles remain more individually distributed.

Hence, provided oxidation of the raw material is avoided, the results presented here confirm that high density ZrB_2-based materials can be produced by pressureless sintering with addition of boron carbide and carbon.

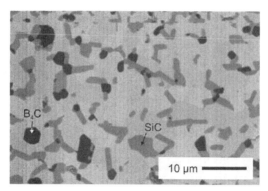

Figure 5. Backscattered electron micrograph of polished cross-sections of ZrB_2 with 20 vol% SiC, 4 wt% B_4C and 1 wt% carbon after pyrolysis in argon at 400 °C for 2 hrs and sintering at 2000 °C for 2 hrs. As indicated, the black particles are B_4C, the grey particles are SiC and the bright background are ZrB_2 grains.

Hardness measurements

Figure 6 reports hardness measurements over a range of applied loads for dense ZrB_2 with 20 vol% SiC and for 90% dense ZrB_2. Also shown in the figure is that the range of hardness values observed here (10-25 GPa) is consistent with the variation reported in the literature. Moreover the variation in reported hardness values is so similar to the variation in hardness with load for a single material that the differences in the literature values for the hardness of ZrB_2 based materials are the result of the different loads used by different authors rather than due to differences in material quality.

For the lowest applied loads, the hardness is found to be independent of the applied load. For higher applied loads the hardness decreases as the load is increased until eventually at the highest applied loads, the hardness is found to become constant again. Over the entire range of applied loads, the hardness of the dense ZrB_2-SiC composite is higher than the hardness of the 90% dense ZrB_2 monolith. Considering that for the lowest load indentations, cracking does not occur, and that making small indents into a relatively coarse microstructure allows the indents to be made far away from the nearest pore, it is proposed that the constant hardness in the low load regime is a measure for the resistance to plastic deformation of the material being indented[14-16] whereas the hardness at the highest loads incorporates the influence of porosity on hardness[17] and other damage mechanisms such as cracking and densification, which are known to have a marked influence on the hardness[14, 18]. The observation that the hardness starts to decrease at lower loads for the ZrB_2 monolith, which is only 90% dense, than for the dense ZrB_2-SiC composite is consistent with this interpretation as for the same pore size, a higher porosity means that the distance between pores is reduced and therefore pores are closer to any indent in 90% dense material than in nearly fully dense material. Hence, the hardness as a measure for the resistance to plastic deformation of ZrB_2 can be estimated to be of the order of 20-22 GPa.

Observation of the smallest indents in the ZrB_2-SiC composite in the scanning electron microscope revealed that the indents were so small that many were entirely contained inside a single particle of the phases making up the microstructure. Therefore, the hardness data was separated into hardness values for the composing phases of the microstructure (ZrB$_2$, SiC and B$_4$C). Figure 7 illustrates that using this procedure the hardness of B$_4$C, SiC and ZrB$_2$ is estimated as 33±1 GPa, 27±2 GPa and 21±2 GPa respectively. The latter agrees very well with the value obtained from the ZrB$_2$ monolithic sample (20±4 GPa), thereby confirming the interpretation that the constant hardness at low loads is due to the resistance to plastic deformation of the material being indented alone. Reported hardness values for hot pressed B$_4$C vary between 29 and 46 GPa[19], for dense SiC the reported hardness varies between 30 to 36 GPa[20-22], and hence the values estimated for the component phases are somewhat low. The fact that the hardness did not vary with load for SiC suggests that this is not a consequence of the softer ZrB$_2$ matrix.

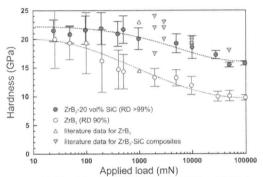

Figure 6. Hardness versus applied load for dense ZrB$_2$-20 vol% SiC and 90% dense ZrB$_2$ and literature data for the hardness of ZrB$_2$[6] and ZrB$_2$-SiC composites[9-12]. Only literature data for relative densities in excess of 97% have been included.

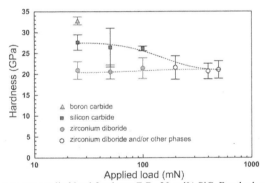

Figure 7. Hardness versus applied load for dense ZrB$_2$-20 vol% SiC. For the lowest applied loads, imaging the indents in SEM made it possible to attribute the indentations to single phase particles, whereas from 200 mN onwards, indentations were found to cover multiple phases and hence only an average is shown.

Rather it is thought that the SiC and B$_4$C phases might contain small scale porosity. The size of the SiC and B$_4$C particles in the microstructure is much larger than the size of the starting powders, and they can therefore not have formed from single powder particles. Since full densification of B$_4$C and SiC at 2000 °C requires hot pressing or liquid phase sintering, it is very likely that the SiC and B$_4$C might contain some residual porosity. Since these pores would be much finer than the pores between the ZrB$_2$ grains, the effect on the hardness will be noticeable at even the lowest loads.

The fact that the hardness is reduced much less for the composite compared to the monolith is attributed to the difference in porosity although some influence of the harder particles might come into play at higher loads as well.

The variation of the hardness with temperature for a sample of dense ZrB$_2$ with 20 vol% SiC, is compared with historical data in Figure 8. Our results confirm that the hardness decreases strongly with temperature. A linear decrease in hardness with temperature is quite common[7, 23] and is indeed expected when thermal activation aids in overcoming the Peierls' stress or lattice resistance[24, 25].

Extrapolating the decrease in hardness with temperature to the service temperature is not possible: it can be expected that as the lattice resistance vanishes other microstructural features such as the presence of grain boundaries and second phase particles as well as the dislocation density will become determining factors in the resistance to plastic flow[26]. Nevertheless, it is clear that at the service temperature, the lattice resistance, which controls the hardness at room temperature, will be very limited in magnitude, and zirconium diboride based materials might therefore to some extent be ductile or show much higher toughness at elevated temperatures. Further work at higher temperatures is needed to clarify this further, but this might explain why near application tests with ZrB$_2$ based materials have proven to be so successful[27].

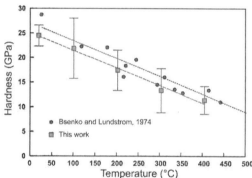

Figure 8. Hardness of ZrB$_2$- 20 vol% SiC with addition of 4wt% B$_4$C and 1 wt% carbon. The applied load was 50 mN for all indentations and the loading was applied in 20s.

CONCLUSIONS

In agreement with claims in the literature, pressureless sintering of ZrB$_2$-based materials with addition of B$_4$C and carbon yields fully dense materials. To achieve such high densities, oxidation of the powders during processing must be limited as much as possible. Where charring is carried out ex-situ, i.e. in a different furnace, it is advisable to limit the charring temperature to 400 °C to avoid oxidation of the boron carbide.

The hardness of ZrB$_2$ depends strongly on the applied load, decreasing from 20-22 GPa for small loads (~ <1 N) to 10-16GPa for loads up to 50 N. The hardness values obtained from low load indentations are determined by the resistance to plastic flow, whereas for higher loads the resistance to cracking and the influence of porosity have a dramatic effect on the measured hardness. The range of reported hardness values in the literature for nearly dense materials (>97% relative density) appears to be more a consequence of the differences in applied loads by different authors rather than due to differences in material quality.

Due to the relatively coarse nature of the microstructures, indents made with small applied loads, can be used to determine the hardness of the different phases in the microstructure (ZrB_2, SiC, B_4C). The hardness of the B_4C and SiC was found to be somewhat lower than literature values, suggesting these phases might not be fully dense. The hardness of dense ZrB_2 was again confirmed to be 20-22 GPa.

The hardness of ZrB_2 with 20 vol% SiC decreases rapidly with temperature, decreasing to 11 GPa at 400 °C in accordance with historical data for the hardness of ZrB_2. This indicates that at the intended service temperatures, the lattice resistance, which controls the hardness at room temperature, will be very small, which could lead to some ductility or increased toughness at those temperatures, but further work at higher temperatures is needed to clarify this.

ACKNOWLEDGEMENTS

Jianye Wang would like to thank the UK Engineering and Physical Sciences Research Council (EPSRC) for a PhD scholarship through the funding for the UK Centre for Structural Ceramics.

REFERENCES

1. M.M. Opeka, I.G. Talmy, and J.A. Zaykoski, Oxidation-based materials selection for 2000 °C plus hypersonic aerosurfaces: Theoretical considerations and historical experience. *Journal of Materials Science*, **39**, 5887-5904 (2004).
2. W.G. Fahrenholtz, G.E. Hilmas, I.G. Talmy, and J.A. Zaykoski, Refractory diborides of zirconium and hafnium. *Journal of the American Ceramic Society*, **90**, 1347-1364 (2007).
3. E. Opila, S. Levine, and J. Lorincz, Oxidation of ZrB_2- and HfB_2-based ultra-high temperature ceramics: effect of Ta additions. *Journal of Materials Science*, **39**, 5969-5977 (2004).
4. F. Monteverde, Ultra-high temperature HfB_2-SiC ceramics consolidated by hot pressing and spark plasma sintering. *Journal of Alloys and Compounds*, **428**, 197-205 (2007).
5. S.Q. Guo, Densification of ZrB_2-based composites and their mechanical and physical properties: a review. *Journal of The European Ceramic Society*, **29**, 995-1011 (2009).
6. A.L. Chamberlain, W.G. Fahrenholtz, and G.E. Hilmas, Pressureless sintering of zirconium diboride. *Journal of the American Ceramic Society*, **89**, 450-456 (2006).
7. A.G. Atkins and D. Tabor, Hardness and deformation properties of solids at very high temperatures. *Proc. R. Soc.*, **A292**, 441-459 (1966).
8. L. Bsenko and T. Lundström, The high-temperature hardness of ZrB_2 and HfB_2. *Journal of the Less Common Metals*, **34**, 273-278 (1974).
9. Y. Zhao, L. Wang, G. Zhang, W. Jiang, and L. Chen, Effect of holding time and pressure on properties of ZrB2-SiC composite fabricated by the spark plasma sintering reactive synthesis method. *International journal of refractory metals & hard materials*, **27**, 177-180 (2009).
10. W. Li, X. Zhang, C. Hong, J. Han, and W. Han, Hot-pressed ZrB_2-SiC-YSZ composites with various yttria content: Microstructure and mechanical properties. *Materials science & engineering A, Structural materials*, **494**, 147-152 (2008).
11. S.C. Zhang, G.E. Hilmas, and W.G. Fahrenholtz, Pressureless sintering of ZrB_2-SiC ceramics. *Journal of the American Ceramic Society*, **91**, 26-32 (2008).
12. A. Rezaie, W.G. Fahrenholtz, and G.E. Hilmas, Effect of hot pressing time and temperature on the microstructure and mechanical properties of ZrB_2-SiC. *Journal of Materials Science*, **42**, 2735-2744 (2007).

13.   G.B. Raju, B. Basu, N.H. Tak, and S.J. Cho, Temperature dependent hardness and strength properties of TiB$_2$ wtih TiSi$_2$ sinter-aid. *Journal of The European Ceramic Society*, **29**, 2119-2128 (2009).

14.   B.R. Lawn and D.B. Marshall, Hardness, toughness, and brittleness: an indentation analysis. *Journal of the American Ceramic Society*, **62**, 347-350 (1979).

15.   D. Tabor, The physical meaning of indentation and scratch tests. *Br. J. Appl. Phys.*, **7**, 159-166 (1956).

16.   L.J. Vandeperre, F. Giuliani, and W.J. Clegg, Effect of elastic surface deformation on the relation between hardness and yield strength. *Journal of Materials Research*, **19**, 3704-3714 (2004).

17.   R.W. Rice, Porosity of ceramics. Materials Engineering, New York: Marcel Dekker Inc. 539, (1998).

18.   G.D. Quinn, P. Green, and K. Xu, Cracking and the indentation size effect for knoop hardness of glasses. *Journal of the American Ceramic Society*, **86**, 441-448 (2003).

19.   J. Deng, J. Zhou, Y. Feng, and Z. Ding, Microstructure and mechanical properties of hot-pressed B$_4$C/(W,Ti)C ceramic composites. *Ceramics international*, **28**, 425-430 (2002).

20.   A. Slutsker, A.B. Sinani, V.I. Betekhtin, A.G. Kadomtsev, and S.S. Ordan'yan, Hardness of microporous SiC ceramics. *Technical physics*, **53**, 1591-1596 (2008).

21.   M. Balog, P. Sajgalik, M. Hnatko, Z. Lences, F. Monteverde, J. Keckes, and J.-L. Huang, Nano-versus macro-hardness of liquid phase sintered SiC. *Journal of the European Ceramic Society*, **25**, 529-534 (2005).

22.   T. Hirai and K. Niihara, Hot hardness of SiC single crystal. *Journal of Materials Science Letters*, **14**, 2253-2255 (1979).

23.   J.H. Westbrook, The temperature dependence of hardness of some common oxides. *Rev. Hautes Temper. et Refract.*, **3**, 47-57 (1966).

24.   W.S. Williams, Influence of temperature, strain rate, surface condition, and composition on the plasticity of transition metal carbide crystals. *Journal of Applied Physics*, **35**, 1329-1338 (1964).

25.   L.J. Vandeperre, F. Giuliani, S.J. Lloyd, and W.J. Clegg, The hardness of silicon and germanium. *Acta Materialia*, **55**, 6307-6315 (2007).

26.   G.E. Dieter, Mechanical Metallurgy. SI Metric Edition ed, London: McGraw-Hill. 751, (1988).

27.   F. Monteverde and R. Savino, Stability of ultra-high-temperature ZrB$_2$-SiC ceramics under simulated atmospheric re-entry conditions. *Journal of the European Ceramic Society*, **27**, 4797-4805 (2007).

NANO-INDENTATION HARDNESS MEASUREMENTS AS A CHARACTERIZATION TECHNIQUE OF SiC- AND PYROLYTIC CARBON LAYERS OF EXPERIMENTAL PBMR COATED PARTICLES.

I.J van Rooyen[a, b], E. Nquma[a], J. Mahlangu[a] and J. H. Neethling[b]
[a] Pebble Bed Modular Reactor (Pty) Ltd, Centurion, South Africa
[b] Department of Physics, Nelson Mandela Metropolitan University, Port Elizabeth, South Africa

ABSTRACT

The Pebble Bed Modular Reactor (PBMR) fuel consists of Tri-Isotropic (TRISO) coated particles in a graphite matrix. The containment of fission products inside the TRISO coated particles (CPs) is dependent on the integrity of the three layer system namely Inner Pyrolytic carbon (IPyC)-SiC-Outer Pyrolytic Carbon (OPyC). The dependency of the fission product transport mechanisms on the properties of SiC is considered in the design of the PBMR Fission Products Transport Mechanism Experimental Program, therefore experiments are developed to study the combined effect of temperature and irradiation properties on SiC properties. Nano-indentation hardness measurements were identified as a possible characterization technique to show changes due to SiC decomposition and PyC becoming anisotropic[1, 2]. The nano-indentation technique is briefly discussed in this article, followed by a discussion of recent nano-indentation results from five experimental PBMR coated particle batches annealed at a temperature range of 1000°C to 2100°C. The effect of long-duration annealing on these nano-indentation hardness values will also be demonstrated by comparing results from 1 hour; 5 hours and 100 hours annealing of coated particles at a temperature of 1600°C.

INTRODUCTION

PBMR fuel is based upon a proven, high quality German fuel design consisting of low enriched uranium triple coated particles (TRISO) contained in moulded graphite spheres. The fuel inside the PBMR reactor core consist of approximately 450 000 fuel pebbles of which each one contains approximately 15 000 TRISO coated fuel particles. The PBMR Fission Product Transport Mechanism Experimental Program is developed to study the combined effect of temperature and irradiation properties on SiC and PyC properties. The SiC layer of the PBMR fuel is deposited by chemical vapour deposition (CVD) using methyltrichlorosilane (MTS) as precursor. The SiC deposited from MTS under the correct conditions produces silicon carbide with density of approximately 3.2 g/cm^3 (nearly 100% its theoretical density). Nano-indentation hardness measurements were identified as a possible characterization technique to show changes due to SiC decomposition and PyC becoming anisotropic[1, 2]. A previous study by Van Rooyen et al.[1] showed the importance of temperature on the PyC microstructure and also showed a tendency of decreasing nano-indentation hardness measurement of IPyC with increasing temperature of the three experimental batches studied. Further research work reported in this article focuses specifically on the very high temperature nano-indentation properties of SiC as SiC being the primary fission product barrier.

In this study a CSM Nano-indentation Hardness tester was used for the hardness and elastic modulus measurements. One advantage of the CSM Nano-indentation hardness Tester (NHT) is that it does not require a stabilization time and nano-indentation measurements can be started immediately in a common laboratory environment. This feature is due to the surface referencing procedure that allows the elimination of the thermal drift in nano-indentation experiments[3]. The typical load range is 0.1 to 500 mN (0.04 μN resolution) up to a depth of 200 μm with a depth resolution of 0.04 nm^3. In this experimental work, all reported nano-indentation results are done on polished cross sectioned pieces of the PBMR coated particles with a load of 100 mN and maintained for 15 seconds before unloading. All experimental batches contain zirconia kernels to enable the measurements to be done in non-nuclear regulatory facilities.

NANO-INDENTATION HARDNESS MEASUREMENT AS A CHARACTERIZATION TECHNIQUE
Principles

The development of measuring the hardness and elasticity of the samples during nano indentation testing (NHT) is well known and described by Oliver-Pharr[4] and Sneddon[5]. An indentor tip with a known geometry is applied for a specific time and the indent measured after the force has been removed[3]. Hardness is expressed by the ratio between the applied load and the contact area. (See Figure 1) In practice:

$$I_{IT} = \frac{F_m}{A_p}$$
(1)

where F_m is the maximum load, and A_p is the contact area between the indenter and the specimen at the maximum depth and load. F_m Being easily measured (checking on the load/unload curve), A_p is needed to determine the hardness. For an ideal indentor, the areas function:

$$A_p = C_o h_c^2$$
(2)

can be used to provide a first estimate of the contact area, where the constant C_o depends on the indenter geometry (C_o=24.5 for a Berkovich tip). However, tips are never ideally sharp and this can be taken into consideration by using a more general function due to blunting of the tip which is written as:

$$A_p = C_o h_c^2 + C_1 h_c + C_2 h_c + C_2 h_c^{1/2} + ... + C_8 h_c^{1/16}$$
(3)

Where constants $C_o...C_8$ can be determined by data measurement using a bulk fused silica and the method outlined by Oliver and Pharr[4]. This full tip function is necessary if an accurate value of nano-indentation hardness is to be determined.

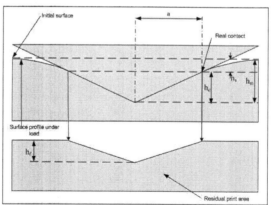

Figure 1: Schematic presentation of the indentation process after unloading[4]

Precautions

Surface roughness is extremely important in indentation testing, because the mechanical properties of the material tested are calculated on the assumption that the surface is flat and smooth to avoid a miscalculation of the hardness[3].

SiC NANO-INDENTATION HARDNESS MEASUREMENT BENCHMARK

A benchmarking literature survey of SiC nano-indentation properties was undertaken and summarized in Table I, to compare the recent nano-indentation values obtained from the SiC layer of the PBMR TRISO coated particles. It was also of interest to identify if phase changes were previously identified by nano-indentation measurements. This survey suggested that the nano-indentation measurements showed no significant difference between the values obtained for hexagonal- and cubic phases of SiC. The interpretation and comparison between the nano-indentation properties of SiC as reported in Table I need to be viewed with care as it is clear that the manufacturing conditions play a significant role on the values obtained. The work reported by López-Honorato et al.[9] also showed the importance of the representative samples by comparing the results obtained from polished flat samples with those of cross sectional polished samples from coated particles manufactured under the same conditions. The importance of the location (cross sectional or surface) of nano-indentation measurements are demonstrated by the work of Tan et al.[11] where differences of approximately 50 GPa are observed with the lower values obtained from the cross sectional samples. Katoh et al.[12] showed that the nano-indentation elastic modulus measurements after irradiation is not indicating the same significant tendency as measured by flexural strength, arguing that the effect of irradiation are decreased by the surface polishing necessary for nano-indentation.

Table I. Measured SiC nano-indentation properties benchmarking

Material	Surface	Hardness (GPa)	Elastic Modulus (GPa)	Deposition Temperature	References
Single crystal 3C-SiC	Smooth surface	31.2 ± 3.7	433 ± 50	1375°C	[6]Reddy et al.
Polycrystalline 3C-SiC	Polished	33.5 ± 3.3	457 ± 50	1375°C	[6]Reddy et al.
Single crystal 6H-SiC	Polished	30-32	Not available	Not available	[7]Ling
Polycrystalline 6H-SiC	Polished	30-32	Not available	Not available	[7]Ling
Polycrystalline 3C SiC from coated particles (Al_2O_3 kernels)	Polished	36.3 ± 0.8	378 ± 13.6	1500°C	[8]López-Honorato et al.
Polycrystalline 3C SiC from coated particles (Al_2O_3 kernels)	Polished	36.5 ± 0.5	424.5 ± 17.5	1300°C without any additions	[8]López-Honorato et al.
Polycrystalline 3C SiC from coated particles (Al_2O_3 kernels)	Polished	42	448	1300°C with propylene additions	[9]López-Honorato et al.
Polycrystalline 3C SiC from coated particles (Al_2O_3 kernels)	Polished	39	415	1300°C with argon additions	[9]López-Honorato et al.
Polycrystalline 3C SiC from coated particles (Al_2O_3 kernels)	Polished	Not available	434.94 ± 1.97	1300°C with propylene additions	[10]López-Honorato et al.
Polycrystalline 3C SiC	Polished	Not available	442.5 ± 13.3	1500°C	[11]Tan et al.

from coated particles (Al₂O₃ kernels)			(external surface). 391.1 ± 12.9 (cross-section).		
PBMR coated particles (ZrO₂ kernels) a. Batch A b. Batch B c. Batch C	Polished	a. 33.38 ± 3.2 b. 37.24 ± 3.5 c. 40.50 ± 2.8	a. 289 ± 20 b. 357 ± 21 c. 427 ± 11	Not available	[1]Van Rooyen et al.
Polycrystalline 3C SiC	Polished	40.6 ± 1.1	456.5 ± 10.4	Not available	[12]Y. Katoh and L.L. Snead

EXPERIMENTAL DESIGN

Five experimental batches were produced by PBMR Fuel Development Laboratories at Pelindaba in two different coater facilities. Three batches were manufactured in the advanced coater facility (ACF) namely G118, G146 and G169 whereas batches B10 and B14 were manufactured in the research facility (RCF). Annealing of the 1600°C long duration (100h, vacuum), batches G146 and G169 at 1600°C (1h, argon), as well as the annealing of all batches at 2000°C (30 min, argon) and 2100°C (15 min, argon), were performed in a resistance heated Webb 89 vacuum furnace supplied by R.D. Webb Company USA. The samples were loaded in graphite or ceramic holders at room temperature and heated to the required temperature at a rate of 25°C per minute. After completion of annealing at the required holding time and temperature, the samples were furnace-cooled to room temperature. Most of the short duration heating below 1980°C was carried out at the PBMR Fuel Development Laboratories at Pelindaba as described previously[1]. The experimental research plan for this work is shown in Figure 2.

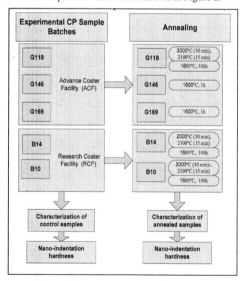

Figure 2: Experimental research plan.

RESULTS AND DISCUSSION

The initial nano-indentation measurements done on the reference (unannealed) samples as well as on the short duration annealed CP samples were done on a selected small area only. The nano-indentation measurements on the long duration (100 h) annealed samples of batch B10 at 1600°C was done at three positions on three CPs each. The purpose of this exercise was to establish if there is a significant difference in measured values between the CPs. These measurements revealed no significant difference in values between the CPs manufactured in the same batch and annealed in the same furnace load. The measurements for batches G118 and B14 were only done on one particle but still at three different positions as indicated in Figure 3 and a total of 27 measurements per batch were measured. Figures 4 to 9 were based on the average values of the results obtained.

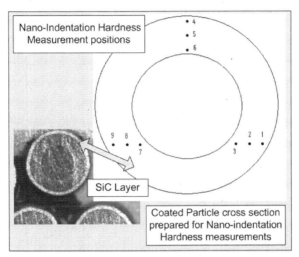

Figure 3: Indentation positions for CP samples after long duration (100 h) annealing at 1600°C

The effect of high temperature annealing on the nano-indentation properties of polycrystalline SiC from PBMR coated particles are schematically presented in Figures 4 to 7. The nano-indentation hardness results of batches G118, B14 and B10 after 30 minutes annealing at 2000°C showed a significant decrease of 30.2%, 40.1% and 56.3% respectively if compared with the unannealed values. From Figures 4 and 5 it is further observed that the results after annealing of batches G118 and B10 showed a significant decrease (sharp drop) between 1980°C and 2000°C. This specific trend is not observed for batch B14, as a more gradual decrease (no sharp drop between 1980°C and 2000°C results) in nano-indentation hardness is observed from unannealed to 2100°C.

Figures 6 and 7 showed that no significant differences in the nano-indentation hardness measurements between the different coater facilities are observed although specifically the two RCF coater batches (B14 and B10) yield higher average values than those of batch G118 of the ACF coater. The elastic modulus of the batch G169 is significantly lower than those of batch G118 and both of these batches are manufactured in the same ACF coater facility. It is noted that the elastic modulus for the RCF batches as manufactured showed a significant higher value if compared to those of the three ACF batches.

It is therefore important to investigate the differences in manufacturing conditions between these two batches as well as to compare the other metallurgical properties of samples manufactured in these two coaters.

The effect of long duration of high temperature annealing on the nano-indentation properties of cubic polycrystalline SiC from PBMR coated particles are presented in Figures 8 and 9. The effect of long duration annealing was more prominent on batches G118 and B10 by a significantly decrease in both the nano-indentation hardness (35.2% and 49.0% respectively) and elastic modulus (45.3% and 64.0% respectively). No trend is observed for batch B14 and it is further observed that the elastic modulus standard deviation of this batch are larger (72.3) if compared with those of the other two batches namely G118 (46.5) and B10 (27.5).

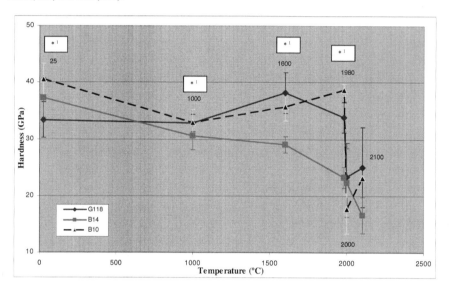

Figure 4. SiC Nano-indentation Hardness Results as a function of high temperature short duration annealing*[1].

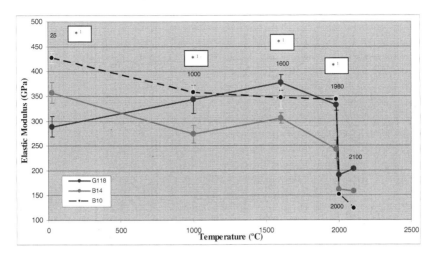

Figure 5. SiC Nano-indentation elastic modulus results as a function of high temperature short duration annealing*[1].

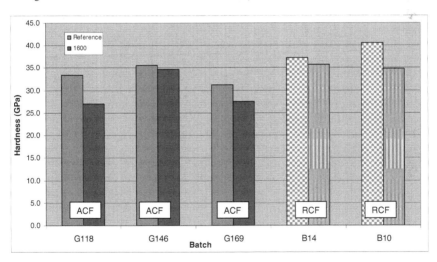

Figure 6. The effect of coater facility on the SiC Nano-indentation hardness results of five experimental PBMR TRISO coated particle batches as a function of short duration annealing at 1600°C*[1].

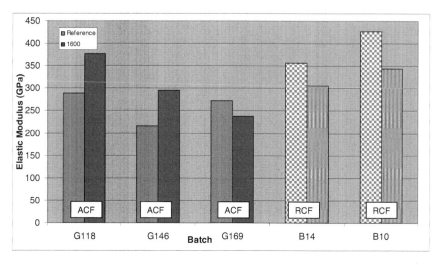

Figure 7. The effect of coater facility on the SiC Nano-indentation elastic modulus results of five experimental PBMR TRISO coated particle batches as a function of short duration annealing at 1600°C*[1].

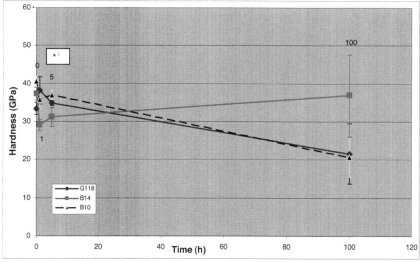

Figure 8. SiC Nano-indentation hardness results as a function of annealing time at 1600°C*[1].

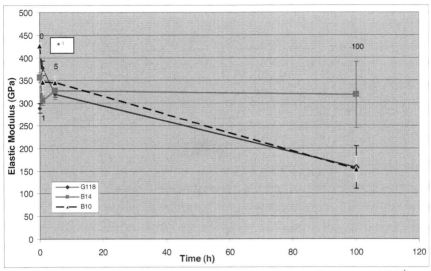

Figure 9. SiC Nano-indentation elastic modulus as a function of annealing time at 1600°C*[1].

CONCLUSION

The nano-indentation measurements of this study showed significant trends prior to irradiation as a function of annealing temperature and therefore this type of characterization is valid for a comparative characterization technique between PBMR batches. The effect of long duration annealing on the nano-indentation results also indicate a useful comparative technique between different batches. It is further concluded that no significant difference in nano-indentation hardness results is observed to differentiate between the two coaters, but that higher elastic modulus for the RCF batches, is suggested.

This study also revealed that the interpretation and comparison between the nano-indentation properties of SiC as reported by different researchers need to be viewed with care as it is clear that the manufacturing conditions and sample preparation and orientation play a significant role on the values obtained.

It is recommended that these results be interpreted in conjunction with detail analysis of transmission electron microscopic (TEM) examination as no microstructural and phase changes are observed from the preliminary TEM analysis after annealing which is supportive of these trends described above. It is further recommended that a full analysis also be done to compare the specific manufacturing conditions, other metallurgical measurements like grain size, defect density and SiC layer density, to establish the reason for the changes observed.

ACKNOWLEDGEMENTS
The authors are grateful for the support of PBMR and NMMU laboratory staff namely William Goosen, Jacques O'Connell and Jaco Olivier.

FOOTNOTES

[*]Nano-indentation hardness results for annealing temperatures from 1000°C to 1980°C graphically presented previously by Van Rooyen et al.[1] and are referenced in this article to show the effect of the annealing done at even higher temperatures of 2000°C and 2100°C. These results are also further analyzed and discussed in perspective of the higher temperature annealing and % changes observed.

REFERENCES

[1]I. J. van Rooyen, J. H. Neethling and J. Mahlangu, Influence of Temperature on the Micro- and Nanostructures of Experimental PBMR TRISO coated particles: A Comparative Study, *Proceedings of the 4th Internation Topical Meeting on High Temperature Reactor Technology, 28 Sept 2008 Washington DC USA*, HTR 2008-58189 (2008)

[2]D. Helary, O. Dugne, and X. Bourrat, Advanced characterization techniques for SiC and PyC coatings on high temperature reactor fuel particles, *Journal of Nuclear Materials*, **373**, 150-156 (2008)

[3]CSM Instruments for Advanced Mechanical Surface Testing: Handbook on Instrumented Indentation 1-85 (2007)

[4]W.C. Oliver, and G.M. Pharr, An improved Technique for determining Hardness and Elastic moduli using load and displacement sensing Indentation experiments, *Journal of Material's Research*, **7**, 1564-1583 (1992).

[5]I.N. Sneddon, The relation between load and penetration in the axisymmetric bousinesq problem for a punch of arbitrary profile. *Int.J Eng.Sci.*, **3**, 47 (1965))

[6]J. D. Reddy, A. A. Volinsky, C. L. Frewin, C. Locke and S. E. Saddow, Mechanical Properties of 3C-SiC films for MEMS Applications, *Mater.Res.Soc.Symp.Proc.*, **1049** (2008)

[7]Y. Ling, Low-Damage Grinding/Polishing of Silicon Carbide Surfaces, *SIMTech Technical Report* (PT/01/001/PM)

[8]E. López-Honorato, P.J. Meadows, J. Tan, and P. Xiao, Control of stoichiometry, microstructure, and mechanical properties in SiC coatings produced by fluidized bed chemical vapor deposition, *J. Mater. Res.*, **23**, No. 6, (2008)

[9]E. López-Honorato, P. J. Meadows, J. Tan, Y. Xiang and P. Xiao, Deposition of TRISO Particles With Superhard SiC Coatings and the Characterization of Anisotropy by Raman Spectroscopy, *Journal of Engineering for Gas Turbines and Power*, **131** 042905-1 (2009)

[10]E. López-Honorato, J. Tan, P.J. Meadows, G. Marsh, and P. Xiao, TRISO coated fuel particles with enhanced SiC properties, *Journal of Nuclear Materials* **392**, 219–224, (2009)

[11]J. Tan, P.J. Meadows, D. Zhang, Xi Chen, E. López-Honorato, X. Zhao, F. Yang, T. Abram and P. Xiao, Young's modulus measurements of SiC coatings on spherical particles by using nano-indentation *Journal of Nuclear Materials*, **393**, 22–29(2009)

[12]Y. Katoh and L.L. Snead, Mechanical properties of cubic silicon carbide after neutron irradiation at elevated temperatures. *Journal of ASTM International*, **2**, no. 8, paper ID JAI12377, (2005)

HIGH TEMPERATURE MECHANICAL LOSS OF NANOSTRUCTURED YTTRIA STABILIZED ZIRCONIA (3Y-TZP) REINFORCED WITH CARBON NANOTUBES

Mehdi Mazaheri, Daniele Mari, Robert Schaller
École Polytechnique Fédérale de Lausanne, Institute of Physics of the Condensed Matter, Station 3, 1015 Lausanne, Switzerland
mehdi.mazaheri@epfl.ch and mmazaheri@gmail.com

Zhijian Shen
[2] Department of Inorganic Chemistry, Arrhenius Laboratory, Stockholm University, S-10691 Stockholm, Sweden

ABSTRACT

High temperature mechanical spectroscopy measurements were conducted on carbon nanotube-reinforced fine-grained 3Y-TZP ceramics processed by conventional (CS, grain size ~ 350 nm) and spark plasma (SPS, grain size ~ 100 nm) sintering. The mechanical loss spectra are composed of a peak and an exponential background appearing at low frequencies or high temperatures. The SPS samples showed a much higher level of internal friction and lower creep resistance, which can be attributed to the easier grain boundary sliding in nanosize-grained specimens. The addition of carbon nanotubes resulted in a decrease in damping with respect to the high purity of zirconia powder and possibly in a reduction of creep.

INTRODUCTION

Toughness improvement of ceramics, even the more classical ones, is still nowadays a subject of interest[1, 2]. Large increase in strength and fracture-toughness is frequently documented for ultra-fine ceramic/metallic structures; a phenomenon that is explained in part by small grain sizes using the well-known Hall–Petch relationship[3]. Grain refining is a promising route for simultaneous increase of mechanical strength and fracture toughness. Such improvement can be attributed to the fact that grain boundaries act as obstacles against deformation. For instance, application of two-step sintering method resulted in processing nanostructured cubic stabilized zirconia[4], which exhibits up to ~ 96% increase in the fracture toughness (i.e. from 1.61 to 3.16 MPa m$^{1/2}$), when the grain size is reduced from ~ 2.15 μm to ~ 295 nm. In addition to mechanical advantages, using nanopowder for fabrication of nanocrystalline parts, considerably, enhances sinter ability at lower temperatures rather than those of micrometric grains.

Spark Plasma Sintering (SPS) is a promising technique for production of nanostructured ceramics. SPS is a newly developed sintering process that combines the use of mechanical pressure and microscopic electric discharge between the particles. The enhanced densification in this process has been attributed to localized self-heat generation by the discharge, activation of the particle surfaces, and the high speed of mass and heat transfer during the sintering process[5]. As a result, samples can rapidly reach full density (in a few minutes) at relatively low temperatures. This process has been used to prepare a large variety of nanograined ceramics, including 3Y-TZP, Al_2O_3 and $BaTiO_3$.

According to the above introduction, processing of nanostructured ceramics (instead of micrometric grains), can increase fracture toughness at room temperature. On the other hand, high temperature mechanical properties of ceramics and especially fine-grained 3 mol% yttria stabilized tetragonal zirconia polycrystals (3Y-TZP) depend highly on grain boundary (GB) properties[6]. GB sliding is an important mechanism of plastic deformation for fine-grained ceramics. It is mostly attributed to the fact that, when nanostructure is concerned, GB sliding can be activated at much lower temperatures than in the case of coarser grain size[6, 7]. Hence, it seems that application of nano-structured ceramics at room temperature is reasonable. But, the domination of GB sliding at high temperature will deteriorate creep resistance. According to the general equation of creep for ceramics, the creep rate ($\dot{\varepsilon}$) is inversely proportional to the grain size:

$$\dot{\varepsilon}(\sigma, T) := A \left(\frac{\sigma - \sigma_o}{G}\right)^n \left(\frac{b}{d}\right)^p D \tag{1}$$

where σ is the applied stress, σ_o the threshold stress (which depends on the nature of the GBs, but is zero in many cases), G the shear modulus, n the stress exponent, b is the Burger's vector, d the grain size, p the grain size exponent and A a material constant. D is the diffusion coefficient expressed as $D := D_o exp(Q/kT)$, with D_o the pre-exponential factor, Q the activation energy which accounts for the underlying controlling mechanism and kT has the usual meaning.

In this work, CNTs were added to nanostructured zirconia samples in order to pin the grain boundaries and reduce GB sliding. As GB sliding is a source of energy dissipation, mechanical loss measurements are well suited to study such a mechanism. The mechanical loss spectrum of fine-grained 3Y-TZP reinforced with CNTs, processed by spark plasma sintering (SPS, grain size ~ 100 nm) and by conventional sintering (CS, grain size ~300 nm) are presented in the current investigation.

EXPERIMENTAL PROCEDURE

Pure commercial yttria-stabillized zirconia (3Y-TZP, Tosoh, Japan) and multiwall carbon nanotube (CNT, synthesized by chemical vapor deposition method) were used as raw materials. The powder with an amount of 3 wt% CNTs was first mixed by attrition milling for 24 h using zirconia grinding balls. Then the mixture was processed using two methods: conventional sintering and spark plasma sintering. Using the conventional method, the mixture was initially cold pressed under 100 MPa, followed by sintering at 1673 K for 3 h in Ar atmosphere. Spark plasma sintering was carried out in Stockholm University SPS apparatus at 1523 K for 5 min under the pressure of 50 MPa. The density of the sintered bodies was measured by Archimedes method in distilled water and using an accurate (10^{-4} g) balance. For scanning electron microscopy (SEM, Philips XL30, Netherlands), the samples were cut, mechanically polished and thermally etched. The grain size of the sintered samples was determined by multiplying the average linear intercept by 1.56^8. For each specimen, 50 line segments were taken into account.

Mechanical spectroscopy measurements were carried out in an inverted forced torsion pendulum, working in a subresonant mode[9]. Samples of the size 25 * 1 * 4 mm^3 were excited in torsion and the deformation of the sample was detected by an optical laser cell. The mechanical loss, tan (ϕ), and the shear modulus, G, were measured from phase lag and the amplitude ratio between stress and strain signals, respectively. The measurements were performed under a high vacuum (10^{-3} Pa) as a function of temperature (at a fixed frequency of 1 Hz) in the range of 300-1600 K with a heating rate of 1 K min^{-1}. The curves obtained are defined as mechanical loss spectra. Compressive creep tests were also preformed on parallelepiped samples (3 * 3 * 8 mm^3) under stress of 8 MPa at 1350 K.

RESULTS AND DISCUSSION

A typical mechanical loss and shear modulus spectrum of a high purity nanostructured 3Y-TZP, processed by SPS, is shown in Figure 1 as a function of temperature. In the measured domain, two regions of special interest may be observed:

a) At about 400 K, a mechanical loss peak is observed, which is associated with a decrease in the shear modulus. The peak has been interpreted as due to the reorientation of elastic dipoles of the type "oxygen vacancy – yttrium cations" under the influence of the applied stress [6].

b) At higher temperatures (>1200 K), the spectrum consists of a monotonic increase of the high temperature damping background and a steep decrease in the material stiffness. The amplitude of the applied cyclic stress during measurements was of the order of 8 MPa. High temperature plastic deformation of polycrystalline ZrO_2 is related to grain boundary (G.B.) sliding, and consequently it seems reasonable to link the mechanical loss spectrum with G.B. sliding as a source of energy dissipation. Lakki[6, 7] has developed a theoretical model in order to interpret the high temperature mechanical loss of 3Y-TZP due to the relative sliding of hexagonal grains separated by an intergranular glassy phase (viscous layer). A dissipative force related to the viscosity of the layer and a restoring force corresponding to the elasticity of the neighboring grains (which limits the sliding at the triple point junctions) are two forces playing a key role in the relaxation process. In the spectrum shown in Figure 1, at temperatures higher than 1200 K, no peak but an exponential increase in the mechanical loss is observed, while, the abovementioned model (Lakki's model [6, 7]) accounts for a Debye peak. This behavior can be explained by considering that the restoring force in triple junction is decreasing drastically with temperature, due to large relative movement of grains (creep). When the restoring force vanishes, grain boundary sliding is no more limited. Consequently, the mechanical loss increases exponentially with temperature and one can consider this exponential background in the spectrum as the signature of creep.

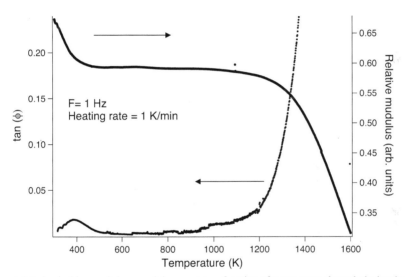

Figure 1. Mechanical loss and shear modulus spectra as function of temperature through the heating of a high purity 3Y-TZP processed via SPS.

Figure 2 shows the mechanical loss spectrum and the relative shear modulus of pure zirconia sintered conventionally in furnace and in spark plasma sintering apparatus. SEM micrographs of both grades of 3Y-TZP are presented in Figure 3 (a) and (b). The grain size of the samples sintered in SPS is about 3 times less than in the conventionally sintered ones (~100 nm for SPS versus ~ 350 nm for conventionally sintered sample). As a consequence, grain sliding in SPS sample starts at lower temperatures, and increases faster than in conventionally sintered samples. In addition, the shear modulus drops at temperature around 1100 K for SPS sample, while this temperature for conventionally sintered sample was observed at ~ 1300 K. It can be attributed to the easier grain boundary sliding in nanosize-grained specimens.

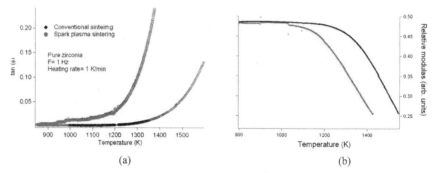

(a)   (b)

Figure 2. Mechanical loss spectra (a) and relative shear modulus (b) of high purity 3Y-TZP samples sintered via SPS and conventional methods.

(a)   (b)

Figure 3. SEM micrographs of pure 3Y-TZP specimens sintered by conventional sintering (a) and spark plasma sintering (b).

For comparison, creep measurements at low stress regime (8 MPa) and 1350 K were carried out in SPS and CS samples. Figure 4 shows that the creep strain in SPS sample is much higher than in CS one. This result is in agreement with mechanical spectroscopy (Fig. 2) and also with the power law creep equation (Eq. 1). A higher level of the mechanical loss at high temperature for SPS specimen can be interpreted by a worse creep resistance of nanosized grain sample.

Figure 5 compares the mechanical loss spectrum of pure (un-doped) and 3wt % CNTs doped 3Y-TZP both processed via SPS, with a mean grain size in the range of 100 nm. Doping the grain boundaries with carbon nanotubes results in a decrease in damping with respect to high purity zirconia. Daraktchiev et al.[10] have interpreted the decrease in damping at high temperature as due to the presence of carbon nanotubes (CNT) on the grain boundaries, which would reduce the grain boundary sliding process drastically. Ionascu et al.[11] have also shown that 3Y-TZP reinforced with CNTs or SiC exhibits a lower exponential background in the mechanical loss spectrum and a lower creep strain than pure zirconia. In these new composites, grain boundary sliding is more difficult and consequently a better creep resistance is observed. In other words, addition of CNTs can provide pinning centers on the grain boundaries, which are stable at high temperature. SEM micrograph of 3Y-TZP reinforced with CNT (Figure 6) is proved that CNTs dispersed homogenously in the microstructure and they can provide new pining centers, which could be against the grain-boundary sliding.

The mechanical loss spectrum of a 30 h annealed sample (reinforced 3Y-TZP) at 1600 K is added to Figure 3. In this case, the decrease in the level of mechanical loss is probably due to grain growth during annealing at 1600 K.

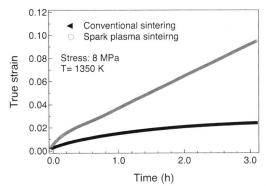

Figure 4. True strain as a function of time during creep test for pure and CNTs doped 3Y-TZP at 1350 K and stress of 8 MPa.

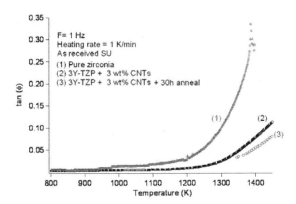

Figure 5. Mechanical loss spectra of high purity, CNTs doped 3Y-TZP, and annealed for 30 h at 1600 K.

Figure 6. SEM micrographs of 3Y-TZP reinforced with 3 wt% CNTs sintered by spark plasma sintering under vacuum.

CONCLUSION

In the present study, high temperature mechanical spectroscopy measurements were performed in zirconia: pure 3Y-TZP and 3Y-TZP reinforced with carbon nanotubes. Two types of samples with different grain size of 100 and 350 nm were processed via conventional and spark plasma sintering methods, respectively. SPS sample showed a much higher level of internal friction, which can be attributed to the easier grain boundary sliding in nanosize-grained specimens. The addition of carbon nanotubes resulted in a decrease in damping with respect to the un-doped zirconia which can be interpreted as a better resistance to creep of doped specimens.

ACKNOWLEDGMENTS

This work was financially supported by the Swiss Science Foundation.

REFERENCES

1.   Peigney, A., Composite materials: Tougher ceramics with nanotubes. *Nature Materials* **2003,** 2, (1), 15-16.
2.   Zhan, G. D.; Kuntz, J. D.; Wan, J.; Mukherjee, A. K., Single-wall carbon nanotubes as attractive toughening agents in alumina-based nanocomposites. *Nature Materials* **2003,** 2, (1), 38-42.
3.   Krell, A.; Blank, P., The Influence of Shaping Method on the Grain Size Dependence of Strength in Dense Submicrometre Alumina. *Journal of the European Ceramic Society* **1996,** 16, (11), 1189-1200.
4.   Mazaheri, M.; Hesabi, Z. R.; Golestani-Fard, F.; Mollazadeh, S.; Jafari, S.; Sadrnezhaad, S. K., The effect of conformation method and sintering technique on the densification and grain growth of nanocrystalline 8 mol% yttria-stabilized zirconia. *Journal of the American Ceramic Society* **2009,** 92, (5), 990-995.
5.   Munir, Z. A.; Anselmi-Tamburini, U.; Ohyanagi, M., The effect of electric field and pressure on the synthesis and consolidation of materials: A review of the spark plasma sintering method. *Journal of Materials Science* **2006,** 41, (3), 763-777.

6.   Lakki, A. Mechanical spectroscopy of fine-grained zirconia, alumina and silicon nitride, Ph.D. Thesis No. 1266, EPF-Lausanne, Switzerland, 1994.

7.   Lakki, A.; Schaller, R.; Bernard-Granger, G.; Duclos, R., High temperature anelastic behaviour of silicon nitride studied by mechanical spectroscopy. *Acta Metallurgica Et Materialia* **1995,** 43, (2), 419-426.

8.   Mendelson, M. I., Average Grain Size in Polycrystalline Ceramics. *Journal of the American Ceramic Society* **1969,** 52, (8), 443-446.

9.   Gadaud, P.; Guisolan, B.; Kulik, A.; Schaller, R., Apparatus for high-temperature internal friction differential measurements. *Review of Science Instruments* **1990,** 61, (3), 2671-2675.

10.   Daraktchiev, M.; Moortèle, B. V. D.; Schaller, R.; Couteau, E.; Forró, L., Effects of carbon nanotubes on grain boundary sliding in zirconia polycrystals *Advacned Materials* **2005,** 17, (1), 88-91.

11.   Ionascu, C. High temperature mechanical spectroscopy of fine-grained zirconia and alumina containing nano-sized reinforcements, Ph.D. Thesis No. 3994, EPF-Lausanne, Switzerland, 2008.

THE INFLUENCE OF NANOSIZE CARBON CONCENTRATION ON MECHANICAL PROPERTIES OF RBSIC

Cristiane Evelise Ribeiro da Silva*
1Universidade Federal do Rio de Janeiro – PEMM/COPPE/UFRJ
Rio de Janeiro, RJ, BRAZIL
*Presently INMETRO – Xerem, RJ, BRAZIL

Rosa Trejo
Oak Ridge National Laboratory (ORNL)
Oak Ridge, TN, USA

Sanghoon Shim
Oak Ridge National Laboratory (ORNL)
Oak Ridge, TN, USA

Edgar Lara-Curzio
Oak Ridge National Laboratory (ORNL)
Oak Ridge, TN, USA

Celio A. Costa
Universidade Federal do Rio de Janeiro – POLI/COPPE/UFRJ
Rio de Janeiro, RJ, BRAZIL

ABSTRACT
    The use nanosize carbon on the processing of RBSiC was studied. The materials were prepared in two different compositions, 28/72 and 50/50 wt % of nanosize carbon (C_N) and SiC, respectively. Molten silicon was infiltrated at 1550 °C in vacuum, resulting in 19-14 wt% of free silicon and about 74 wt% of SiC. When 50 wt% of C_N was used, there was nucleation and growth of SiC in the metallic Si phase, while for the 28 wt% C_N the nucleation was not observed and the growth might have happened in the SiC of the preform. Four point flexural tests were conducted at room temperature and 400, 800 and 1200 °C, and both compositions showed the same behavior for all tested temperatures, especially with respect to the stiffness. Nanoindentation tests were conducted, given a measured hardness of 24 GPa and an elastic modulus of 294 and 282 GPa for the 50/50 and 28/72 wt % compositions, respectively. The microstructure showed potential for exhibiting high strength at high temperature, but improvement in processing conditions must be done.

INTRODUCTION
    Reaction Bonded Silicon Carbide (RBSiC) ceramics combine very attractive features for a variety of applications. For instance, it is a near net-shape process conducted at low temperatures (1450-1700 °C) when compared to solid state SiC (2000-2200 °C), also, it has good tribological properties, high strength and excellent thermal shock resistance. Those characteristics make them suitable to applications such as combustion chambers, gas turbines, heat-exchangers, pump-sealing, welding-nozzles, etc [1,2].
    RBSiC is a composite material processed by infiltration of melted silicon in a porous preform of SiC and carbon resulting in a dense composite. The molten silicon reacts with the free

carbon and a new SiC is formed. The RBSiC is formed by particles of the old SiC and new SiC, metallic silicon (around 5 to 30 vol %), carbon and pores [3, 4, 5]. In the literature, these ceramics are usually found to be brittle. However, many efforts are been done to understand and improve the mechanical properties of the RBSiCs [6, 7].

The present study analyzes the applicability of nanosize carbon on the processing of RBSiC and how the mechanical properties depend on its amount. Thus, nanosize carbon in the amount of 28 wt% and 50 wt% and α-SiC were used to process the preform. The RBSiC properties were analyzed by Scanning Electronic Microscopy (SEM), Arquimedes density, X-Ray Diffraction (XRD) - Rietveld's method. Four point flexural tests were conducted at room temperature and 400, 800 and 1200 °C. The hardness and elastic modulus were determined by instrumented nanoindentation.

EXPERIMENTAL PROCEDURE

The green body (preforms) were prepared using an α-SiC (Saint-Gobain - Brazil), nanosize carbon (Degussa – Brazil) and metallic silicon (RIMA – Brazil). The physical properties of the raw materials are shown in Table I.

Table I- Properties of the raw materials used.

Raw Material	Mean Particle Size (d$_{50}$) (μm)	S (m^2/g)	Density (g/cm^3)
SiC	75.65	0.33	3.22
Nanosize Carbon (C$_N$)	Not measured	105.72	2.03
Si	420	Not measured	2.55

Two chemical compositions were tested and the major difference was the amount of nanosize carbon present, as shown Table II. The compositions were individually wet ball-milled as follows: the milling media was composed of bi-distilled water, PVA as binder (2 wt%), dispersant and alumina balls. The mixtures were ball-milled for 12 hours and afterwards, dried in an oven and then sieved through a 60 mesh screen. The powders were uniaxially pressed with 30 MPa and the resulting preform had the dimensions of 58.7x 65.2 x 5 mm. The processing occurred at 1550 °C in a 10^{-3} torr vacuum for 30 min. The silicon metal was placed beneath the preform and was infiltrated by capillarity.

Table II- Chemical composition and sample identification.

Preform Identification	Composition (wt%)	
	Nanosize Carbon (C$_N$)	SiC
50C$_N$/50SiC	50	50
28C$_N$/72SiC	28	72

The density of RBSiC materials were measured by Archimedes' method, and at least two samples of each plate were used. The volume fraction of the residual silicon was determined using the Rietveld's method. XRD analysis was done using X´Pert PRO Diffractometer (Philips, Panalytical) (CuKα radiation – 1.5418 Å, 30kV/15mA), the scanning area was 2θ from 10° to 100°, and scanning rate was 0.05° s⁻¹. The microstructural observations were performed using Scanning Electron Microscopy (SEM, Jeol JSM-6460LV).

The instrumented nanoindentation technique was used to measure the Bercovick hardness (BVH) and the elastic modulus (Oliver and Pharr equation). The tests were conducted at constant depth of 1000 nm and the needed force to reach this depth was then measured. The average BVH for each material was calculated over 50 indentations.

The mechanical strength were measured at room and high temperatures (400, 800 and 1200 °C), following ASTM 1161 [8] and 1211 [9], respectively. The tests were conducted in four-point bend, sample size of 4x3x45 mm and cross-head speed of 0.5 mm.min⁻¹. The average room temperature flexure strength was obtained from at least six samples, while the high temperatures strength used three samples.

RESULTS AND DISCUSSION

The infiltration process of both preforms resulted in materials with identical density and very similar amount of total SiC, even though the amount of silicon was different by 5%, as shown in Table III. The amount of SiC obtained in the 50C$_N$/50SiC preform was 76.8 wt%, more than 50% increase in the original amount of SiC in the composition. On the other hand, the preform 28C$_N$/72SiC ended up with a marginal increase of 1.7%. The final density and wt% of each phase are in the range of RBSiC found in the literature, which shows a very broad range of values, since the raw material, additives and processing vary from one author to another. For instance, Fernández et al. [10] reported the volume fraction of SiC in Refel, Cerastar RB and Cerastar RX being 89%, 80 and 74%, respectively. Paik et al [11] processed RBSCs with densities varying from 2.7 to 2.85 g.cm⁻³, while Wang et al. [12] found densities from 2.81 to 3.07 g.cm⁻³.

Table III - Archimedes density and phase percentage after infiltration

Preform Identification	Average Archimedes Density (g/cm³)	Phase percentage after infiltration (Rietveld) Si / SiC wt%
50C$_N$/50SiC	2.91 ± 0.05	14.4 / 76.8
28C$_N$/72SiC	2.91 ± 0.05	19.5 / 73.7

The microstructure and the hardness of each material will be discussed together, and both will aim to understand the effect of the C$_N$ in the present process. Figure 1 and Figure 2 show the microstructure and the nanoindentation performed on the 50C$_N$/50SiC and 28C$_N$/72SiC, respectively.

In Figure 1 it is observed the large original SiC grains surrounded by the Si metal and a very large amount of very small SiC particles distributed in the Si metal phase. These small particles were nucleated and grew from the combinations of C$_N$ and the silicon metal. This reaction consumed both elements and explain the phases quantified by the Rietveld method (Table III), where the amount of SiC formed was 26.8 wt% and the residual Si was 14.4 wt%. It is possible that a longer shelf time would increase the amount of new SiC formed and a network between the SiC grains would be created, which could improve the high temperature properties. The density measured reflects the presence of defects and they can be seen in the upper and lower left corner of this Figure. The Bercovick hardness measured by nanoindentation was 23 GPa. When Figure 2 is analyzed (28C$_N$/72SiC), the large original SiC grains and the Si metal can be clearly observed, but the small SiC particles are almost not seen, which shows that nucleation/growth corresponded to only 1.7 %wt. The average BVH hardness was 24 GPa, showing that both composition reached similar values and they are in the range reported in the literature[13] , compared to Vickers hardness.

The microstructure observation combined to Rietveld analysis and BVH help to understand why both materials possessed similar mechanical behavior. The average mechanical properties do represent an average of the phases present in the microstructure; consequently, a similar mechanical behavior should be expected since both compositions had around 74 wt% of SiC.

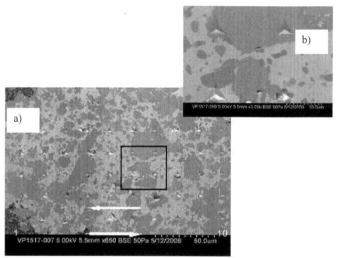

Figure 1. Microstructure and indentations of the 50C$_N$/50SiC composition: a) low magnification (650 X) and b) high magnification (3000 X). The presence of defects can be observed in the upper and lower left corners.

Figure 2. Microstructure and indentations of the $28C_N/72SiC$ composition: a) low mag. (650 X) and b) high mag. (3000 X). The presence of defects can be observed in the middle of micrograph.

The instrumented hardness tests were conducted on different phases of both compositions as can be seen by indentations on both Figures 1 and 2. When the indentation was done on the ceramic phase (SiC) alone, cracks emanated from their corners; on the other hand, when it was done on the metallic Si phase alone, no cracks were observed due to the plastic deformation. When two phases were presents, there was a mixed behavior. The dependence of the mechanical behavior to the microstructure was better demonstrated by the measure of the elastic modulus, as shown in Figure 3. For both compositions, the elastic modulus of the pure SiC phase was about 400 GPa, which is the common value for high dense solid state sintered SiC. As penetration goes below 100 nm, the elastic modulus decreases probably due it is a problem with the accuracy to which the tip of the indenter can be characterized. The rate of this decrease was faster for the $28C_N/72SiC$ than for $50C_N/50SiC$, and it might be related to the microstructure surrounding the crack propagation path. When pure metallic Si was indented, the measured elastic modulus was about 150 GPa, and there seems to be an strain hardening process, which is more pronounced for the $50C_N/50SiC$ composition (in the $28C_N/72SiC$ the behavior looks like elastic perfectly plastic). This behavior can be explained by the microstructure of the $50C_N/50SiC$ composition, since it has many small SiC particles nucleated and grown in the metallic Si phase, which act as barriers to dislocations movement therefore strain hardening the Si phase. When the indentation was carried out in two phase region the behavior was in between pure SiC and Si, and the $50C_N/50SiC$ composition was more homogenous than the $28C_N/72SiC$.

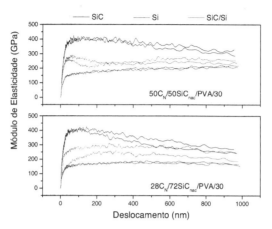

Figure 3 - Elastic modulus versus displacement of the $50C_N/50SiC$ and $28C_N/72SiC$ compositions.

The instrumented nanohardness test is able to measure properties on specific phase, as shown in Figure 3. However, the materials property is an average of the weight percent (or volume fraction) of the phases present. Two plates were processed for the $50C_N/50SiC$ composition and three plates for the $28C_N/72SiC$; the number of indentation was over 50 for each plate. The average elastic modulus (E) and the standard deviation are shown in Table IV. It can be observed that the $50C_N/50SiC$ composition had a much lesser scatter of data, which is a direct result from microstructure, since the amount of metallic Si is 25% lower than in the other composition (see Table III). When the overall E average is compared, both materials have similar elastic modulus, reflecting the identical density and the similar amount of SiC of both materials.

Table IV. Elastic modulus (E) measured by instrumented nanoindentation (NI)

Preform ID	Elastic Modulus by NI (GPa)	E average (GPa)
$50C_N/50SiC$ (plate 1)	289 ± 85	294
$50C_N/50SiC$ (plate 2)	298 ± 88	
$28C_N/72SiC$ (plate 1)	276 ± 98	282
$28C_N/72SiC$ (plate 2)	260 ± 99	
$28C_N/72SiC$ (plate 3)	309 ± 101	

The four point flexural strength of both compositions was measured at room and high temperatures (400, 800 e 1200 °C). The typical curves are shown in Figure 4. Basically, both compositions showed the same behavior for all tested temperatures, especially with respect to the stiffness. It shall be pointed out that linear behavior occurred throughout the room temperature tests and up to 0.75 mm of the LVDT displacement for all other temperatures, where plastic deformation started to take place. The flexure strengths are summarized in Table IV. It can be observed that the values are very similar for both compositions at all temperatures, with the exception of $28C_N/72SiC$ tested at room temperature, which showed a value of roughly half of the other measurements. This lower value cannot be attributed to the higher amount of metallic Si in the microstructure, since the high temperature strength was very similar and the amount of metallic Si was the same; instead, it was the result of processing defects, which were found in both compositions, as shown in Figure 5. These defects can be the responsible for the low strength values measured here, which were from 18% to 45% lower than reported in the literature. For instance, Martinez-Fernandez et al. tested a RBSiC (88 wt% of SiC) and measured the flexure strength to be 168 ± 30 and 245 ± 20 MPa [14]. However, in the present study, the strength was not reduced as the temperature increased, as noted by Chakrabarti and Das [15], with the difference that they tested at 1300 °C and higher. The ability to keep the strength at high temperature maybe explained by the SiC nucleated/grown in the metallic Si, but further testes need to be done to confirm it.

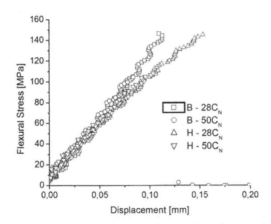

Figure 4 – Typical Flexural stress-displacement curves for both compositions at different temperatures.

Table IV. Flexural stress at room and high temperature (400, 800, 1200°C). For 25 °C tests, average was calculated from 6 samples, while for the others temperatures only from three samples.

Code	Flexural Strength (MPa)			
	25°C	400°C	800°C	1200°C
$28C_N/72SiC$	75 ± 7	135 ± 6	117 ± 32	131 ± 17
$50C_N/50SiC$	124 ± 34	112 ± 11	118 ± 14	139 ± 102

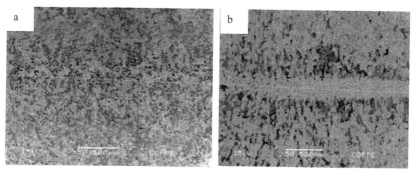

Figure 5. SEM Microstructure of the polished surfaces of (a) 28C_N/72SiC and (b) 50C_N/50SiC compositions.

So far, this study demonstrated that nanosize carbon as precursor for SiC formation is dependent on the amount of carbon (wt %) used. When 50 wt% of C_N was used, there was nucleation and growth of SiC in metallic Si phase, while for the 28 wt% C_N the nucleation was not observed and the growth might have happened in the previous SiC of the preform. This behavior is corroborated by the lower amount of metallic Si in 50C_N/50SiC composition and microstructure observation. Change in wet processing and/or in the sintering Schedule may increase the amount of SiC formed, which would be a good achievement.

CONCLUSIONS

The processing and infiltration schedule resulted in identical density for both compositions and there seemed to be a saturation limit to the formation of new SiC, which might be changed. The mechanical properties (hardness, elastic modulus and flexure strength) were again very similar for both compositions, which reflect the total amount of SiC in the materials. However, the microstructures were different and at high temperatures (800 and 1200 °C) the elastic modulus and strength were less affected, when compared to RBSiC reported in the literature. It is necessary to improve the processing conditions in order to improve density and strength.

ACKNOWLEDGMENT

This work was supported by CNPq, FINEP (contract 01.05.0966.00), PETROBRAS, ANP-PRH 35.

REFERENCES

[1]M. WILHELM., M. KORNFELD and W. WRUSS, Development of SiC–Si composites with fine-grained SiC microstructures, *Journal of the European Ceramic Society*, **19**, 12 2155-2163, (1999).
[2]M. K. AGHAJANIAN, B. N. MORGAN, J. SINGH, R. J. MEARS and R. A. Wolffe, A new family of reaction bonded ceramics for armor applications, *In Ceramic Armor Materials by Design, Ceramic Transactions*, **134**, 527-539 (2002).

[3]M. WILHELM, and W. WRUSS, Influence of annealing on the mechanical properties of SiC–Si composites with sub-micron SiC microstructures. *Journal of the European Ceramic Society*, **20**, 1205-1213, (2000).

[4]S. HAYUN, N. FRAGE, and M. P. DARIEL, The morphology of ceramic phases in BxC-. SiC-Si infiltrated composites, *J. Solid Chem*. **179**, 2875-2879, (2006).

[5]YU. P. DYBAN, Structuring of multiphase compacts in the SiC-Carbon System. I. Structuring in Green Blanks*, Powder Metallurgy and Metal Ceramics*, Ukraine, **40**, 1-2 (2001).

[6]YU, P. DYBAN, Structuring of multiphase compacts in the SiC-Carbon System. II. Structuring During Sintering, *Powder Metallurgy and Metal Ceramics*, Ukraine, **40**, 5-6 (2001).

[7]I. V. GRIDNEVA , Yu. V. MIL, G. G. GNESIN, S. I. Chugunova, Yu. P. Dyban and Z. V. Sichkar, Effect of crystallite size on the mechanical properties of self-bonded silicon carbide, *Powder Metallurgy and Metal Ceramics*, **20**, 137-141, (1981).

[8]STANDARD ASTM, Standard Test Method for Flexural Strength of Advanced Ceramics at Ambient Temperature, Designation: C 1161 - 02, West Conshohocken, *American Society for Testing Materials*, (2007).

[9] STANDARD ASTM, Standard Test Method for Flexural Strength of Advanced Ceramics at High Temperature, Designation: C 1211 - 02, West Conshohocken, *American Society for Testing Materials*, (2007).

[10] Fernández, J. M.; Muñoz, A.; López, A. R. A.; Feria, F. M. V.; Rodrígues-Domínguez, A.; Singh, M.; Microstructure-mechanical properties correlation in siliconized silicon carbide ceramics; *Acta Materialia* 51 (2003) 3259-3275.

[11] Paik, U.; Park, H-C.; Choi, S-C.; Ha, C-G.; Kim, J-W.; Jung, Y-G.; Effect of particle dispersion on the microstructure and strength of reaction-bounded silicon carbide; *Materials Sci and Eng.* A334 (2002) 267-274.

[12] Wang. Y.; Tan, S.; Jiang, D.; The effect of porous carbon perform and the infiltration process on the properties of reaction-formed SiC; *Carbon* 42 (2004) 1833-1839.

[13]F. GUTIERREZ-MORA, K. C. GORETTA, et al., Indentation Hardness of Biomorphic SiC, *International Journal Refractory Metals & Hard Materials*, **23**, 369-374, (2005).

[14]P. MARTÍNEZ-FERNÁNDEZ, et al., Biomorphic SiC: A New Engineering Ceramic Material, *International Journal of Applied Ceramic Technology*, **1** ,56–67, (2005).

[15]O. P. CHAKRABARTI and P. K. Das; High temperature load-deflection behavior of reaction bonded SiC (RBSC), *Ceramics International*, **27**, 559-563, (2001).

SI/SIC AND DIAMOND COMPOSITES: MICROSTRUCTURE-MECHANICAL PROPERTIES CORRELATION

S. Salamone, R. Neill, and M. Aghajanian
M Cubed Technologies, Inc.
1 Tralee Industrial Park
Newark, DE 19711

ABSTRACT

Composites containing diamond are desirable because of the unique physical properties of diamond, e.g., high hardness and high thermal conductivity. These properties are invaluable in traditional wear products, thermal management systems and any application that can take advantage of these distinctive characteristics. However, because diamond is a meta-stable compound that can convert to graphite when exposed to high temperatures and low pressures, processing diamond composites with conventional methods can be difficult.

A method was developed to fabricate reaction bonded SiC + diamond composites that exhibit increased density and Young's modulus, depending on the diamond concentration. The reaction bonding process has a key advantage that enables successful fabrication of diamond containing composites, i.e., processing temperatures are relatively low. This makes it possible to include lower stability compounds (e.g., diamond) into the microstructure without significant thermal degradation.

Successful compositional modifications include the incorporation of various diamond particle sizes into preforms (average particle size 22-65 μm) and increasing the amount of diamond particles (0-70 volume%) in the preforms. The effect of the final silicon metal content on properties is also reported. Microstructures of the various composites are correlated with the measured properties.

INTRODUCTION

Composites containing diamond particles are of interest in many fields due to the high hardness and high thermal conductivity of diamond. Engineering applications as diverse as wear products, such as mining, drilling, cutting and grinding, to electronic packages controlling thermal flow make use of these unique properties.[1-2] The processing techniques commonly employed usually involve sintering at high temperatures and high pressures with various reactive infiltration alloys and diamond particle sizes.[3-6] In contrast, the reaction bonding process developed and discussed here results in a Si/SiC and diamond composite fabricated at lower processing temperatures with a low weight percent of residual metal.

The reaction bonding process has two key steps. First a preform is made that consists of ceramic particles (e.g., SiC and Diamond) and carbon. Second, the preform is infiltrated with molten Si. During the infiltration process, a Si + C → SiC reaction occurs. Thus, in the case where the starting preform consists of SiC and diamond particles, and carbon, the finished ceramic contains four constituents, namely the original SiC and diamond particles, reaction formed SiC and residual Si. The relative fractions of the constituents are controlled by factors such as the SiC:C ratio and the packing of the starting preform.

The ability to manipulate the microstructure and composition, thereby tailoring the physical properties, is the key to the success of any materials system. The RBSC and diamond composites in the present work are an example of how variables (diamond particle size, volume percent, SiC grain size and carbon content) can be used to increase the complexity of the microstructure and manipulate physical properties.

EXPERIMENTAL PROCEDURES

Powder compacts consisting of 50 μm SiC particles, and 76, 65, 33 or 22 μm diamond particles, adjusted depending on the desired microstructure, were combined with specific levels of additional carbon. Preforms were consolidated using these SiC/diamond/C mixtures, and then infiltrated with molten Si to yield a Si/SiC and diamond ceramic.

The physical properties of the infiltrated composites were measured using several common techniques summarized in Table I. All the microstructures were characterized by examining fracture surfaces using a JEOL JSM-6400 Scanning Electron Microscope. The scanning electron microscope (SEM) images were taken in Back-Scattered Mode to differentiate the phases present (compositional differences) - e.g., Si metal (brightest phase), SiC (intermediate/gray phase), and Diamond (dark phase). All images are static fracture surfaces due to the difficulty of polishing diamond containing specimens.

Samples were sent to an independent laboratory to identify residual silicon metal content using inductively coupled plasma atomic emission spectroscopy (ICP-AES)

Table I: Summary of Properties and Techniques Used to Quantify the Various Composites.

Property	Technique	Standard
Density	Immersion	ASTM B 311
Elastic Modulus	Ultrasonic Pulse Echo	ASTM D 2845
Silicon Content	ICP-AES	
Phase Analysis	Powder XRD	

RESULTS AND DISCUSSION

Diamond is an excellent reinforcement for ceramics made by reaction bonding. It has extremely high hardness and Young's modulus[7-9] so small additions will have a large impact on properties. Moreover, diamond is transitioning from a specialty to a commodity raw material due to its growing use in broad applications (cutting tools, oil drilling, polishing, etc.). Therefore, future availability and cost concerns are becoming diminished. (Although the price is still significantly higher than silicon carbide powder). Diamond is also a meta-stable compound (i.e., it converts to graphite at high temperatures and low pressures)[10], however, it is stable under reaction bonding processing conditions.

In addition to the aforementioned advantages of diamond, it is also a very high density form of carbon. During the reaction bonding process, some of the diamond (e.g., the surfaces of the particles) will react with the infiltrating Si alloy to form β-SiC.[10] Due to the high density of diamond, the volume of SiC formed during this reaction is very high relative to the volume of diamond that is reacted (Table II), which allows the production of a reaction bonded ceramic with very low free Si content in the microstructure.

Table II: SiC Volume Created by Si + C → SiC Reaction

Form of Carbon	Density (g/cc)	Volume of 1 Mole of Carbon (atomic weight/density) (12/density)	Volume of 1 Mole of SiC Created by Reaction with Infiltrated Si (atomic weight/density) (40/3.21)	Volume Increase (%)
Carbon Black	1.70	7.0	12.5	78%
Graphite	2.25	5.3	12.5	136%
Diamond	3.52	3.4	12.5	268%

Thermal Stability

Experiments with diamond containing preforms were conducted at an isothermal temperature for several hours. This temperature range was initially used to achieve complete infiltration and still maintain the chemical and physical properties of the diamond particles. When processed in the preform and before infiltration, the diamond particles are protected with an in situ formed carbon coating to ensure that they survive the infiltration with silicon metal. However, in the interest of defining the processing parameters, uncoated and unprotected diamond powder was exposed to the same infiltration temperature/vacuum cycle as they would experience in an actual infiltration experiment. This experiment was considered the "worst case" scenario for diamond stability. Table 3 summarizes the results after the powder was sent for X-Ray diffraction analysis to determine the phases that were formed.

Table III: XRD Analysis of Diamond Powder After Exposure in Vacuum at Temperature

Average Particle Size (μm)	Weight Fractions (wt%)		
	Diamond	Graphite	SiC
65	95.5	3.1	1.4
33	91.3	3.9	4.8
22	86.0	8.5	5.5

Table III shows that as the diamond particles decrease in average size, they are more susceptible to instability and reaction with Si vapor. This reaction is most likely due to the increased surface area of the finer particles.

Final Silicon Content as a Function of Initial Carbon Content

The presence of residual silicon in RBSC materials has been shown to decrease the desired high stiffness and high hardness properties of this class of materials.[11] Increasing the carbon content prior to infiltration with molten Si allows more SiC to form at the expense of Si. This reaction creates a composite with an increased ceramic phase. With that in mind, silicon content has been measured for a number of infiltrated materials currently under investigation. This experiment is intended to give a better understanding of how processing can affect (and preferably minimize) the residual silicon left in any newly developed materials. Table IV contains the residual silicon content of 50 μm SiC + 14% 65 μm Diamond composites, as a function of carbon weight gain, as measured by wet chemistry analysis (ICP-AES). There is a clear inverse correlation between the carbon content of the preform and the residual silicon content. Also, with small changes in the density of the composites, a correspondingly large change in Young's modulus and hardness is evinced. This density/modulus correlation was also seen in biomorphic silicon carbide structures where increased density corresponds to a large increase in elastic modulus.[12] Therefore, by simply altering the initial carbon content can be used as a mechanism to tailor the density and Young' modulus of Si/SiC with diamond composites over a wide range.

Table IV: Residual Silicon Content of Infiltrated 50 μm Matrix Materials as a Function of Carbon Content of Preform

SiC Size (μm)	Dia. Size (μm)	Density (g/cc)	Young's Modulus (GPa)	Carbon Addition to Preform (%Wt Gain)	Silicon (Wt %)
50	65	3.13	449	0	8.78
50	65	3.14	451	2.66	8.00
50	65	3.16	464	4.49	6.70
50	65	3.17	482	5.26	5.83
50	65	3.18	477	5.72	4.70

Microstructure

Based on the results discussed above for infiltrated Young's modulus/density and measured silicon content, one would expect to be able to see a progressive change in microstructure with changing carbon content. Figure 1 consists of SEM images of the 50 μm SiC + 14% 65 μm Diamond material, highlighting the effect of the lowest and highest carbon contents, at two levels of magnification. In each of these micrographs, the darkest phase is the diamond, the gray phase is SiC and the white (or lightest) phase is residual silicon.

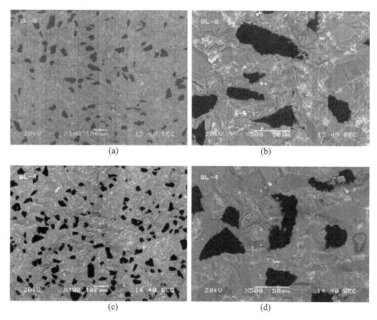

(a) (b)

(c) (d)

Figure 1: Fracture surfaces of 50 μm SiC + 14% 65 μm diamond infiltrated cross sections at various carbon contents and magnifications (a) no added carbon @ 100X (b) no added carbon @ 500X (c) 5.72 wt% added carbon @ 100X (d) 5.72 wt% added carbon @ 500X

At both magnification levels, one can detect a progressive diminishing of the silicon phase in favor of the SiC phase. This finding is consistent with the density, Young's modulus and chemical analysis results.

Diamond Content
In conjunction with the study of increased carbon content on properties, the effect of the volume percent of second phase has been studied. It is important to try to quantify the performance benefit for including additional percentages of diamond particles to an existing preform. At some empirically determined quantity, the diamond properties should start to dominate the bulk response of the composite. There are several reasons why this is a worthwhile endeavor:

- The cost associated with substituting SiC particles with diamond particles becomes very important. The SiC is approximately $6 per lb versus the nominally $100 per lb (current pricing) for diamond particles.
- As the volume percent of diamond is increased, the preform becomes more difficult to infiltrate.
- Weight reduction is always a very important parameter and knowing the change in performance as a function of volume percent diamond could translate to lighter materials.

Preforms containing diamond particles ranging from 7 volume percent to 70 volume percent were consolidated and infiltrated. The average size of the diamond particles was 76 μm and the average particle size of the SiC powder used was 50 μm. Figure 2 is a graph of the density and Young's modulus of the various samples compared to the reaction bonded silicon carbide containing no diamonds. The Young's modulus ranges from approximately 400 GPa for typical RBSC materials to 700 GPa for highly loaded diamond and Si/SiC composites.

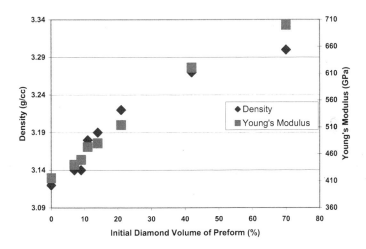

Figure 2: Density and Young's modulus as a function of initial diamond volume percent

The microstructures are shown in Figure 3. The increased concentration of diamond particles is clearly visible and pronounced. The microstructure of the sample with 70 volume percent initial diamond content is obviously different than the others. This is a bi-modal distribution of 76 and 33 μm particles consolidated to increase the packing of the preform. Since no SiC powder was used in the consolidation process, the grey phase is assumed to be reaction formed β-SiC.

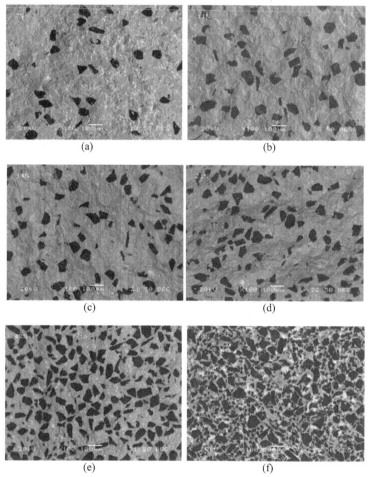

(a) (b)

(c) (d)

(e) (f)

Figure 3: Microstructures of Si/SiC and diamond composites containing (a) 7 volume percent, (b) 11 volume percent (c) 14 volume percent (d) 21 volume percent (e) 42 volume percent (f) and 70 volume percent initial diamond concentration

Young's Modulus & Density

A systematic study of the effect of diamond content on the final silicon content was conducted using the 50 μm SiC + 65 μm Diamond material. The density and Young's modulus were measured and reported in Table V as a function of the increasing volume percent of diamond contained in the preform. Both the density and Young's modulus increase as more diamond particles are added to the preform. Not accounting for packing behavior and reaction with silicon, the trends behave as expected. The density difference between the diamond and SiC is about 10 percent, so the increase should be modest as the SiC is systematically replaced with diamond. However, the Young's modulus for diamond is about two and half times greater than that of SiC. As the diamond content is increased, the Young's modulus should change significantly. This change is evident from the approximately 26 percent increase in Young's modulus seen in the sample sets.

The increase in diamond content also decreases the final silicon content. As described in previous sections, there is a reaction that takes place involving the diamond surface and the silicon alloy. More surface area available for reaction translates to a lower silicon metal content in the final composite.

Table V: Comparison of Physical Properties of RBSC Ceramics with Varying Diamond Volume Percent

Average SiC Particle Size – 50 μm			
Diamond Particle Size – 65 μm			
Diamond (vol%)	Density (g/cc)	Young's Modulus (GPa)	Silicon (wt%)
7	3.16	452	6.2
14	3.19	488	4.4
17.5	3.21	506	4.3
21	3.21	520	3.4

Microstructure

A comparison of the microstructures of the Si/SiC + diamond RBSC material with varying diamond content is provided in Figure 4. The commonality among all the fracture surface images is that the prevalent mode of fracture appears to be one of transgranular behavior.

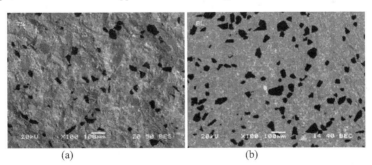

(a)   (b)

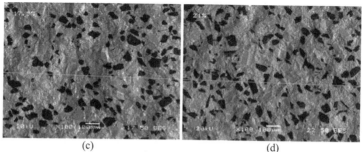

(c)   (d)

Figure 4: Fracture surfaces of 50 µm RBSC with 7 volume percent (a), 14 volume percent (b), 17.5 volume percent (c) and 21 volume percent (d) additions of the 65 µm diamond particles

Diamond Size Effect

Another parameter to explore in this series of composites was the effect of the size of the diamond added to the RBSC matrix at a constant volume percent of the second phase particle. A series of samples was made using 65 µm, 33 µm and 22 µm diamond at 14% diamond in the final product. This sampling was done for a 50 µm average particle size SiC matrix. A summary of the most significant results is provided in Table VI.

In overview, the data demonstrate several important points. First, as the diamond particle size decreases, the weight percent silicon remaining also decreases. The diamond dispersion throughout the RBSC matrix should vary with size, with smaller diamonds being more evenly dispersed than larger ones. However, there is some visual evidence of particle clustering in the 22 µm microstructure. Different diamond particle size distributions substituted into a base SiC matrix also contributes to a variation in overall packing of the preforms. The density and Young's modulus is affected by increasing reactivity of the diamond with the infiltration alloy as the particle size decreases (increasing surface area for reaction).

Table VI: Comparison of Properties of RBSC Ceramics with Varying Diamond Particle Sizes

SiC (µm)	Diamond (µm)	Density (g/cc)	Young's Modulus (GPa)	Silicon (wt%)
Diamond Content – 14 vol%				
50	22	3.21	487	2.3
50	33	3.19	479	3.9
50	65	3.19	488	4.4
RBSC Control (No Diamond)				
50	N/A	3.12	417	7.4

Young's Modulus & Density

Modulus and density of infiltrated tiles were measured for each composition above. The results give an indication of the reactivity for the different diamond sizes. In theory, the smaller the diamonds within the matrix, the more diamond surface area is available to react with the infiltration alloy. This increased reaction with decreasing diamond particle size should be reflected in a decrease of modulus and increase in density (due to Si to SiC conversion) as more diamond should be consumed during infiltration, thus changing the relative contents of diamond and SiC in the resultant matrix. As the diamond size decreased, the density increased but the Young's modulus was essentially the same. The value varied less than 2 percent which is within the error associated with the measurement.

Silicon Content

The residual silicon content of each of these compositions was tested using wet chemistry analysis. The increased reactivity of finer diamond is consistent with the trend of the measured silicon contents (i.e., finer diamond size resulted in lower Si-content, presumably due to increased reaction between the diamond (C) and the Si used to infiltrate the preforms). This reactivity also correlates positively with the measured densities of the infiltrated samples, where higher density compositions have lower measured Si-contents.

Microstructure

Figure 5 is comprised of microstructures comparing the fracture surfaces of the 50 μm RBSC with different average diamond particle sizes, at a constant volume fraction of 14 percent. The most noticeable difference in the images, at this scale, can be seen in the dispersion of the diamond particles and the prevalence of large Si metal regions. As the diamond particles decrease in size they are more uniformly distributed within the given area. In images (a) and (b), there are also regions of Si metal on the scale of the SiC matrix particles. Although present, these regions do not appear to be as large in the 22 μm (c) microstructure.

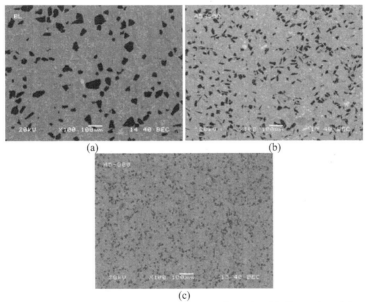

(a)   (b)

(c)

Figure 5: Representative microstructures of the 50 μm RBSC matrix with diamond particles averaging (a) 65 μm (b) 33 μm and (c) 22 μm

SUMMARY

In this study it was found that the physical properties of the Si/SiC + diamond composites can be greatly influenced by the addition of carbon to the preform. The density and Young's modulus of the samples increase and the final silicon percentage in the structure decreases with increasing amounts of carbon. In addition, the diamond content was shown to have a larger impact on Young's modulus than on density. There also appears to be a size/surface area effect of the diamond particle on the final silicon percentage, most likely due to increased reactivity with decreasing particle size. Knowing and understanding these effects can lead to a wide range of physical properties, making this class of materials extremely tailor-able.

ACKNOWLEDGEMENTS

This work was partially supported by the US Army Natick Soldier RD&E Center under contract number W911QY-06-C-0041.

REFERENCES

1.   I. E. Clark and P. A. Bex, "The Use of PCD for Petroleum and Mining Drilling," Industrial Diamond Review, 43-49 (1999).
2.   W. Zhu, G. P. Kochansky, S. Jin, "Low-Field Electron Emission from Undoped Nanostructured Diamond", Science 282 1471 (1998).
3.   Y. S. Ko, T. Tsurumi, O. Fukunaga, and T. Yano, "High Pressure Sintering of Diamond-SiC Composite," Journal of Material Science, vol. 36, 469-475, (2001).
4.   J. Qian, G. Voronin, T. W. Zerda, D. He, and Y. Zhao, "High-Pressure, High-Temperature Sintering of Diamond-SiC Composites by Ball-Milled Diamond—Si Mixtures," J. Mater. Res., vol. 17, No. 8, 2153-2160, (2002).
5.   S. K. Gordeev, S. G. Zhukov, L. V. Danchukova, and T. C. Ekstrom, "Low pressure Fabrication of Diamond-SiC-Si Composites," Inorganic Materials, vol. 37, No. 6, 579-583, (2001).
6.   G. Bobrovnitchii, A. Skury, A. Osipov, R. Tardim, "SiC Infiltrated Diamond Composites" *Materials Science Forum Vols.*, 591-593, 654-660, (2008).
7.   *Engineered Materials Handbook, Vol. 4, Ceramics and Glasses* (ASM International, Metals park, OH, 1991).
8.   NIST Structural Ceramics Database (SCD), Citation No. Z00225.
9.   H. Sumiya and T. Irifune, "Indentation Hardness of Nano-Polycrystalline Diamond Prepared from Graphite by Direct Conversion," *Diamond and Related Materials*, 13 [10] 1717-76, (2004).
10.   C. Pantea, G. A. Voronin, T. Zerda, J. Zhang, L.Wang, Y. Wang, T. Uchida, Y. Zhao, "Kinetics of SiC formation during high P–T reaction between diamond and silicon", Diamond & Related Materials 14, 1611 – 1615, (2005).
11.   S. Salamone, P. Karandikar, A. Marshall, D. D. Marchant, M. Sennett, "Effects of Si:SiC Ratio and SiC Grain Size on Properties of RBSC", in Mechanical Properties and Performance of Engineering Ceramics and Composites III, *Ceram. Eng. Sci. Proc.*, **28**, E. Lara-Curzio et al. editors, 101-109, (2008).
12.   M. Singh and J. A. Salem, "Mechanical properties and microstructure of biomorphic silicon carbide ceramics fabricated from wood precursors", *J. Euro. Ceram. Soc.*, **22**, 2709-2717 (2002).

MECHANICAL PROPERTIES AND FAILURE CRITERION OF SILICON-BASED JOINTS

L. M. Nguyen [1] [2] [3], **O. Gillia** [1], **Emmanuelle Rivière** [2], **Dominique Leguillon** [3]

(1) CEA/LITEN/DTH, F-38054 Grenoble, France, (33) 4 38 78 03 83, leminh.nguyen@cea.fr;
(33) 4 38 78 62 07 olivier.gillia@cea.fr;
(2) CNES/DCT/TV/SM SER Structures et Mécanique, France, (33) 5 61 28 30 88,emmanuelle.riviere@cnes.fr;

(3) Institut JLRA – CNRS UMR 7190, Université Pierre & Marie Curie (Paris 6), 4 place Jussieu, 75252 PARIS CEDEX 05 – France, (33)01 44 27 53 22, leguillo@lmm.jussieu.fr;

ABSTRACT

Brazing of material is a well-known technique widely used to assemble structural parts. Special brazing alloys are required for joining ceramics and ceramic matrix composites (CMC), such as BraSiC® silicon based brazes. In the present study, the BraSiC® process has been chosen to join sintered silicon carbide parts. This joining technology is advantageously used for applications in optical mirror for space, in high temperature heat exchangers and in chemical reactors.

In order to develop a fracture criterion of the assembly, detailed in part 6 of this article, the mechanical properties of the joint material must be measured within the assembly. We have therefore measured the Young modulus and the fracture toughness of the joint material. The second objective of this paper is focused on the determination of joint strength data using SiC-BraSiC®-SiC specimens tested in four point bending tests. The samples consist of two SiC beams butt-jointed by silicon-based braze. The effect of chemical composition and thickness of joint on mechanical properties of silicon-based joints were explored and correlated with SEM morphological characterization.

The third part is devoted to a theoretical model of fracture relying on an asymptotic approach of the stress field which avoids time consuming Finite Element computations carried out on strongly refined meshes. The predictions agree fairly well with the experiments.

1. INTRODUCTION

In recent publication, brazing of material is a well-known technique widely used to bulk up structures [1-3]. Several problems may appear. Brazed joints may have different thermal expansion coefficient with the ceramic which results in stress concentration in the joint area. The reaction between the braze and the substrate material may lead to the formation of fragile interfaces which weaken the strength of the assembly. The use of pure silicon joint associated with a strengthening silicide phase allow the creation of a non reactive but adhesive strong braze joint [1,2]. The BraSiC® process is based on a non-reactive silicon based braze. It has been chosen to joint sintered silicon carbide parts for our study. It is unique in terms of producing joints with tailorable microstructures wherein the composition and the resulting thermomechanical properties can be varied and controled. This has application in joining SiC parts to use at high temperature for heat exchangers and for chemical reactors. The behaviour at lower temperature (from cryogenic to 200°C) is generally addressed for spatial applications in optical mirror [3].

In order to determine the strength of the assembly, various testing configurations have been used: 4-point butt joint bending test, asymmetry bending test, tensile test, but poor data in literature was discussed on the base of a failure criterion. In this paper, one failure criterion is presented based on mechanical properties of joint material and strength fracture of the assembly. Prior to presentation and application of the criterion, the mechanical characterisation of the properties of the joint material (elastic parameters and fracture toughness) is detailed. The measure of the strength of the assembly is then presented in 4 points bendeing tests. The effect of chemical composition and thickness of joint on mechanical properties of silicon-based joints is also shown.

2. JOINT MORPHOLOGY & MATERIALS

The micrograph on Fig.1 shows the BraSiC® brazed joint between two substrates of silicon carbide after cutting and polishing cross-sections. The BraSiC® process is a non reactive brazing, id. no interfacial reaction is visible. No supplementary phase is created between the joint and SiC substrates. Mechanical and thermal properties of ceramic parts are therefore unaltered [1]. The material of the joint is a two-phase material. The test specimens prepared for this study were made of α-SiC substrates joined by using two new kind of BraSiC® named CEA1, and CEA2. The joint microstructure can be described by a distribution of silicon (grey) phase in an XSi_2 (white) phase (Fig.1). The size and the shape of the XSi_2 phase are randomly distributed.

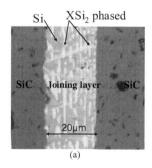

(a)

Fig.1. Microstructure of a joint brazed sample (Optical micrograph)

In aforementioned applications, the thickness of joints is generally situated between 2µm to 200µm. The image in Fig.2 obtained by tomography technique shows the microstructure obtained for different joint thicknesses. The images, at least for 90µm and 190µm thick joint, reveal a bi-phased dendritic solidification morphology where silicide whiskers like structures develop isotropically in 2D plane joint thickness.

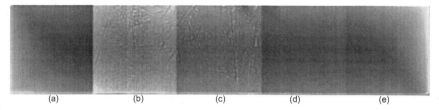

(a)   (b)   (c)   (d)   (e)

Fig.2. Tomography images through the thickness of a brazed joint layer of thickness from 3 to 190µm: (a) 3 µm; (b) 190µm; (c) 90µm; (d) 19 µm; (e) 3µm. For 3µm and 19µm joints the spatial tomography resolution is not sufficiently precise to obtain an image through the joint thickness

3. EXPERIMENTAL PROCEDURE
3.1 Micro and nano-indentation

The identification of BraSiC® joint material properties is generally difficult because of their small thickness. The use of BraSiC® bulk samples, like ingots, is not considered reliable because the solidification structure is different from the one it has in thin joints.

To determine the Young modulus and fracture toughness of brazed joints and silicon carbide, the micro and nano-indentation methods were applied using MTS Micro and Nano Indenter XP at different areas on perpendicular surface cuts of the joint. With these techniques, the fracture toughness is evaluated by measuring the crack length from the centre of the indent to the end of crack using an optical microscope or scanning electron microscopy (SEM) after indentation [4]. The value of fracture toughness is given by:

$$K_{IC} = a \frac{F}{C_0^{3/2}} \left(\frac{E}{H} \right)^{1/2} \tag{1}$$

where, K_{IC} is the fracture toughness, F the applied load, H the hardness, E the Young modulus and C_0 the crack length.

Load of 0.2 N was chosen for the nano-indentation method of the brazed joint, Figu.4. Loads between 3-5N were chosen for the micro indentation of the sintered silicon carbide. Concerning the brazed joint, micro-indentation is non applicable for joints smaller than 30μm, the indentation mark is too large compared to the joint thickness.

The Young elastic modulus is calculated from a part of the loading curve F(h) where F is the applied load and h is the penetration heigh of the indentor. The stiffness S writes:

$$S = \frac{dF}{dh} = \frac{2\beta}{\sqrt{\pi}} E_r \sqrt{A} \tag{2}$$

with A being the projected surface of the indentor imprint. The corresponding E was directly calculated through the expression:

$$E^r = \left(\frac{1-v^2}{E} + \frac{1-v_i^2}{E_i} \right) \tag{3}$$

where E, v and E_i, v_i are the Young modulus and Poisson's of the joint and the indentor respectively

3.2 Four point bending test

According to NF EN 843-1 and ASTM C1211-02 [5-6] for flexural monolithic ceramics, we used a specially designed fixture to avoid friction problem and geometrical defects of the experimental rig (Fig. 3). The load rate was set to 0.5 mm/min [7] and data collection was performed using Testworks software. The peak load reported by the software was used to calculate the flexure strength.

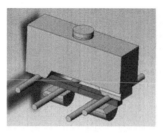

Fig.3. Partial view of the four point jig according to standard NF EN 843-1. Cylindrical supports allow remedial of geometric default of the experimental rig of specimen.

Specimens for bending tests were machined from butt joined plates. The joint is located in the middle of the flexure bars. Dimensions of specimens were respectively 4×3×46 mm3. The thickness of the joint is varied from 3±1 μm to 190 ±10 μm with the help of small steps introduced when brazing plates assemblies. All surfaces of samples are grinded along longitudinal direction and small chamfers are made on edge of the tensile surface during bending test. Apply stress was calculated from analytical equation (4) of classical bending beam submitted to homogenous 4-point bending tests. The equation is:

$$\sigma = \frac{3 * F(L_{EXT} - L_{INT})}{2bh^2}$$

(4)

where F is the maximum load (N), L_{EXT} is the outer span of upper support (mm), L_{INT} is the inner span of the lower support (mm), b is the beam width (mm), h is the beam thickness (mm). The outer and the inner span lengths were 40mm and 20mm respectively.

A minimum of seven joined specimens were tested at each thickness. Testing condition was at room temperature in air. For the as-received SiC materials, at least seven to fifteen specimens were tested at room temperature. After testing, fracture surfaces were examined by optical and scanning electron microscopy to identify the failure origins and fracture mechanism was evaluated on fracture surface

Fracture strength can be considered as a statistical problem due to the scattered distribution of flaws in the assembly. The probability of fracture in volume of specimens test can be given by Weibull

$$P_R = 1 - \exp\left[-\frac{1}{V_0} \int \left(\frac{\sigma_r - \sigma_u}{\sigma_0}\right)^m dV\right]$$

(5)

where m is the Weibull modulus, σ_0 is a scale parameter and σ_u is the threshold stress define as lowest of fracture stress below which fracture probability has a zero value. In the case of 4-point bending tests for ceramic, we admit σ_u is zero, the probability of fracture becomes:

$$P_R = 1 - \exp\left[-\frac{1}{V_0} \frac{V}{2(m+1)} \left(\frac{\sigma_r}{\sigma_0}\right)^m\right]$$

(6)

4. RESULTS AND DICUSSION

The results of fracture toughness are $K_{IC}(\text{SiC})$ = 2.5±0.4 MPa.m$^{1/2}$, $K_{IC}(\text{CEA1})$= 0.69±0.13 MPa.m$^{1/2}$ from micro-indentation and $K_{IC}(\text{CEA1})$= 0.73±0.16 MPa.m$^{1/2}$ from nano-indentation. The nano-indentation gives slightly higher fracture toughness than micro-indentation. The silicon carbide exhibits better cracking resistance than the joining layer. The joint is thus more fragile than the silicon carbide part.

The Young modulus (E) of the CEA1 joint obtained through nano-indentation test is 164±8.4 GPa. The joint exhibits a lower Young's modulus than the substrate material SiC which is 420 GPa [1].

Fig.4. Nano-indentation imprints in brazed joint CEA1 sample

The room-temperature (RT) flexural strengths of joined sintered silicon carbide are shown in Fig. 5.

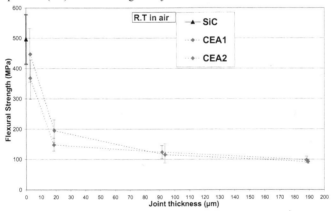

Fig.5. Flexural strengths of joint brazed at RT. Standard deviation is also reported for each point.

The room-temperature flexural strength of thin joint specimens CEA1 (3±1μm) is 446±91 MPa. It demonstrates slightly lower value than the fracture strength of bulk SiC structure which is around 496±91 MPa, but fracture does not always occur in the joint as discussed later. Fig.5 shows also the thickness dependence of fracture strength for CEA1. There is big significant change in the strength of joints as a function of thickness. A rapid decrease of the bending strength to 194±42 MPa was recorded when thickness varies from 3μ to 19μ. For the joint thickness of about 93μm, 190μm flexural strength decreases less dramatically to 115±42 MPa, 91±7 MPa respectively.

Under similar test conditions, the joined CEA2 specimens with different thickness of joint of 3μm, 18.6μ, 91μm and 190 μm has bending strengths of 368±72 MPa, 148±23 MPa, 123±24 MPa and 98±17 MPa respectively. The CEA1 joint shows a better strength than the CEA2 joint, for low thicknesses.

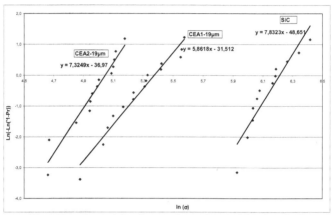

Fig.6. Experimental data-base to determine module of Weibull

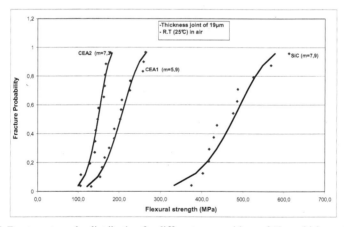

Fig.7. Fracture strengths distribution for different compositions of 19μm thickness joints

In Fig.7, a high dispersion of flexural strength for CEA1 and CEA2 joints were confirmed. The Weibull modulus was determined by a method of least square linear regression in a ln(-ln(1-P_r)) as a function of ln(σ), Fig.6.

The photography in Fig. 8 shows the effect of thickness on the failure surface after testing. Failure principally occurs in the brazing joint for a thick joint. This confirms lower fracture properties of the joint compared to SiC material.

For thin joints ($3\pm1\mu m$), there are five and eleven among fourteen tested specimens for which failure occurred in the brazing joint, respectively for CEA1 and CEA2 brazes. We called failure in the brazing joint when the joint cracked along interface or in the joint. The other failures occurred between the loading rollers, close or under the loading roller. Resistance of thin joints is almost as strong as SiC, the standard deviations overlap, thus possibility of breakage in the joint, or in SiC material depends where the critical defect is.

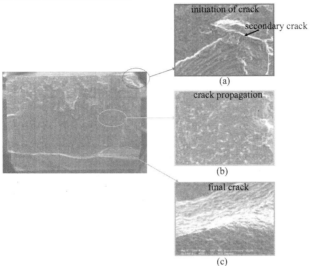

Fig.9. Fracture surface analysis of sample CEA1 testing; (a) region 1; (b) region 2; (c) region 3

In the BraSiC joint materials, the fractographies show typical brittle behavior. A cleavage plane and secondary crack zones can be observed in the area near the tensile surface in Fig.9 (a). We suspect that this is where the crack starts. It seems that the origin of the crack is located near the chamfer of the structure. In the next region, Fig.9 (b), fracture surface is very homogeneous, the crack extends in the centre of joint perpendicularly to principal stress direction. The direction of crack changes slightly to SiC at the final stage of propagation Fig.9 (c).

Analysis of SEM observation on rupture surface (see Fig. 8) shows that for thin joints, some small fragments of bonded SiC are observed on the brazed joint indicating good bonding between these materials, the rupture is not interfacial, id. not adhesive. The grain shape in the thin joint is more of granular shape whereas for thick joints, it is more like plates with large cleavage planes. The grain size in thick joint is larger than in thin joint, which may be related to the difference in term of mechanical performance.

SiC

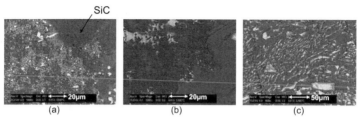

(a)   (b)   (c)

Fig.8. Typical surface fracture of CEA1 sample at different thickness (a) 3μm; (b) 19μm; (c) 90μm

5. THE THEORETICAL AND MATCHED ASYMPTOTIC PROCEDURE

A criterion using both strength and toughness parameters to predict crack onset at a sharp notch in homogeneous brittle materials has been presented in a previous paper [8]. It involves two related asymptotic expansions. One in the far field (outer expansion) takes into account the geometry of the structure but ignores the small joint thickness. The other in the near field (inner expansion) takes into account the joint layer but ignores the global structure geometry.

This two scale approach allows evidencing some significant parameters which intervene in the fracture process: the tensile stress to failure and the energy release rate [9]. Note that it is not a straightforward application of [8], it requires a modification of the expansion procedures [9].

5.1 Nucleation of a crack and tensile stress prior to failure

As observed in the experiments and as can be derived from the existence of singular points, the end of the adhesive layer is a privileged site for crack nucleation. A crack can appear either at the interface between the layer and the substrate (adhesive fracture, figure 10a) or inside the layer (cohesive fracture, figure 10c) or even inside the substrate (cohesive fracture, figure 10b) at a short distance of the interface. We assume that it's the crack length ℓ at initiation is of the same order or smaller than the layer thickness e. In the stretched domain the dimensionless crack length is $\xi = \ell / e$. Otherwise, if ℓ is far much larger than e, the expansions should be carried out first with respect to ℓ (and then with respect to $1/\xi = e / \ell$).

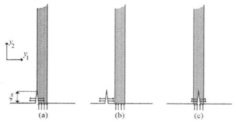

(a)   (b)   (c)

Fig 10. Adhesive (a) and cohesive (b and c) failures and the corresponding loading in the near field

The normalized tensile stress can be computed along the 3 presupposed crack paths, prior to crack initiation. The tensile stress prior to failure is not a decreasing function of the distance to the free boundary [9]. In terms of fracture, it means that if the actual minimum value is larger than the tensile strength of the material where the crack is supposed to grow, then the stress criterion holds true whatever the distance to the free boundary

5.2 The energy release rate

The change in potential energy δW^P between the two states (i.e. prior to and following the crack nucleation) can be written using a path independent integral [8]

$$G = -\frac{\delta W^P}{l} = e.g(\xi)T^2 + ... \tag{7}$$

The function $g(\xi)$ (MPa^{-1}) depends on the geometry of the perturbation, i.e. the length and the location of the crack. T is the tensile stress component. Nevertheless, it is observed that the three $g(\xi)$ curves almost merge into a single function close to a linear one as shown in figure 11 [9]

$$g(\xi) = 0.0044 * \xi \tag{8}$$

The line corresponding to the crack in the middle of the layer deviates slightly from this ideal line for small values of ξ.

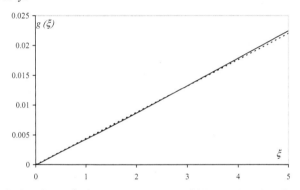

Fig.11. The function $g(\xi)$ for the matrix crack (solid line) and the best fit (dot line).

It must be noticed that the dimensionless crack length ξ is unknown, it is usually determined by a compatibility condition between the energy and stress inequalities [8]. This procedure cannot be used in the present case due to the special form of the stress field which implies that the fracture is completely governed by the energy condition

$$e\, g(\xi)\, T^2 \geq G_c \quad \Rightarrow \quad T \geq \sqrt{\frac{G_c}{e\, g(\xi)}} \tag{9}$$

The tensile stress at failure depends on the layer thickness as $1/\sqrt{e}$. The Log-Log diagram in Fig.12 shows the best fit with a linear function of the form

$$\ln(T) = -\frac{1}{2}\ln(e) + b \tag{10}$$

where b is the only degree of freedom. l

The experimental points come from 2 families of tests conducted on various grades of solder: CEA1 and CEA2. Fig.13 shows the corresponding approximation on the actual measured values. Of course all the curves must be bounded from above by the tensile strength of the matrix (SiC).

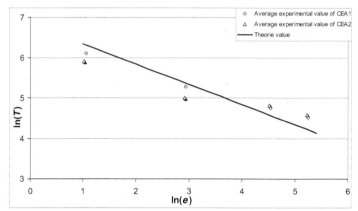

Fig.12. The least square fit with the measures according to eq. (10)

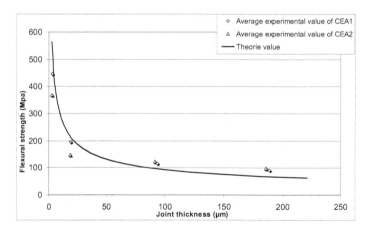

Figu.13. The approximation of the tensile stress at failure following eqn.(9).

6. CONCLUSION

The BraSiC® process is capable of producing joints with good mechanical properties. Young's modulus and fracture toughness were measured. These joint properties can be applied in a failure criterion of brazed structure in further work. Flexural strength of joint is affected by the joint thickness of the tested specimens, demonstrating that the thiner the joint, the stronger the assembly. The theoretical approach well traduces this evolution. Besides, a more homogeneous microstructure and composition have been observed when the joint thickness is reduced. Furthermore, to understand fracture mechanisms more precisely, the effect of testing atmosphere on the joint strength and influence of residual stresses are presently evaluated.

From a theoretical point of view, the fracture mechanism is governed mainly by the incremental energy release rate. It appears that the tensile stress at failure depends on the inverse square root of the brazed joint thickness. The results show that this prediction agrees fairly well with the actual values measured experimentally.

Acknowledgments

The authors would like to thank Boostec Society for providing SiC materials and EADS-ASTRIUM Company for helpful discussions and samples machining.

REFERENCES

[1] Matthieu Puyo Pain, Comportement mécanique d'assemblage de composite 2D SiCf/SiC brasé par un joint à base Silicium: Mesures de champs par corrélation d'images numériques en conditions extrêmes, Thèse Université Bordeaux, 2004
[2] M. Singh, Joining of sintered silicon carbide ceramics for high-temperature application, Journal of material science letters 17 (1998) 459-461
[3] Bath David A., Spain David Ness Eric, Williams Steve, BOUGOIN Michel Evaluation of segmented and brazed mirror assemblies, Optical materials and structures technologies. Conference No2, San Diego, CA, Etats-Unis (01/08/2005), vol 5868, pp 586805.1 – 586805.8
[4] Dylan J. Morris and Robert F. Cook, In Situ Cube-Corner Indentation of Soda-Lime Glass and Fused Silica, J. Am. Ceram. Soc., 87 [8] 1494–1501 (2004)
[5] NF EN 843-1, Mechanical properties of monolithic ceramics at room temperature-Determination of flexural strength, 2007. French Standard, 2007
[6] Standard ASTM C 1211 2, Standard Test Method for Flexural Strength of Advanced Ceramics at Elevated Temperatures, ASTM International
[7] George D.Quinn, Roger Morrrell, Design data for engineering ceramics: a review of flexure test, J. Am. Ceram. Soc. 74 191 2037-66 (1991)
[8] Leguillon D. (2002), Strength or toughness ? A criterion for crack onset at a notch, Eur. J. of Mechanics – A/Solids, 21. Ceramic based material, 61-72.
[9] LM. Nguyen, Comportement mécanique d'une jonction brasée SiC/BraSiC/SiC – critère de dimensionnement, PhD thesis, University Paris 6, in preparation

EFFECT OF VARIOUS SnAgTi-ALLOYS AND LASER INDUCED TEXTURING ON THE
SHEAR STRENGTH OF LASER BRAZED SiC-STEEL-JOINTS

I. Südmeyer, M. Rohde, T. Fürst
Karlsruhe Institute of Technology, Institut für Materialforschung I, Karlsruhe, Germany

ABSTRACT
    A pressureless sintered Silicon carbide has been joined to commercial steel using SnAgTi-brazing alloys. The cylindrical ceramic-steel-compounds were heated up to brazing temperatures of 950°C by a CO_2-laser system ($\lambda = 10.64$ μm) in an Argon stream. The braze pellets were dry pressed on the basis of commercially available powders and polished to a thickness of 300 μm. With the aim of improving the wetting and brazing behavior on silicon carbide, the compositions with different SnAgTi-fractions were tested. Also, the effect of Nd:Yag-laser texturing ($\lambda = 1064$ nm) on ceramic and metal joining surfaces on the compound strength was investigated.
    A homogenous wetting of the different braze fillers could be observed on the untextured and textured joining surfaces. However, only a partial formation of Ti-rich reaction zones was detected. The ceramic-steel joints were tested in a shear arrangement and the fracture stress values were evaluated according to Weibull. The results have shown that the shear strength improved with an increasing amount of Sn up to values of $\tau_0 = 20$ MPa. Furthermore the compound strength could be enhanced by surface textures to a characteristic shear strength of $\tau_0 = 24$ MPa.

INTRODUCTION
    The excellent thermal and mechanical properties of silicon carbide make this ceramic a material of high potential for a wide range of high performance applications. In order to profit from these properties, the integration of the ceramic into the metal surrounding is often necessary.
Those applications, that neither allow form nor force closure, can be solved by adhesive bonding in some cases. But adhesive bonding is limited in terms of the shearing capacity, so that brazing can be an adequate joining technique.
    Traditionally brazing of ceramics was only realizable by a pre-metallization. In the 1970s the joining technique was simplified by introducing active filler materials, which allow a direct brazing of ceramics. Especially successful vacuum brazing of alumina-metal joints with AgCuTi-fillers was well investigated and documented [1-4]. The wetting of oxide ceramics exhibited to be less difficult than that of non oxide ceramics as for example silicon carbide, because of the extreme affinity of active filler alloys to oxides [1,2,7,8]. In spite of the difficulties, successful active brazing of e. g. silicon carbide and silicon nitride to metals have been proved [2,5,6,9,10].
    Although oven brazing of different ceramic-metal-braze-systems have been well explored in the last years, it is still an issue of research due to several aspects. One aspect is the influence of design. The strength and reliability of a brazed ceramic-metal joint depend on the design of the the geometrical component. A further influencing factor is the individual wetting behaviour of each ceramic. Our investigation has shown for example, that although the brazing of silicon carbide with AgCuTi-filler was proved in literature, does not mean that it will work for all SiC ceramics.
    A wide range of investigations treat the reduction of residual stresses, which are created by the mismatch of thermal expansion and lead to a reduction of compound strength. The application of an interlayer is the most promising and successful method for reducing residual stresses [10-13]. Among others a rather new approach are the use of structured interlayer or textured joining faces in order to improve the compound strength [14,15].
    In addition to design methods the reduction of residual stresses can also be achieved by optimization of process parameters. The process variables of oven brazing are limited to heating and

119

cooling gradient, brazing temperature and dwell time. The process of laser brazing offers the additional possibility of a locally limited heat input, so that thermal induced stresses can be decreased by focussed heating. The selective warming up of the joining zone minimises the residual stresses as well as metal distortion.

But similar to reactive air brazing, laser brazing of ceramic-steel joints is not operated under vacuum or inert atmosphere. A reasonable laser brazing process can only be realised in an argon stream. For that reason results known from vacuum brazing experiments cannot be simply transferred to a laser brazing process, but need to be investigated individually.

2 EXPERIMENTAL PROCEDURES

2.1 Materials and geometry

A commercial steel (Mat. 1.1191) was brazed to pressureless sintered silicon carbide (Ekasic-F, ESK Ceramics). Two commercially available brazing foils on AgCuTi-basis were tested for brazing: The CB4 (Brazetec) and the Incusil-ABA (Morgan Chemicals). Additionally different SnAgTi-fractions were mixed and pressed on the basis of conventional powders (Alfa Aesar) and used for brazing. All materials composition and properties are listed in Table I.

Table I. Properties of the component materials

material	company	composition	strength	E	CTE	brazing temperature T	average grain / particle size
			MPa	GPa	10-6/K	°C	μm
SiC-PLS	ESK Ceramics	SiC 100%	350	410	4.7	-	1.9
C45E	-	Matno. 1.1191	620	212	11.1	-	-
CB 4	Brazetec	Ag-Cu-Ti 70.5-26.5-3	230	72	18.95	900-950	-
Incusil-ABA	Morgan Chem.	Ag-Cu-In-Ti 59-27.2-12.5-1.25	338	76	18.2	850-900	-
50SnAg2Ti	KIT , IMF I	Sn-Ag-Ti 50-48-2	-	68	-	950	Sn: 11 Ag: 70 Ti: 380
40SnAg2Ti		Sn-Ag-Ti 40-58-2	-	69	-		
30SnAg2Ti		Sn-Ag-Ti 30-68-2	-	70	-		

Cylindrical samples with an outer diameter of 16 mm were chosen for the brazing experiments. The metal carriers were prepared with a height of 16 mm and a drilling hole (d = 12 mm, h_i =14 mm) for heating up, figure 1 a). The drilling hole was sandblasted and a graphite layer was applied for increasing the absorption of the laser beam. The thickness of the ceramic parts was 4 mm. All joining faces were polished with 600 grain.

Cone shaped textures (d = 400 μm, h = 1600 μm) were applied on a selection of ceramic and metal joining faces with a Nd:YVO$_4$ laser, figure 1 b).

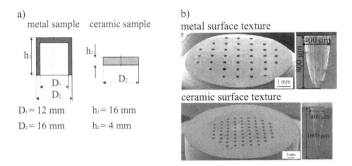

Figure 1. a) sample geometry and laser induced textures on b) metal surface and ceramic surface

2.2 Brazing Process

The brazing foil or pellet was placed between the joining partners and the compound was fixed in a screwing joint, so that a pressure of 1 MPa could be applied. The sample holder was positioned in an insulation box in order to reduce heat loss and thermal shock.

The laser beam ($\lambda = 10.6$ µm, d = 12 mm) was directed to the metal drilling hole and during brazing an Argon stream was aligned with the joining zone for avoiding oxidation reactions. During brazing the temperature was measured in joining zone by a pyrometer, as illustrated in figure 2 a).

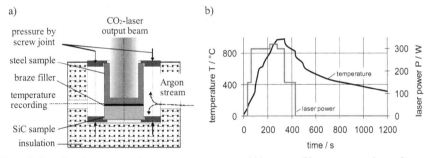

Figure 2. laser brazing process: a) process arrangement and b) course of laser power and sample temperature

The compound was heated up by a gradual increase of laser power until the brazing temperature was achieved. The brazing temperature of the SnAgTi-alloys was at 950°C. For successful brazing a dwell time of 120 s was necessary. The laser power was subsequently reduced to a temperature of about 400°C followed by natural cooling in the insulation box down to room temperature (figure 2b)).

2.3 Characterization of joints

The compound quality was evaluated by microsections of the different joints, which were analysed by scanning electron microscopy (SEM) and X-ray spectroscopy (EDX).

The compound strength of the brazing joints was determined by shear testing. The load was applied with a velocity of v = 0.005m/s and shear stress evaluated on the basis of the fracture load. Ten samples of each series were tested and the fracture shear stress was statistically evaluated according to the Weibull theory, so that a characteristic fracture shear stress τ_0 at a fracture probability of 63,2 % and Weibull modulus m was calculated as a reliability parameter.

3 RESULTS

3.1 Microstructure

The investigations started with brazing the silicon carbide with commercial brazing alloys on AgCuTi basis (CB4, Incusil-ABA, see table I). Only a view SiC-AgCuTi-steel-compounds could be investigated, because in most cases no wetting of silicon carbide was achieved. Despite a distinctive Ti- rich reaction zone only fragmentary wetting of SiC was achieved as visible in figure 3 a). The EDX-analysis did not provide an indication of reasons for that wetting behaviour.

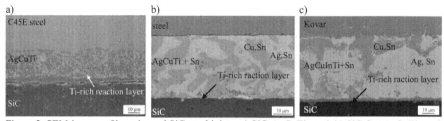

Figure 3. SEM-images of laser brazed SiC-steel joints: a) SiC-AgCuTi-steel, b) SiC-Sn+AgCuTi-steel and c) SiC-Sn+AgCuInTi-Kovar

The wetting of silicon carbide was improved by positioning a thin Sn-foil (Goodfellow, 0.25 μm) between ceramic and braze (figure 3 b)). However, a bonding between ceramic and braze was achieved by this interlayer, the ceramic joining partner fractured right after cooling. The use of additional interlayers between braze and metal, e.g. Cu, Mo or Ni or Kovar (figure 3 c)), still resulted in fractured ceramics after joining. The SEM-images figure 3 b) and c) show cross-sections of these fractured samples.

Further investigations using SnAgTi-filler for brazing based on the assumption, that Sn influenced the wetting of the carbide ceramic positively. According to literature [16,17] elements like Sn, that approve a lower surface energy in comparison to the filler materials, Ag and Ti, improves the wetting on ceramic surfaces. Moreover the residual stresses, which are induced during cooling by the ceramic/metal mismatch of thermal expansion, should be reduced according to the low melting temperature of Sn.

Wetting experiments with different SnAgTi-composites resulted in a sufficient wetting (wetting angle below 30°) for Sn-fractions of above 25 Mass %. In the following the brazing results of SnAgTi-alloys are presented and compared: 30Sn68Ag2Ti, 40Sn58Ag2Ti and 50Sn48Ag2Ti.

The SEM images of the polished cross-sections brazed with these SnAgTi-fillers, as pictured in figure 4 a) to c), exhibited an equally homogenous wetting for these three filler fractions. However an explicit, constant Ti-rich reaction zone was not detected along the ceramic-braze interface. Instead, Ti-particles in the middle of the brazing zone were recognized.

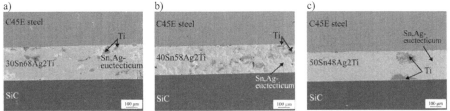

Figure 4. SEM-images of laser brazed SiC-steel joints: a) SiC-30Sn68Ag2Ti-steel, b) SiC-40Sn58Ag2Ti-steel and c) SiC-50Sn48Ag2Ti-steel

The inspection of the joints with cone textured ceramic surfaces resulted in a partial infill at a brazing temperature of 950°C and a complete infill of the brazing alloy at 1000°C. Those joints, which were brazed with textured steel surfaces, exhibited irregular and incomplete filling of the textures. Only view cones were completely filled with the filler material. Still, also for the textured compounds Ti-rich reaction layers were recognizable at the surface of silicon carbide and in the cone textures.

3.2 Shear strength and fracture surfaces

The values of fracture shear stresses and corresponding Weibull lines of non textured SiC-steel joints are illustrated in figure 5 a). It becomes clear, that an increase of compound strength was achieved by a rising Sn-fraction. A higher Sn-fraction obviously improved the bonding between ceramic and braze, although no major differences were detectable in the SEM-images. The results of shear testing are summarized in table II. The characteristic fracture shear stress was increased from $\tau_0 = 15.7$ MPa, that was achieved by 30Sn68Ag2Ti-joints, to $\tau_0 = 22.1$ MPa (40Sn58Ag2Ti) respectively $\tau_0 = 19.1$ MPa (50Sn48Ag2Ti) with Sn-fractions of 40wt% and 50wt%. Thereby a superior Weibull modulus of m = 4.9 was achieved with 50Sn48Ag2Ti-fillers in comparison to the 40Sn58Ag2Ti-samples (m = 3.1). As the value for reliability is almost as important as the shear strength, the 50Sn48Ag2Ti-alloy was used for further investigations on laser textured SiC-steel joints.

Table II. Characteristic fracture shear strength τ_0 and Weibull modulus m of laser brazed SiC-steel joints

SiC-steel joints	τ_0	m
braze filler	*MPa*	-
30Sn68Ag2Ti (non textured)	15.7	2.7
40Sn58Ag2Ti (non textured)	22.1	3.1
50Sn48Ag2Ti (non textured)	19.1	4.9
50Sn48Ag2Ti (textured ceramic)	24.5	5.5
50Sn48Ag2Ti (textured steel)	20.1	3.1

The texturing of joining surfaces raised the degree of shear strength. In case of the textured metal joining surface, a slight increase of shear strength was achieved without improving the reliability ($\tau_0 = 20.1$ MPa, m = 3.1, figure 5 b)). The cone textures on the ceramic joining surface provoked a more evident enhancement of shear strength up to $\tau_0 = 24.5$ MPa. At the same time the reliability of the compound was slightly advanced to a Weibull modulus of m = 5.5.

a)   b)

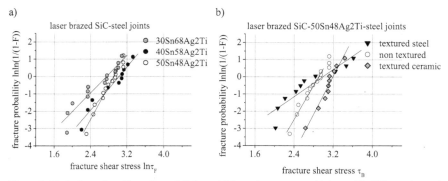

Figure 5. Weibull graphs of laser brazed SiC-steel joints: a) non textured joints with different SnAgTi-fillers and b) textured joints brazed with 50Sn48Ag2Ti

Looking at the fracture surfaces it becomes obvious that even for the good bonding of the non textured 50Sn48Ag2Ti-joints only a fragmentary strong adhesion between SiC and SnAgTi-filler was achieved. For samples, which fractured at lower shear stresses, the fracture line completely ran along the ceramic-braze interface, figure 6 a). In contrast for higher fracture loads, small ceramic fragments adhered to braze after fracture, figure 6 b).

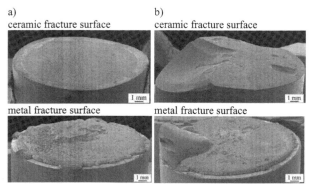

Figure 6. Fracture surfaces of non textured SiC-50Sn48Ag2Ti-steel joints: a) fracture stress $\tau_F \approx 15$ MPa and b) $\tau_F \approx 20$ MPa

A similar relationship between the fracture surface and the compound strength was observed for the textured compounds, figure 7 a), b).

a)
textured ceramic fracture surface

b)
ceramic fracture surface

metal fracture surface

textured metal fracture surface

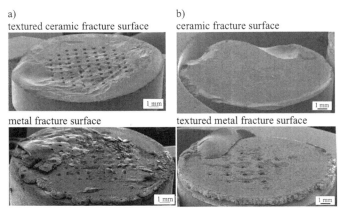

Figure 7. Fracture surfaces of textured SiC-50Sn48Ag2Ti-steel joints: a) textured ceramic surface ($\tau_F \approx 25$ MPa) and b) textured metal surface ($\tau_F \approx 25$ MPa)

4 CONCLUSIONS

Joining SiC to steel with commercial active braze fillers on AgCuTi-basis by a laser brazing process (CO_2-laser) resulted only in poor or no wetting of the ceramic joining surface. For that reason SnAgTi-filler with different Sn/Ag-fractions were developed and applied for the laser brazing process. The investigations have shown that a successful brazing of SiC with SnAgTi-alloys could be achieved for Sn-fractions of above 25 wt%.

Three different Sn:Ag fraction were used for brazing: 30Sn68Ag2Ti, 40Sn58Ag2Ti and 50Sn48Ag2Ti. Whereas the microscopic analysis did not exhibit differences between those three brazing alloys, the shear testing resulted in an increase of shear strength with a rising Sn-fraction. In that manner the joints brazed with SnAgTi-fillers containing a 30 wt% Sn achieved a characteristic shear strength of $\tau_0 = 15.7$ MPa and samples, which were joined with 50Sn48Ag2Ti, accomplished a strength of $\tau_0 = 19$ MPa. At the same time the Weibull modulus as parameter for the reliability was also improved from m = 2.7 (30Sn68Ag2Ti) to m = 4.9 (50Sn48Ag2Ti).

A further increase of the characteristic shear strength and the Weibull modulus was accomplished by laser texturing of the joining surfaces using the 50Sn48Ag2Ti-filler. The ceramic as well as the metal textured samples achieved higher shear stress values than the non textured compounds. A characteristic shear strength of 24 MPa was achieved by cone textures on the ceramic joining surface and a Weibull modulus of m = 5.5.

Still, the microsections of the laser brazed SiC-SnAgTi-steel-joints exhibited only thin, non continuous Ti-rich reaction layers. This way the residual stresses are reduced, but also the strength of the SiC-steel joints is limited. The inhomogenous Ti-rich reaction layer could be related to large Ti-particles, which were found in interface distal regions and anticipated the creation of a reaction layer. Though the use of finer Ti-powder could result in a more homogenous reaction zone, first investigations have shown, that the high reactivity of fine Ti-particles lead to undesirable pre-brazing oxidation, which evokes non-wetting of SiC. Further experiments with a variation of particle size and distribution are needed in order to optimize the wetting and joining behaviour of our SnAgTi-filler.

REFERENCES

[1] R. M. do Nascimento, A. E. Martinelli, A. J. A. Buschinelli, Recent advances in metal-ceramic brazing, Ceramica, 49, 178-198(2003).

[2] K. Suganuma, Y. Miyamoto, M. Koizumi, Joining of ceramics and metals, Annual Review of Materials Science, 18, 47-73 (1988).

[3] A. Kar, A. K. Ray, Characterization of Al_2O_3-304 stainless steel braze joint interface, Materials Letters, 61, 2982-2985 (2007).

[4] W. B. Hanson, K. I. Ironside, J. A. Fernie, Active Metal Brazing of Zirconia, Acta Materialia, 48, 4673-4676 (2000).

[5] L. Huijie, F. Jicai, Q. Yiyu, Microstructure of the SiC/TiAl joint brazed with Ag-Cu-Ti filler metal, Journal of Materials Science Letters, 19,1241-1242 (2000).

[6] J. K. Boadi, T. Yano, T. Iseki, Brazing of pressureless-sintered SiC using Ag-Cu-Ti alloy, Journal of Materials Science, 22, 2431-2434 (1987).

[7] L. Grimillard, E. Saiz, V. R. Radmilovic, A. P. Tomsia, Role of titanium on the reactive spreading of lead-free solders on alumina, Journal Materials Research, 21 (12), 3222-3233 (2006).

[8] M. G. Nicholas, T. M. Valentine, M. J. Waite, The wetting of alumina by copper alloyed with titanium and other elements, Journal of Materials Science, 15, 2197-2206 (1980).

[9] P. Prakash, T. Mohandas, R.P. Dharma, Microstructural characterization of SiC ceramic and SiC-metal active metal brazed joints, Scripta Materialia, 52, 1169-1173 (2005).

[10] H. Chang, S.-W. Park, S.-C., Choi, T.-W. Kim, Effects of residual stress on fracture strength of Si_3N_4/stainless steel joints with Cu-interlayer, Journal of Materials Engineering and Performance, 11 (6), 640-644 (2002).

[11] H. Hao, Y. Wang, Z. Jin, Y. Wang, The effect of interlayer metals on the strength of alumina ceramic and 1Cr18Ni9Ti stainless steel bonding, Journal of Materials Science, 30, 4107-4111 (1995).

[12] R. A. Marks, D. R. Chapman, D. T. Danielson, A. M. Glaeser, Joining of Alumina via Copper/Niobium/Copper Interlayer, Acta Materialia, 48, 4425-4438 (2000).

[13] J.-W. Park, P. F. Mendez, T. W. Eagar, Strain energy release in ceramic-to-metal joints by ductile metal interlayer, Scripta Materialica, 53, 857-861 (2005).

[14] A. A. Shirzadi, Y. Zhu, H.K.D.H. Bhadeshia, Joining ceramics to metals using metallic foams, Materials Science and Engineering A, 496, 501-506 (2008).

[15] H. Xiong, C. Wan, Z. Zhou, Increasing the Si_3N_4/1.25Cr.-0.5Mo steel joint by using the method of drilling holes by laser in the surface of brazed Si_3N_4, Journal of Materials Science Letters, 18, 1461-1463 (1999).

[16] R. Standing, M. G. Nicholas, The wetting of alumina and vitreous carbon by copper-tin-titanium alloys, Journal of Materials Science, 13, 1509-1514 (1978).

[17] M. G. Nicholas, T. M. Valentine, M. J. Waite , The wetting of alumina by Cu alloyed with Ti and other elements, Journal of Materials Science, 15, 2197-2206 (1980).

[18] M. Brochu, M.D. Pugh, R.A.L. Drew, Joining silicon nitride ceramic using a composite powder as active brazing alloy, Materials Science and Engineering A, 374, 34–42 (2004).

Acknowledgements

These studies were supported by the Deutsche Forschungsgemeinschaft (DFG) in context with the Sonderforschungsbereich 483 ''High performance sliding and friction systems based on advanced ceramics''.

The authors wish to thank M. Beiser and P. Bähr for preparation of the laser textured samples as well as F. Lutz for manufacturing the braze pellets.

CHARACTERIZATION OF POLED SINGLE-LAYER PZT FOR PIEZO STACK IN FUEL INJECTION SYSTEM

Hong Wang,[*] Tadashi Matsunaga, and Hua-Tay Lin
Ceramics Science and Technology Group, Materials Science and Technology Division, Oak Ridge National Laboratory[†]

ABSTRACT

Poled single-layer PZT has been characterized in as-extracted and as-received states. PZT plate specimens in the former were extracted from a stack. The flexure strength of PZT was evaluated by using ball-on-ring and 4-point bend tests. Fractography showed that intergranular fractures dominated the fracture surface and that volume pores were the primary strength-limiting flaws. The electric field effect was investigated by testing the PZT in open circuit and coercive field levels. An asymmetrical response on the biaxial flexure strength with respect to the electric field direction was observed. These experimental results will assist reliability design of the piezo stack that is being considered in fuel injection system.

INTRODUCTION

The mechanical strength and integrity of PZT (lead titanate zirconate) layers continue to be a major concern for the use of PZT fuel injector in heavy duty engine.[1] These PZT layers are subjected to mechanical loading while acting as functional elements in external electric field. The level of the loading can be substantial, regarding that the actuator is desired to generate more than 2,500 N of axial force to activate a typical control valve.[2] It is well known that the PZT is prone to any format of fracture due to its low fracture toughness (~ 1.0 MPa·m$^{1/2}$). Hence, axial splitting would be induced readily should a large enough flaw exist.[3] The eccentric loading from misalignment contributes to the fracture also.[1] Secondly, strain mismatch is inherently involved with the actuator that has an inter-digital electrode configuration.[4] The mismatch in the transitional zone between the active and inactive regions tends to induce cracks; this issue has been addressed in modern PZT actuator design by using a structure of stacked multilayer actuator plates.[5] This strategy has only limited the magnitude of tensile stress, but not eliminated potential cracking. Finally, any load on a layered structure, either mechanical or electric, is generally subjected to dynamic amplification. This is characteristically associated with the response time of < 100 μs used in a fuel injection cycle.[1]

The failure of PZT materials has been studied extensively using setups like 3-point bend[6] and ring-on-ring (RoR) flexure.[7] Studies on the relevant fracture of PZT materials have also been conducted in a variety of efforts. Indentation fracture,[8,9] compact tension,[10] and single-edge prenotched beam flexure[11,12] were all attempted. Particularly, behaviors of characteristic crack systems have been focused including those oriented normal and parallel to the poling.[13] However, these methods were originally developed for examining large artificial cracks and cannot be trusted to provide effective characterization of naturally occurring flaws.

[*] Corresponding author. Tel: 1-865-574-5601; email: wangh@ornl.gov.
[†] This manuscript has been authored by UT-Battelle, LLC, under Contract No. DE-AC05-00OR22725 with the U.S. Department of Energy. The United States Government retains and the publisher, by accepting the article for publication, acknowledges that the United States Government retains a non-exclusive, paid-up, irrevocable, world-wide license to publish or reproduce the published form of this manuscript, or allow others to do so, for United States Government purposes.

As a result, efforts have been made recently by the authors to characterize the mechanical strength and electric effects using ball-on-ring (BoR) tests;[14,15,16] fractography integrated in this approach enables the characterization of fracture behavior of materials based on natural flaws. These studies were carried out on the PZT fabricated by tape cast; the electric field did exhibit substantial effects on the biaxial strength of poled PZT-5A. A similar study is expected to be extended to piezoceramics used in the piezo stack considered for systems of interest and fabricated with other methods. This paper will describe our experimental studies that address some of these issues.

EXPERIMENTAL TECHNIQUE

Material and Specimen Preparation

A commercial piezoceramic, PZWT (Kinetic Ceramics, Inc., Heyward, CA), was examined in this study. This material is similar to PZT-5A and is currently being considered as one of the candidates for piezo stack in the fuel system of diesel engine.

Two groups of specimens were tested and evaluated. The first group was prepared from a stack through dissolving bonding agent using a chemical solution (Dynasolve 185, DYNALOY, Indianapolis, IN) at 150°C. The extraction usually took 30 to 90 minutes to have individual plates separated and detached from the stack. These plates were examined using an optical microscope afterwards to see if the surfaces were free of bonding agent. Those with residuals then went through another cycle. The stack for the extraction was obtained through Cummins, Inc. (Columbus, IN). The second group of specimens was supplied by Cummins, Inc. These specimens were electroded and poled by the manufacturer (Kinetic Ceramics, Inc.). The surfaces of as-received plates were covered with visible scratches. A laser profilometer (Rodenstock RM600, Rockford, IL) was used to quantify the condition. The R_a (arithmetical mean deviation) was measured from 0.8 to 1.4 μm and R_z (mean roughness depth) was from 12 to 20 μm. The plate specimens of both groups had same geometric dimensions. They were circle-shaped with two diametric ends cut into parallel flats with the same length. The nominal diameter was 15.40 mm, the flat-to-flat distance was 12.60 mm, and the thickness was 0.500 mm.

Test Method

Ball-on-ring test
A BoR set-up with the capability of electric loading[14,15,16] was used in tests. Two combinations of loading ball and supporting ring were examined. The first combination was a polymer-ceramic pair consisting of a 10 mm polymer ring and a 2 mm Si_3N_4 ball. The second combination was a steel-steel pair consisting of a 9.5 mm steel ring and a 6.35 mm steel ball. The latter has been proposed because it enables the application of an electric field. The electric loading method was same as before.[15,16] High voltage was provided by an amplifier (Trek 609E-6, Medina, NY) whose two monitor channels (voltage and current) were used in monitoring the contact and loading process. Mechanical loading was achieved through a displacement-controlled mode; most of the rates were set at 0.01mm/s and a small part of the rates was at 0.001 mm/s when a positive electric field was involved. A 10 lb load cell was used to monitor the mechanical load.

4-point bend test

A semi-articulated 4-point bend fixture was incorporated into the same test frame as in the BoR. In the literal sense, this 4-point means the 4 contact lines between the rollers and the plate under test. The loading and supporting spans were designed as 3.175 mm and 6.35 mm, respectively. A 100 lb load cell was utilized in measuring mechanical load. The cross-head speed remained at 0.01 mm/s in all of the tests.

Piezodilatometer test

Characterization of single-layer PZT necessitates the investigation on the response of PZT under a characteristic electric field; namely, coercive field (E_c). However, the datasheet did not provide any relevant information. A piezodilatometer developed at ORNL was used for this purpose. A 0.1 Hz triangular wave was employed to drive the PZT plate in thickness direction (x_3). The mechanical displacement in the plate plane (x_1) was measured using a linearly varying capacitance (LVC) sensor and the electric displacement (charge) in the thickness (x_3) was monitored with a modified Sawyer-Tower circuit.

Test Procedure

The *as-extracted* specimens were tested using both BoR and 4-point bend set-ups. In the BoR, two combinations of ball and ring as mentioned above were used to study the effect of contact on the strength response of PZT. The tests including both setups all were conducted without external electric field. The tensile side of plate specimen under flexing with respect to the polarity of electrode was randomly assigned since positive and negative sides of extracted plates were not tracked. The effect of which side of the flexure plate is tensioned has been demonstrated to be negligible for tape cast PZT in the open circuit (OC).[15,16] and it is assumed that the same holds for the pressed PZT. The *as-received* specimens were examined first using the piezodilatometer to determine the E_c of the PZT material. Then BoR tests were carried out under four electric conditions: open circuit, $E = \pm E_c$, and $E = + 2E_c$.

Failure of PZT specimens in the BoR exhibited multiple peaks on the load curve before the failure. The load at the first peak has been identified to be the fracture load[15,16] and used in the evaluation of flexure strength.[17] The estimate of disk radius (R) accounting for the irregular geometrical shape of the specimen in this evaluation was based on a recommended approach.[18] The load curve in the 4-point bend was linear to the failure. Flexure strength of PZT based on 4-point bend was evaluated according to the bending plate theory.[19] In all of these analyses, the electroded PZT layer was modeled as a monolithic layer regardless of electrode effect. The loading rate was about 24 MPa/s when the cross head speed was set at 0.01 mm/s; this rate was typical of fast fracture and the strength measured can be interpreted as representative of material's inert strength.

The probability of failure (P_f) as a function of failure stress (σ_f) can be expressed using a 2-parameter Weibull distribution as follows,

$$P_f = 1 - \exp[-(\sigma_f / \sigma_\theta)^m]$$ (1)

where σ_θ is the characteristic strength and m is the Weibull modulus. The characteristic strength is related to the material scaling parameter. Thirteen to nineteen specimens were tested under each load condition. The strength data sets were analyzed using Weibull statistics software (WeibPar, Connecticut Reserve Technologies, Inc., Strongsville, OH). The characteristic strength and Weibull modulus were estimated along with their 95% confidence intervals according to ASTM Standard C1239-00.[20] Finally, fractographical studies were conducted on selected failure specimens by following ASTM Standard C1322-96a.[21] Both optical microscopy (Nikon Nomarski Measure Scope MM-11, Tokyo, Japan) and SEM (Hitachi S4800 field emission scanning electron microscope, San Jose, CA) were used to examine specimens.

EXPERIMENTAL RESULT

Flexure Strength of As-Extracted PZT

The biaxial strength obtained by BoR tests was insensitive to the selection of contact. This can be seen from the overlaid estimate intervals of Weibull parameters in two contact conditions. The estimate intervals of the characteristic strength (σ_θ) and Weibull modulus (m) for the steel-steel pair were (64 MPa, 71 MPa) and (7, 14), respectively; those for the polymer-ceramic pair were (64 MPa, 77 MPa) and (4, 9). The results obtained with the steel-steel pair are given in Fig. 1 as a Weibull plot along with 95% confidence ratio ring. Fractography on selected specimens revealed that intergranular fractures prevailed over fracture surface and that volume pores appeared to be the type of strength-limiting flaw. A representative fracture surface with the fracture origin is given in Fig. 2. The Weibull strength distribution and confidence ratio ring for the 4-point bend are shown in Fig. 3. The flexure strength had values significantly lower than those obtained with the BoR flexure (Fig. 1). This was expected as the 4-point bend sampled a larger stressed volume and surface than the BoR flexure. A variety of strength-limiting flaws was also obtained that featured volume pore, porous region, agglomerate, and surface void. Pores and porous regions comprised more than 50% of sampled strength-limiting flaws.

Flexure Strength of As-Received PZT

The loops of mechanical strain (S_{31}) – electric field (E_3) and charge density (D_{33}) – electric field (E_3) are shown in Fig. 4 for the as-received single-layer PZT. Three plate specimens were tested. The E_c appeared to be quite stable in each test and their average was 1.05 kV/mm. BoR tests at the level of $+2E_c$ resulted in dielectric breakdown. As a result of that, the subsequent discussion will focus on the OC and $\pm E_c$ conditions only. Varying the cross-head speed from 0.01 to 0.001 mm/s in the field of 1.05 kV/mm did not result in any significant difference in flexure strength. These two sets of data were pooled and Weibull parameters were re-estimated. The 0.001 mm/s will be implicitly included when the rate is generally referred as 0.01 mm/s in the following pertaining to this electric field case.

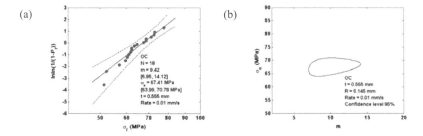

Fig. 1. (a) Weibull plot and (b) confidence ratio ring (95% confidence level) for flexure strength of single-layer PZT; ball-on-ring setup was with 9.5 mm steel supporting ring and 6.35 mm steel loading ball. PZT plate specimens were extracted from a stack.

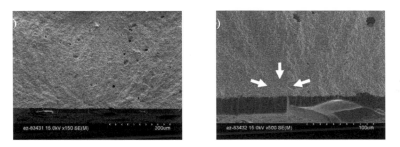

Fig. 2. (a) Fracture surface and (b) fracture origin featuring a volume pore in as-extracted PZT. Ball-on-ring setup was with 10 mm polymer supporting ring and 2 mm steel loading ball; the rate was 0.01 mm/s and failure stress was 60 MPa.

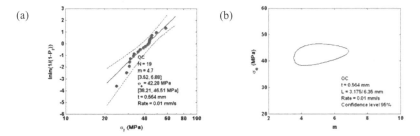

Fig. 3. (a) Weibull plot and (b) confidence ratio ring (95% confidence level) of flexure strength for single-layer PZT obtained using semi-articulated 4-point bend setup with 3.175 mm of loading span and 6.35 mm of supporting span; specimens were extracted from a stack.

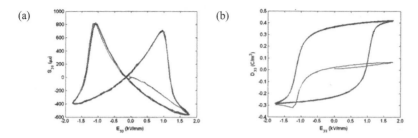

Fig. 4. (a) Mechanical strain (S_{31}) – electric field (E_3) loop and (b) charge density (D_{33}) – electric field (E_3) loop for as-received single-layer PZT. A 0.1 Hz triangular wave was used and data are based on piezodilatometer tests.

The PZT material exhibited an asymmetrical response of flexure strength with respect to the electrical load. The σ_θ and m in the OC were found to be 92 MPa and 20, respectively. These values are higher than with the as-extracted PZT specimens under similar loading condition. The electric loading at $E = 1.05$ kV/mm did not cause significant difference in strength value as shown in Fig. 5. The electric loading at $E = -1.05$ kV/mm resulted in a flexure strength appreciably lower than in the OC; the Weibull parameters were 89 MPa and 35, respectively.

DISCUSSION

Size Scaling of Strength

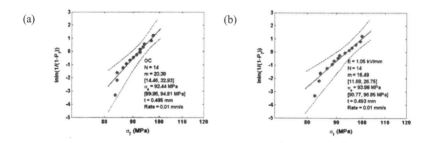

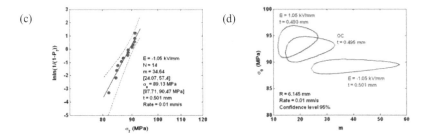

Fig. 5. Weibull plots for flexure strength of as-received single-layer PZT in (a) OC, (b) $E = 1.05$ kV/mm, (c) $E = -1.05$ kV/mm, and (d) confidence ratio rings (95% confidence level). Data were obtained using ball-on-ring setup with 9.5 mm steel ring and 6.35 mm steel ball.

The strength-limiting flaw in all of BoR and 4-point bend tests was identified to be the volume flaw located on or near the PZT plate's tensile surface. A couple of them can be viewed as a surface void; they were inherently volume-distributed flaws intersecting the plate surface. When volume flaws are strength-limiting, the Weibull distribution for strength-scaling can be represented by the following expression,

$$P_f = 1 - \exp[-k_V V (\sigma_f / \sigma_{0V})^m]$$

(2)

where k_V is the loading factor, V is the total volume of the specimen, and σ_{0V} is the material's scaling parameter. The product $k_V V$ is typically referred as the effective volume (V_e). For the studied PZT, the expected failure stress can be size-scaled for the equal failure probability,[22]

$$\sigma_{fBoR} = \left(\frac{V_{e4pt}}{V_{eBoR}}\right)^{1/m} \sigma_{f4pt}$$

(3)

where additional subscripts "BoR" and "$4pt$" refer to the ball-on-ring flexure and 4-point bend, respectively. The effective volumes were estimated as 0.028 and 1.950 mm³ for the BoR and 4-point bend. By further substituting the average value of 7.1 (Figs. 1 and 3) for m into Eq. (3), the expected strength of the BoR is 1.823 times that of the 4-point bend. On the other hand, the observed strength ratio of two cases was actually 1.595. The marginal discrepancy may be due to either the existence of concurrent flaw populations or the shear-sensitivity[23] that was not captured by the idealized flaw. Nevertheless, the amount of size scaling in the strength correlated to the predicted based on Weibull distribution, validating the role of volume flaw, particularly, the volume pore in the failure of this material.

As-Extracted PZT versus As-Received PZT

A comparison on the data between the as-extracted and as-received groups (Figs. 1 and 5, the OC) revealed a higher σ_θ and a higher m in the as-received PZT. Besides batch-to-batch variation in fabrication, different treatments involved with these two groups of specimens were mainly responsible for the difference observed in the Weibull parameters. The single-layer PZT plates typically went through screen printing, laminating, sintering, encapsulating, and poling to be finally imbedded in a stack. These plates experienced same aging and mechanical and electric loading as their parent stack. Lastly, the plates had to be heated at a specified temperature (in a defined period) for the extraction. On the other hand, the as-received PZT plates had a short record as they were only subjected to electroding and poling after sintering. Therefore, the above-observed discrepancy is understandable with regard to such distinct histories between them.

Porosity

Porosity of PZT was characterized using backscattering SEM. A cross section of the as-extracted specimen was prepared for this purpose by grinding and polishing down to 0.25 μm. The backscattering images were sized with 178 μm x 254 μm and area and size distributions of pore were analyzed using image processing software (Image J, NIH, Bethesda, MD). The maximum of pore size was found to be around 20 to 33 μm that, in fact, fell inside a wider size range of strength-limiting flaws revealed by fractography on 4-point bend specimens, 9 to 39 μm. This size range also agreed with the surface roughness depth (12 to 20 μm of R_z) of as-received specimens measured using the profilometer. This is established because the R_z reflects the half size of flaw. It was found that, although their m values were quite close (20 versus 17), the as-received PZT in this study had a smaller characteristic strength, 92 MPa, than the tape cast PZT studied previously, 140 MPa.[15] The consistent observations on the porosity thus partially explained this difference.

Electric Field Effect

A similar asymmetrical response to the electric loading was also observed in another study on PZT using a 3-point bend setup.[6] However, the asymmetry associated with the pressed PZT in this study is different from on the tape cast PZT.[14,15,16] It has been noticed that the degree of the strength decrease in this study was quite low compared to that in the tape case PZT[15,16] and the strength increase in the positive electric field was not statistically significant (Fig. 5). The lowered effect of electric field and sizable volume pores revealed by various approaches suggested that the porosity overweighed the electric field. The larger size and density of flaws impacted material's performance in such a manner that an applied electric would hardly manage to work effectively. On the other hand, the porosity of the tape cast PZT [15,16] was not in such extent where the failure was actually more dictated by surface-located flaws. At the same time, it has been seen that intergranular fractures were predominant across the fracture surface, which was different from that in the tape cast PZT too.[15,16] This indicated that the grain boundary phase in the studied PZT played a unique role. The local electric field can be affected because the boundary phase usually has a different dielectric constant from that of surrounding grains. The degree of this effect on the overall mechanical strength of the pressed PZT under electric field remains to be investigated.

CONCLUSION

A commercial PZT was experimentally investigated in both as-extracted and as-received states. Ball-on-ring and 4-point bend tests were conducted and electric field effect was studied. The following conclusions can be drawn based on the observations:

1. The studied PZT exhibited relatively lower characteristic strength under the BoR flexure compared to that of equivalent PZT but processed by tape cast.
2. Measured flexure strengths were apparently affected by the specimen states. The as-received PZT showed a higher characteristic strength than the as-extracted PZT.
3. Strength-limiting flaws were mainly volume-distributed; volume pore was identified to be the strength-limiter of this PZT.
4. Electric field had a certain effect on the strength of the as-received PZT. This effect was asymmetrical with respect to the direction of applied electric field.

ACKNOWLEDGMENTS

Research sponsored by the U.S. Department of Energy, Assistant Secretary for Energy Efficiency and Renewable Energy, Office of Vehicle Technologies, as part of the Propulsion Materials Program. The authors thank Jesus Carmona-Valdes of Cummins, Inc. for providing the PZT specimens for this study. The authors thank Drs. Hsin Wang and Enis Tuncer for reviewing the manuscript.

REFERENCES

[1] U. Joshi, Y. Kalish, C. Savonen, V. Venugopal, and N. Henein, "Materials for fuel systems," in "Heavy Vehicle Propulsion Materials FY 2003 Progress Report," pp.7–14, 2003.
[2] S. Kim, N. Chung, and M. Sunwoo, "Injection rate estimation of a piezo-actuated injector," *SAE Technical Paper Series*, 2005-01-0911, SAE Int., Warrendale, PA, 2005.
[3] B. Zickgraf, G.A. Schneider, and F. Aldinger, "Fatigue behavior of multilayer piezoelectric actuators," *Proc. ISAF 1994*, University Park, PA, 325-328, 1995.
[4] H. Aburatani, S. Harada, K. Uchino, A. Furuta, and Y. Fuda, "Destruction mechanisms in ceramic multilayer actuators," *Jpn. J. Appl. Phys.* **33**, 3091–2094, 1994.
[5] C. Schuh, T. Steinkopff, A. Wolff, and K. Lubitz, "Piezoceramic multilayer actuators for fuel injection systems in automotive area," in "Smart Struct. Mater. 2000: Active Mater.: Behavior and Mechanics," C. S. Lynch Ed., *Proc. of SPIE 2000* **3992**, 165-175, 2000.
[6] H. Makino and N. Kamiya, "Effect of dc electric field on mechanical properties of piezoelectric ceramics," *Jpn. J. Appl. Phys.* **33**, 5323–5327, 1994.
[7] M.G. Cain, M. Stewart, and M.G. Gee, "Mechanical and electric strength measurements for piezoelectric ceramics: Technical measurement notes," NPL REPORT CMMT (A), 99, March 1998.
[8] K. Mehta and A.V. Virkar, "Fracture mechanisms in ferroelectric- ferroelastic lead zirconate titanate (Zr:Ti = 0.54:0.46) ceramics," *J. Am. Ceram. Soc.* **73**, 567–574, 1990.
[9] A.G. Tobin and Y.E. Pak, "Effect of electric fields on fracture behavior of PZT ceramics," *Proc. SPIE*—Int. Soc. Opt. Eng. **1916**, 78–86, 1993.
[10] S.B. Park and C.T. Sun, "Fracture criteria for piezoelectric ceramics," *J. Am. Ceram. Soc.* **78**, 1475–1480, 1995.

[11] V. Heyer, G.A. Schneider, H. Balker, J.Drescher, and H.A.Bahr, "A fracture criterion for conducting cracks in homogeneously poled piezoelectric PZT-PIC 151 ceramics," *Acta Mater.* **46**, 6615–6622, 1998.

[12] F. Fang and W. Yang, "Poling-enhanced fracture resistance of lead zirconate titanate ferroelectric ceramics," *Mater. Lett.* **46**, 131–135, 2000.

[13] T.-Y. Zhang, M. Zhao, and P. Tong, "Fracture of piezoelectric ceramics," *Adv. Appl. Mech.* **38**, 147–289, 2001.

[14] H. Wang and A.A. Wereszczak, "Effects of electric field on the biaxial strength of poled PZT," *Ceram. Eng. Sci. Proc.* **28**, 57-67, 2007.

[15] H. Wang and A.A. Wereszczak, "Effects of electric field and biaxial flexure on the failure of poled lead zirconate titanate," *IEEE Trans. Ultras., Ferroelect., and Freq. Contr.* **55**, 2559-2570, 2008.

[16] H. Wang, H.-T. Lin, and A.A. Wereszczak, "Strength properties of poled PZT subjected to biaxial flexural loading in high electric field," *J. Am. Ceram. Soc.*, in print.

[17] D.K. Shetty, A.R. Rosenfield, P. McGuire, G.K Bansal and W.H. Duckworth, "Biaxial flexure tests for ceramics," *Ceram. Bull.* **59**, 1193–1197, 1980.

[18] J.E. Ritter, Jr., K. Jakus, A. Batakis, and N. Bandyopadhyay, "Appraisal of biaxial strength testing," *J. Non-Cryst. Solids* **38-39**, 419-424, 1980.

[19] Timoshenko, S. P. and Goodier, J. N., *Theory of Elasticity*, McGraw-Hill Int. Editions, 1970. pp. 288-290.

[20] ASTM Standard C1239-00, *Standard Practice for Reporting Uniaxial Strength Data and Estimating Weibull Distribution Parameters for Advanced Ceramics*, Volume 15.01, March 2002.

[21] ASTM Standard C1322-96a, *Standard Practice for Fractography and Characterization of Fracture Origins in Advanced Ceramics*, Volume 15.01, March 2002.

[22] A.A. Wereszczak, T.P. Kirkland, K. Breder, H.T. Lin, and M.J. Andrew, "Biaxial strength, strength-size-scaling, and fatigue resistance of alumina and aluminum nitride substrates," *Int. J. Micro. Electr. Packag.* **22**, 446-458, 1999.

[23] J. Lamon and A.G. Evans, "Statistical analysis of bending strengths for brittle solids: a multiaxial fracture problem," *J. Am. Ceram. Soc.* **66**, 177-182, 1983.

THERMAL TOMOGRAPHIC IMAGING FOR NONDESTRUCTIVE EVALUATION OF
CERAMIC COMPOSITE MATERIALS

J. G. Sun
Argonne National Laboratory
Argonne, IL 60439

Infrared thermal imaging has been widely used for nondestructive evaluation of ceramic
matrix composite (CMC) materials that are being developed for high-temperature structure and
engine applications. Current thermal imaging methods may measure the gross thermal property
or detect flaws within the CMCs. They however are not well suited to determine the property
and flaw distributions in the depth direction. This paper presents a thermal tomography method
that can resolve the depth distribution of a material property in the tested sample. It utilizes the
one-sided flash thermal-imaging data to construct three-dimensional images of material's
thermal effusivity in the entire sample volume. Because thermal effusivity is an intrinsic
material thermal property, it equals to the square root of the product of thermal conductivity and
heat capacity, thermal tomography results can be used to determine thermal properties as well as
to detect flaws in the material. This paper describes the capabilities and presents typical
experimental results for a CMC sample obtained by the thermal tomography method.

INTRODUCTION
Ceramic matrix composites (CMCs) are advanced materials with superior mechanical
properties at elevated temperatures. They are being developed for various high-temperature
structure and engine applications. Examples of these applications include the reinforced carbon-
carbon composites used in the leading edge of space shuttle's wings [1], CMC thermal protection
systems for space vehicles [2], and CMCs combustor liners for turbine engines [3,4]. Because
CMC components play critical functional and safety roles in those applications, their reliability
must be maintained during the service lifetime. Nondestructive evaluation (NDE) is therefore
normally required to inspect CMC components during the manufacturing process as well as in
service. One critical defect to be detected is delamination between the plies. Other defects such
as cracks, voids, and porosities should also be characterized in terms of their position, size, and
severity.
One-sided pulsed thermal imaging has been widely used for NDE of CMC components.
Compared with other NDE technologies, thermal imaging is noncontact, sensitive to relatively
shallow features, and fast for large-area inspection. Therefore, it is especially suitable for testing
CMC components because they are usually thin-wall structures with large surface areas.
Thermal imaging can detect all types of defects in CMCs, although most of the studies were
focused on detection and characterization of delaminations [5,6]. However, because heat transfer
is a diffusion process, not a wave-propagation process, thermal imaging data are convolutions of
both depth and severity information of a defect as well as heat absorption and diffusion that
becomes more prominent with increased heat propagation distance. Therefore, raw thermal
imaging data are usually not precise and detailed for quantitative interpretation of material's
internal properties. Development of data processing methods to enhance defect features as well
as determine quantitative defect and material parameters are essential for the advancement of this
technology.

Many thermal imaging data-processing methods have been developed. Methods used for image enhancement typically convert the measured surface temperature into nondimensional parameters that display unique characteristics to defects in the resultant images [7,8]. For quantitative property measurement, such as material properties or delamination depth, regression methods based on theoretical models of the material system are usually used [3,4,9]. Although these methods are useful for NDE characterization of CMCs, they do not provide complete information for the detailed material property and defect distributions within the component volume. This paper introduces a new thermal tomography method that can construct 3D thermal effusivity distribution in the entire component volume for direct visualization of its internal structures and properties [10]. Experimental results for a CMC sample are presented to demonstrate its capabilities for detecting delaminations as well as other distributed defects.

PULSED THERMAL IMAGING TECHNOLOGY

Pulsed thermal imaging is based on monitoring the temperature decay on a sample surface after it is applied with a pulsed thermal energy that is gradually transferred inside the sample. A schematic one-sided pulsed-thermal-imaging test setup is illustrated in Fig. 1. When pulsed thermal energy is applied, a thin layer of material on the surface will be heated instantaneously to a high temperature. Heat transfer then takes place from the heated surface to the interior of the sample, resulting in a continuous decrease of the surface temperature. If a delamination exists under the surface, the local heat transfer rate is reduced, resulting in a higher temperature in the surface area above the delamination than in the surrounding areas. This is seen in the thermal images as a local "hot spot." The hot spot appears earlier during the transient if the delamination is shallow and later if the delamination is deep.

Figure 2 illustrates a diagram of a CMC plate sample with seven machined flat-bottom holes to simulate delamination defects of various sizes and depths. The composite consists of eight plies of SiC continuous-fiber cloth and a SiNC matrix [11]. It was not completely densified and contained many defects and distributed porosities. The depths, listed in Fig. 2, refer to the distance from the hole bottoms to the front flat surface where thermal imaging data were taken. The thickness of the sample ranged from 2.3 to 2.7 mm. Its average thermal diffusivity was estimated to be 0.97 mm^2/s [11]. Figure 3 shows two thermal images taken respectively at an early and a later time after the flash. It is seen that Hole A as a hot spot is visible in the early time because the surface material in that region is thinner, while other holes are shown in later times depending on their depths. The images in Fig. 3 also indicate that the bottoms of the holes are inclined; they are thinner at the top-right corner. In addition, a large defect (very thin delamination) is visible at the lower-left corner in the early image, so are many small porosities (bright spots) distributed over the entire sample surface.

Although the raw thermal images (e.g., those in Fig. 3) can be used to detect delaminations and determine their sizes and depths based on the sizes and the initiation times of the "hot spots", the thermal contrast for deeper delaminations can be extremely small and undistinguishable from other thermal features induced by material heterogeneity or nonuniform flash heating on the front surface. One simple method to enhance the delaminations is to calculate the first or second derivatives (in log scale) of the measured surface temperature with time, and construct a series of first- or second-derivative images for analysis [8]. It has been identified that these derivatives exhibit characteristic changes with a fixed contrast around the time when heat flux reaches the delamination depth, so defect signature is not dependent on depth and can be easily detected [5]. However, analysis for these data is not efficient because a large number of images have to be studied and the user must be familiar with the characteristic signatures of different defects. Alternatively, many data reduction methods were developed. These methods may directly produce a depth image showing the distribution of delamination depth and sample thickness based on the assumption of uniform material properties for the test sample [5]. All these methods are normally used for detection of a single dominant defect, not for the distribution of various defects, under each surface location. Further, these methods are essentially designed for detection of subsurface discontinuities and have weak or no sensitivity to material properties, they are not suitable to determine the material property distributions under the surface. These problems are the major reasons that limit the wide use of thermal imaging for scientific and engineering applications. They can be resolved if a material property can be determined in 3D for the entire sample volume; this is achieved by the thermal tomography method discussed in the following.

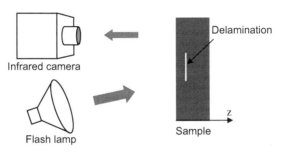

Fig. 1.   Schematics of pulsed thermal imaging test setup.

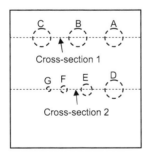

Hole	Diameter (mm)	Depth (mm)
A	7.5	0.25
B	7.5	1.12
C	7.5	0.97
D	7.5	0.87
E	5.0	0.78
F	2.5	0.85
G	1.0	0.85

Fig. 2.   Diagram and list for dimensions of machined flat-bottom holes in a CMC plate.

| $t = 0.007$ s | $t = 0.67$ s |

Fig. 3. Thermal images taken at $t = 0.007$ and 0.67 s after thermal flash from front surface of the CMC sample illustrate in Fig. 2.

THERMAL TOMOGRAPHY METHOD

There are two major steps for the development of the thermal tomography method: to construct a 3D spatial representation of the test sample and to express the thermal-imaging data in the form of a material property. Fortunately, pulsed thermal imaging does acquire a 3D set of data: the 2D surface temperature distribution (thermal images) as a function of time t, i.e., $T_s(x, y, t)$ when the surface plane is considered as the (x, y) plane. At each fixed surface position (x_i, y_j), its surface temperature is a function of time, or a 1D profile in the time domain. When this 1D profile is converted into a profile in the depth z domain and this conversion is repeated for all surface positions, the measured thermal-imaging data in the (x, y, t) domain becomes the 3D data in the (x, y, z) spatial domain for the test sample. Therefore, thermal tomography method can be established by the construction (or deconvolution) of a depth profile from a time profile based on a direct relationship between depth z and time t. This relationship was determined to be [5,10]:

$$z = (\pi \alpha t)^{1/2}, \tag{1}$$

where α ($= k/\rho c$) is thermal diffusivity; ρ is density, c is specific heat, and k is thermal conductivity.

The other step for thermal tomography development is to convert the measured surface temperature into a material parameter of the test sample. One such material property is the thermal effusivity defined as $e = (\rho c k)^{1/2}$. From the thermal imaging data, i.e., the surface temperature $T_s(t)$ at each surface position (x_i, y_j), an apparent thermal effusivity $e_a(t)$ can be determined from [10,12]:

$$e_a(t) = \frac{Q}{T_s(t)\sqrt{\pi t}}, \tag{2}$$

where Q is the energy deposited by the flash on the surface position. It can be shown that for semi-infinite single-layer materials with constant properties, $e_a(t)$ is a constant and equals to the material effusivity $(\rho c k)^{1/2}$. In general, $e_a(t)$ differs from the material effusivity for inhomogeneous materials.

Based on Eqs. (1) and (2), the measured surface temperature profile $T_s(t)$ can be used to construct the material effusivity as a function of depth $e(z)$ at each surface position (x_i, y_j) [10]. When this process is repeated for all surface positions, the thermal tomography method converts the measured surface temperature $T_s(x, y, t)$ into a 3D distribution of the thermal effusivity within the sample volume $e(x, y, z)$. The 3D thermal effusivity data are similar to those obtained by 3D x-ray CT, and can be sliced in planes either parallel or perpendicular to the imaged surface for analysis. However, unlike the x-ray CT data where the diffusion is small and uniform throughout the volume, thermal tomography data exhibit considerable diffusion effect that increases with depth. As a result, image contrast decreases with depth, as to be seen in the data presented below.

EXPERIMENTAL RESULTS

The thermal tomography method was used to construct 3D thermal effusivity data for the flat-bottom-hole sample illustrated in Fig. 2. The 3D data are sliced in plane and vertical cross sections for visualization of the features inside the sample. Based on the experimental settings, the pixel size in the plane direction (x, y) is 0.24 mm, and each constructed depth layer has a thickness of 0.042 mm. Figure 4 shows two plane slices below the imaged flat surface. At the depth of 0.23 mm, only part of Hole A is reached, because it is inclined with a center depth of 0.25 mm. There are many defects distributed at this depth, with a large one at the lower-left corner. All defects are displayed with low effusivity, because they consist of air that has very low effusivity compared to the CMC material. At the deeper depth of 0.86 mm, all flat-bottom holes are visible. It should be noted that Holes B and C are deeper than this depth (see list in Fig. 2); they appear at this depth because of the thermal diffusion effect which makes a sharp boundary to become blurred (this effect is seen more clearly in the cross sectional images in Fig. 5). There are also many defects at this depth, although their contrast is much weaker than that in the shallow depth.

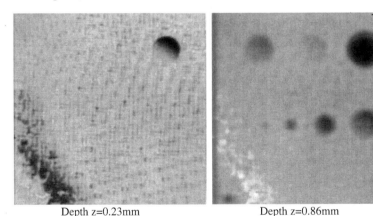

Depth z=0.23mm   Depth z=0.86mm

Fig. 4.   Plane thermal effusivity images at two depths under the front surface of the flat-bottom-hole plate illustrated in Fig. 2.

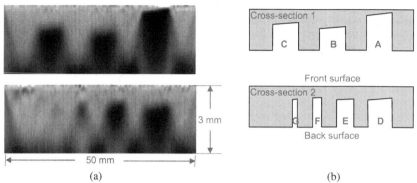

(a)   (b)

Fig. 5.   (a) Cross-sectional thermal effusivity images and (b) diagrams at two cross sections of the flat-bottom-hole plate illustrated in Fig. 2.

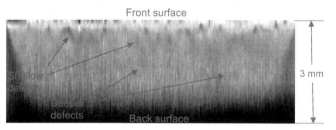

Fig. 6.   Rescaled cross-sectional thermal effusivity image at a middle region of the flat-bottom-hole plate illustrated in Fig. 2.

Figure 5 shows two cross-sectional effusivity images and corresponding cross-sectional diagrams along the two horizontal lines marked in Fig. 2. It is seen that all flat-bottom holes are well defined with detailed depth resolution of their bottom surfaces (note that all holes have inclined bottom surface). The images also show the degradation of spatial resolution with depth in both lateral and depth directions due to the 3D diffusion effect. In addition to the holes, many shallow and deeper defects (darker spots) are detected within the sample. They are better resolved in the rescaled cross-sectional image in Fig. 6 which is located at the middle of the sample without the machined holes. These defects are small voids due to the incomplete densification of the plate.

From the results in Figs. 4-6, it is evident that thermal tomography is a true 3D imaging technology based on one-sided thermal imaging data. The data are quantitative and complete for the entire sample volume, and can be easily interpreted. This is the first 3D thermal imaging method.

CONCLUSION

A thermal tomography method was developed for the construction of 3D thermal effusivity data for the entire volume of a test sample from one-sided pulsed thermal imaging data. Because thermal effusivity is an intrinsic material thermal property that can be unique for different materials within a component, thermal tomography results can be easily interpreted and allow for direct visualization of the internal structures, material variations, as well as defect distributions within inhomogeneous materials. Thermal tomography was used in this study for NDE characterization of a CMC plate sample with internal defects and machined flat-bottom holes to simulate delaminations. It was demonstrated that thermal tomography can clearly resolve the delaminations (bottom surfaces of the holes), the back surface of the plate, and distributed defects within the sample volume. Therefore, thermal tomography is an efficient method for NDE characterization of CMC components and other engineering structures.

ACKNOWLEDGMENT

This work was sponsored by the U.S. Department of Energy, Office of Fossil Energy, Advanced Research and Technology Development/Materials Program, and by the Heavy Vehicle Propulsion Materials Program, DOE Office of FreedomCAR and Vehicle Technology Program, under contract DE-AC05-00OR22725 with UT-Battelle, LLC.

REFERENCES

1. D.J. Roth, N.S. Jacobson, J.N. Gray, L.M. Cosgriff, J.R. Bodis, R.A. Wincheski, R.W. Rauser, E.A. Burns, and M.S. McQuater, "NDE for Characterizing Oxidation Damage in Reinforced Carbon-Carbon Used on the NASA Space Shuttle Thermal Protection System," in Ceramic Engineering and Science Proceedings, Vol. 26, No. 2, pp. 133-141, 2005.

2. C.J. Kacmar, K.J. LaCivita, K.V. Jata, and S. Sathish, "Thermal Characterization of TPS Tiles," in Review of Quantitative Nondestructive Evaluation, Vol. 25, eds. D.O. Thompson and D.E. Chimenti, pp. 1740-1747, 2005.

3. J. G. Sun, J. Benz, W. A. Ellingson, J. B. Kimmel, and J. R. Price, "Nondestructive Evaluation of Environmental Barrier Coatings in CFCC Combustor Liners," in Ceramic Eng. Sci. Proc., eds. A. Wereszczack and E. Lara-Curzio, Vol. 27, no. 3, pp. 215-221, 2006.

4. M. J. Verrilli, G. Ojard, T. R. Barnett, J. G. Sun, and G. Baaklini, "Evaluation of Post-Exposure Properties of SiC/SiC Combustor Liners Tested in the RQL Sector Rig," in Ceramic Engineering and Science Proc., eds. H-T Lin and M. Singh, Vol. 23, Issue 3, pp. 551-562, 2002.

5. J. G. Sun, "Analysis of Pulsed Thermography Methods for Defect Depth Prediction," J. Heat Transfer, Vol. 128, pp. 329-338, 2006.

6. J. G. Sun, "Evaluation of Ceramic Matrix Composites by Thermal Diffusivity Imaging," Int. J. Appl. Ceram. Technol., Vol. 4, pp. 75-87, 2007.

7. X. Maldague, F. Galmiche, and A. Ziadi, "Advances in Pulsed Phase Thermography," Infrared Physics and Technology, Vol. 43, No. 3-5, pp. 175-181, 2002.

8. S. M. Shepard, J. R. Lhota, B. A. Rubadeux, D. Wang, and T. Ahmed, "Reconstruction and Enhancement of Active Thermographic Image Sequences," Optical Eng., Vol. 42, pp. 1337-1342, 2003.

9. J.G. Sun, "Thermal Imaging Characterization of Thermal Barrier Coatings," in Ceramic Eng. Sci. Proc., eds. J. Salem and D. Zhu, Vol.28, no. 3, pp. 53-60, 2007.

10. J. G. Sun, "Method for Thermal Tomography of Thermal Effusivity from Pulsed Thermal Imaging," US Patent No. 7,365,330, 2008.
11. J. G. Sun, C. Deemer, W. A. Ellingson, T. E. Easler, A. Szweda, and P. A. Craig, "Thermal Imaging Measurement and Correlation of Thermal Diffusivity in Continuous Fiber Ceramic Composites," in Thermal Conductivity 24, eds. P. S. Gaal and D. E. Apostolescu, pp. 616-622, 1999.
12. D. L. Balageas, J. C. Krapez, and P. Cielo, "Pulsed Photothermal Modeling of Layered Metarials," J. Appl. Phys., Vol. 59, pp. 348-357, 1986.

A MORE COMPREHENSIVE NDE: PCRT FOR CERAMIC COMPONENTS

By Leanne Jauriqui, Lem Hunter – Vibrant Corporation; Albuquerque, New Mexico

ABSTRACT

Ceramic material advancements have produced many new components to meet the needs of the growing technology requirements. Traditional NDT methods (UT, ET, RT, and PT) often provide incomplete results when used to inspect these components, leading to multiple costly and time-consuming inspections.

Process Compensated Resonant Testing (PCRT) has been demonstrated as an excellent alternative for inspecting ceramic parts that cannot be inspected with traditional NDT. PCRT goes beyond traditional resonance testing by applying pattern recognition algorithms and process control statistics to precise resonant data. PCRT is a trained method that learns which resonance variation is acceptable, and which is unacceptable, while also monitoring in-control processes.

PCRT is used for studying structural integrity and functional performance in a variety of ceramic parts including ceramic balls, seal rings, and armor plates. PCRT is a whole body inspection that detects internal defects such as cracks, voids, and inclusions. PCRT can also perform surface inspection of balls, using Surface Acoustical Waves. PCRT can be performed in real manufacturing time, with inspection times measured in seconds, and computer control eliminates operator error and subjectivity. PCRT is a proven, emerging technique for quality verification and process control.

INTRODUCTION

Ceramic material advancements have produced many new components to meet the needs of the growing engine technology requirements. Ceramic components, including seal elements and bearing components, provide higher stiffness, lower thermal expansion, lighter weight, increased corrosion resistance, and higher electrical resistance than comparable steel products. However, NDE techniques currently applied to these ceramics cannot keep pace with typical rates of manufacturing production. Ceramics are a relatively brittle material and the presence of defects can lead to catastrophic failure. Traditional NDT methods (UT, ET, RT, and PT) often provide incomplete results when used to inspect ceramic components. This leads to a requirement for multiple costly and time-consuming inspections.

Process Compensated Resonant Testing (PCRT) has been demonstrated as an excellent alternative for inspecting ceramic parts that cannot be inspected with traditional NDT. PCRT goes beyond traditional resonance testing by applying pattern recognition algorithms and process control statistics to precise resonant data. PCRT is a trained method that learns which resonance variation is acceptable, and which is unacceptable, while also monitoring in-control processes.

Many ceramic materials, including Silicon Nitride and Zirconia, resonate extremely well, leading to very precise measurements and an exceptional ability to compare samples and populations of samples. In this study, PCRT is used for studying structural integrity and functional performance in a variety of ceramic parts used in aerospace, power generation and armor applications. The parts include ceramic balls with varying diameters, seal rings, and armor plates. PCRT is a whole body inspection and provides an assessment for structural integrity in terms of cracks, voids, inclusions, heat treatments, material properties, etc. as they relate to the ultimate performance of the parts. Cosmetic defects are not rejected. Parts with acceptable structural integrity and performance are separated from parts with manufacturing and structural and/or material deficiencies. PCRT can be performed in real manufacturing time, with inspection times measured in seconds, and is computer controlled for operator-independence. PCRT is limited in that it does not characterize the nature or location of the defect; it simply highlights the part as being statistically 'different' from the acceptable training population.

As an alternative to methods that can study only the surface of ceramic balls (hybrid bearing rolling elements), PCRT is used for studying both the internal structural integrity as well as the surface finish quality. Whole body inspection is performed using lower frequency modes sensitive to internal porosity or inclusions, and to basic material properties. Surface Acoustical Waves are used to evaluate only the outer surface for cracks and other stress-risers. The results from this study show that PCRT successfully sorts acceptable parts from unacceptable parts based on materials/structural integrity, functional performance and defects such as microstructure changes, c-cracks, density variation and inclusions. PCRT has been successfully implemented for inspecting production parts in many applications, and has reduced inspection costs tremendously while other uses have led to improvements in the manufacturing process. The pursuit of progressive improvements in the usage of PCRT, like any other NDT methods, is an on-going process.

PCRT should be used by ceramic part manufacturers to verify the quality of the manufacturing process, and to produce a supply stream of defect-free components. PCRT can also be utilized by assemblers and OEM's to verify the quality of the part stream, or to analyze components in a repair environment.

PROCESS COMPENSATED RESONANCE TESTING (PCRT)

Process Compensated Resonance Testing (PCRT) is a relatively young approach in NDT. The underlying technology was developed in the late 1980's at Los Alamos National Laboratories. PCRT is based on the physics fundamental that any rigid component will resonate at specific frequencies that are a function of its mass, shape, and material properties. Material changes or flaws change the normal resonant pattern. While elementary resonance techniques may suffer from low-precision and a limited frequency range, PCRT utilizes high-precision resonance measurements, across the range of 1 kHz to 15 MHz. Most resonance techniques do not have the ability to distinguish acceptable process variation from the effect of a small defect, as both cause changes in the resonance spectra. PCRT relies on proprietary software algorithms, developed to compensate for normal manufacturing variations, and novel algorithms for monitoring structural changes in a part over its life. These analytical tools combine to significantly reduce field failures of component, increase production yield, and optimize part life.

PCRT is a fundamental shift in NDT philosophy and applications. Current technologies strive to highlight indications that could represent structural deficiency in a component. PCRT accurately measures the structural similarity of a component to known good parts, and is also able to measure the structural changes in a single part throughout its useful life. Proprietary pattern recognition software algorithms, called Sorting Modules, are developed to identify defects while compensating for normal manufacturing variations. The final product is a rapid, accurate, computer-controlled and operator-independent PCRT evaluation. The technology is achieving growing acceptance (over 150,000,000 automotive parts tested with PCRT in 2009) in inspection of manufactured components such as connecting rods, crank shafts, suspension arms, etc.

PCRT has been applied as a viable NDT technique on many different components. It can be used on a wide range of materials, including ceramics, and most metals. Generally speaking, the size and geometry of components does not limit the application, with successful applications ranging from 3/8" diameter ceramic ball bearings up to 200 lb automobile engine blocks.

PCRT applies statistics, comparative analysis, and pattern recognition to precise resonance data. Measurements are taken using 3 contact PZT transducers – 1 used as a 'drive', and 2 used to receive. Swept sine waves are used to stimulate the part through a proprietary signal generator/processor. Resulting measurements have repeatability of .05% (standard deviation) or better.

More basic methods of resonant inspection suffer from both a lack of precision, and the inability to discriminate resonance changes due to acceptable process variation from those due to the defective conditions. Defects in ceramic parts generally result in one of two outcomes – 'shifted' peaks, due to changes to the component's bulk properties / strength, and degenerate, or 'split' peaks, due to breaks in symmetry in a symmetrical component. The PCRT system builds on its precision and statistics to create computer-based comparison algorithms that can sort parts into 'good' and 'defective' categories. The resulting inspection is objective, can be fully automated, requires no part preparation (other than drying), and generates no waste.

To deal with defects that have a significant effect on the bulk material properties, such as density variation, or a crack, in ceramic armor, the patented PCRT software utilizes a Mahalanobis-Taguchi System (MTS) statistical analysis to identify a multi-frequency central tendency or pattern that groups the good parts, and excludes many types of defects. Defects with a lesser effect on the resonant spectra are excluded with a secondary discriminator known as the Bias. This is all determined in 'n-dimensional' space; however, some simplified graphics are shown in Figure 1 (right hand graphic is actually PCRT software output).

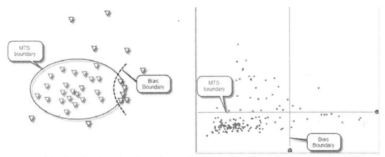

Figure 1 – Graphic showing how MTS evaluates the 'central tendency' of a group, and Bias excludes parts that may also belong to the group, but showing their own similarities (left). A PCRT software output (right), shows the sorting algorithm's 'passing' region at the lower left, with the green dots indicating 'good parts'. Red X's represent the training set's Bad Parts, which fall outside both boundaries, and will be rejected by the PCRT System.

Additionally, many ceramics end-users are concerned with much finer defects, such as surface cracking on a ceramic ball, or chipping of a ceramic seal. These defects do not have a significant effect on the bulk material properties, but can dramatically affect the resonant frequency of degenerate modes, or modes where symmetrical resonances generally occur 'on top' of one another. When a surface crack, or chip, is present, modes that were previously identical are not any longer, and a 'split' is detected (see Figure 4). The size of the resonance split is proportional to the size of the source of the asymmetry. By evaluating these splits statistically, a test can be developed to identify components with more asymmetry than is desired.

Both PCRT methods require a basic 'training set', or input model, to discover the range of 'normal' variation, and the source of differences, both acceptable and unacceptable. Sorting algorithms can use a 'process control' concept to reject samples outside of the range of 'normal' variation, or they can use a targeted 'pattern recognition' strategy to identify parts with the characteristics of specific defects.

APPLICATIONS OF PCRT

Hybrid Bearing Rolling Elements (Ceramic Balls)

New developments in gas turbine engines are driving the requirements of current bearing technology to its design limits in terms of material performance, capability and reliability, as well as affordability. Hybrid ceramic rolling element/metal race bearing technology has proven itself to be a valid candidate to meet the needs of the growing engine technology requirements. Compared to all-steel components, hybrid bearings offer increased load-carrying capability, decreased friction and heat generation, greater stiffness and corrosion resistance, lower coefficients of thermal expansion, and increased thermal stability. Ceramic balls have proven to be robust in use provided that they are free of manufacturing defects. However, NDE techniques applied to ceramics for this class of bearing have proven to be ill-suited for quality-assurance inspections for typical rates of manufacturing production. The slow inspection rate and high inspection costs of the current ceramic NDE methods may prevent hybrid bearings from being widely used in current or future fleets. Figure 2 shows examples of C-spall and crack defects found in ceramic balls. A distinct advantage of the PCRT process is the ability to detect internal defects that cannot be detected by any surface measurement technique.

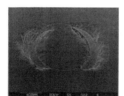

Figure 2 - Examples of C-Spall and Crack Defects in Ceramic Balls[3].

Lower-frequency resonances can be used to evaluate the spherical symmetry and identify bulk-body defects. Higher-frequency resonances, where surface or Rayleigh waves dominate, propagate without dispersion in an isotropic media and therefore, any distortion in the measured Rayleigh wave may be attributed to near surface flaws[1,2]. An example of this distortion is shown in Figure 4. The resonances exhibit peak splits and shifts in frequency and these are key characteristics that the PCRT System uses to classify a part as acceptable or unacceptable.

PCRT tests ceramic balls in a fixture similar to those shown in Figure 3. Inspection does not require any ball-rotation, and the balls can be loaded into the fixture by an automatic parts handling system. As the test does not require any human interpretation, the entire process can be automated. Testing times less than 2 minutes could be easily achieved with these methods.

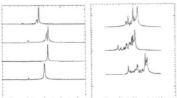

Figure 4 – Resonances of Good Samples (left) and SAW-Like Modes Caused by Surface Defects (right).

0.375-in. SiN Ball

0.5-in. SiN Ball

Figure 3 – Nest Designs for Ceramic Balls.

Relatively low frequency resonances (first 10-15 modes) for the ceramic balls have proved capable of detecting a number of whole body and surface defects such as microstructure differences, surface chemistry / reaction layer defects, metallic inclusions, and density and raw material variation. An example of a reaction layer defect is presented in Figure 5.

Significant surface scuffs and gouges were also detectable by these lower modes. For repeated samples, it was found that PCRT sorting methods could reliably detect a scuff or gouge representing about a .02% volume effect, and/or about a 1% effect on the surface area. Figure 7 shows the increased effect (frequency splitting) as a 'scuff' defect was created and increased in size to approximately 0.130" x 0.170". The size of the frequency split is proportional to the size of the 'asymmetry' imparted to the ball, as shown in Figure 6. Opportunities to detect smaller defects of these types are currently being investigated at higher frequencies.

Resonant Frequency Variation of Good Balls

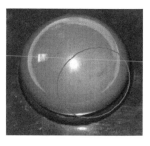

Figure 5 - Resonant peak data for Reaction Layer Defect and photo.

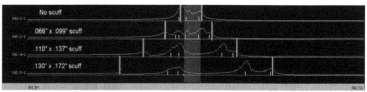

Figure 7 - Increased 'splitting' of 994 kHz resonance due to increased surface scuffing[4].

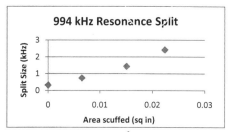

Figure 6 - Plot of Split Size vs. scuffed area (in^2). The split size is proportional to the size of the defect[4].

Smaller surface damage is detectable by higher order surface resonance. Figure 8 (left) shows peak splitting around 8.1 MHz due to a C-spall crack introduced into a 1.125-in. ball (via impact with a 0.375-in. ball). Figure 8 (right) shows peak splitting due to surface damage inflicted by a micro-hardness tester, resulting in chip and crack. These splitting characteristics are expected due to surface acoustic wave behavior, as described by Migliori1, and were evident on many other resonances over the broadband range.

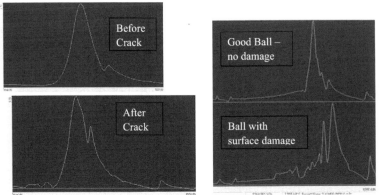

Figure 8 - Resonance Comparisons for balls with no surface damage, and balls with surface damage (C-cracks and/or surface chipping).

In a repeated study, a ball was impacted in a lesser manner to try to induce a 'smaller' c-crack. A ball free of surface defects was measured 15 times to provide a baseline. Although the ball had no defects, some resonance splitting was still observed, due to the difficulty in making a completely symmetrical specimen. Then the ball was impacted with another ceramic ball, with increasing force, to inflict surface damage. A crack was not detected with visual or dye-penetrant inspection until the 'cracked' stage, shown in Figure 9, but perhaps the effect of the initial 'impacts' (which did not produce a visible crack) are detected (see 'impacted' stage in Figure 9). Five resonances between 6 and 9 MHZ were studied in this experiment, with all showing similar results for the 'cracked stage'. Only the highest frequency peak studied (shown above) shows a difference in the 'impacted' stage. These studies continue.

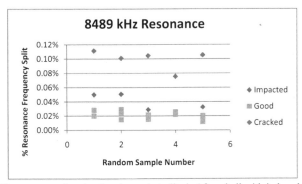

Figure 9 - 8.5 MHz resonance showing the change in 'split size' for a ball with induced c-crack.

Resonant Frequency data can also be used for process control, receiving inspection, and supplier comparison. Figure 10 shows how the frequency of a representative resonance started out tightly controlled, and drifted over manufacturing time. A change to the process imparted additional variation to the samples. PCRT can be used by the manufacturer to identify this type of change, process variables, and produce the most consistent product possible.

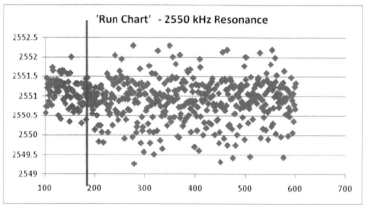

Figure 10 - Run Chart of Resonant Frequencies for a production batch of 1/2" ceramic balls.

Figure 11 shows a similar result for a separate set of balls. This type of variation, seen in low frequency 'bulk resonant modes' is indicative of variation in either the dimension, or base material properties of the balls. As the dimensions are very tightly controlled, it is assumed the variation is due to material property difference. Resonant Ultrasound Spectroscopy (RUSpec) measurements of the Young's Modulus varied noticeably, from a statistical distribution standpoint, but only about 1.5% between the batches.

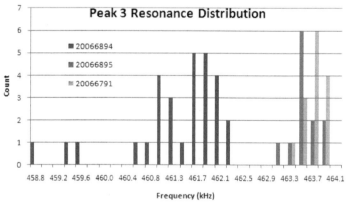

Figure 11 - Resonant Frequency Distributions for multiple batches of ball production.

Evaluation of other sample sets shows similar results. Variation can be seen between batches from the same supplier, and more obviously between balls from different suppliers. Some batches show relatively consistent variation (similar to Batches 2 and 3 in Figure 11), and others show dramatically different results (Batches 1 and 2 in Figure 11). PCRT can be a valuable part of a supplier evaluation program, or in-house process control, identifying trends and changes in the product that can be traced back to process changes and manufacturing conditions.

Ceramic Seals

Vibrant has completed some initial feasibility studies of the use of Process Compensated Resonant Testing for ceramic seal NDE. The seals were evaluated for process control opportunity (batch comparison), chips, and cracks.

PCRT data showed a clear difference between the good parts and the cracked part provided. The crack showed both as a frequency 'shift', due to bulk property changes, and a 'split', due to the break in the ring's symmetry. Nearly every resonant peak was affected, and the defective part was easily identified. Standard PCRT testing does not specify the 'type' or 'location' of defects detected, but with further PCRT study, and/or application of more traditional NDE methods, a comprehensive study of the cause and effect of the defect can be completed.

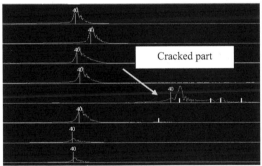

Figure 12 - Cracked part's resonances shift significantly from resonances for good seals and show a split due to break in symmetry.

The chipped parts show only the 'splitting' characteristic in the resonance spectra, because the chip does not have as significant an effect on the overall strength of the part. Vibrant validated these findings by inflicting a chip similar to a production sample, in an otherwise 'good' part, and noting the spectra 'before' and 'after' (Figure 13, Figure 14). Split size has been shown in many applications to correlate well to the size of the defect present. Splits will also identify internal sources of asymmetry, such as porosity, inclusion, and chemistry anomaly. The chip sizes evaluated to date are readily detectable by a PCRT system.

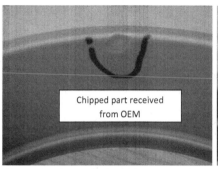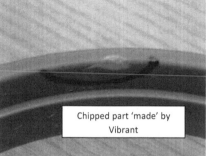

Figure 13 - Photos of chips in Ceramic Seals, both received from the production inspection, and induced by experiment at Vibrant. Both defects show similar, detectable, effects to the resonance spectra.

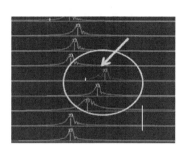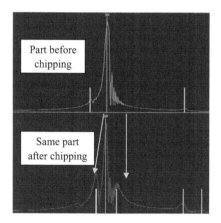

Figure 14 - Stack of Resonance Spectra showing the 'split' evident in the chipped part, as received (left). Higher resolution spectra of part before and after chipping, for experimental part (right). Peak splitting increased significantly due to chip.

The resonant data also make evident the batch-batch variation in the manufacturing process (Figure 15). PCRT was specifically designed to accommodate this type of variation, but additionally, the resonance data could provide some means of process control, monitoring, and quality assurance.

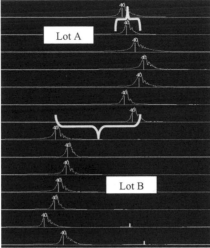

Figure 15 - The variation between production lots is about 4 times greater than the variation within a single production lot. This type of situation requires PCRT's pattern recognition algorithms and statistics, to identify changes that are the result of defects, rather than 'acceptable' production variation.

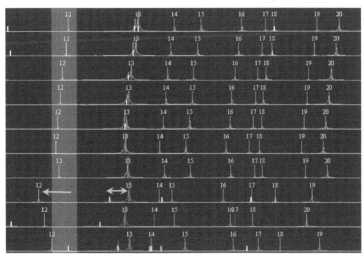

Figure 16 - Resonance pattern changes are evident in these samples of defective armor plates. Peaks shift (peak 12), display splitting (peak 13), and change significantly in pattern (peaks 16-20).

Ceramic Armor

PCRT can also be used to examine ceramic armor plates for production conditions and defects. The Army Research Laboratory has recently commissioned a system to test armor plates for density variation and cracks following a feasibility study that included a successful 'blind' test. The system will be used by BAE for the inspection of multiple part numbers. Density changes produce a proportional shift in resonant frequency, and variation outside of the 'normal' controlled process is easily identified. Cracking leads to significant changes in the resonant spectra, such that the 'normal' resonant frequency patterns are not evident.

PCRT data has also been used by other organizations to compare armor plate suppliers, with notable differences in the stability of production processes. Resonance data for the armor plates is high quality, allowing very precise, repeatable non-destructive evaluation in a matter of seconds. Figure 16 shows a snapshot of such data for a group of good plates, compared to some defective plates. Clear differences in the resonance spectra are evident, attributable to the defects.

Some armor sets evaluated display significant process variation in the 'good' parts. Detecting defective conditions in the presence of this variation will require PCRT's patented pattern recognition techniques. Figure 17 shows the output of the VIPR (Vibrational Pattern Recognition) software that identifies the MTS and Bias discriminators for the same data set show in Figure 16.

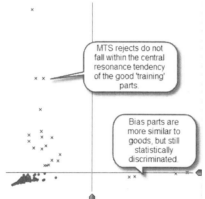

Figure 17 - PCRT's VIPR Output for armor plates. Defects are discriminated while process variation is accommodated.

REFERENCES

[1] A. Migliori, J.L. Serrao, Resonant Ultrasound Spectroscopy, Wiley Interscience, 1997.
[2] G. Rhodes, Non-Destructive Evaluation of Ceramic Bearings, ARPA (DoD) Order No. 5916, 1994.
[3] H. Kolsky, Stress Waves in Solids, Dover Publications, 1963.
[4] "Reprinted with permission from S. Singh, L. Jauriqui and T. Sloan, "NonDestructive Evaluation of Hybrid Bearing Ceramic Rollers Using Process Compensated Resonant Testing (PCRT),"
in Review of Progress in Quantitative Nondestructive Evaluation vol. 29, AIP Conference Proceedings #1211, 2010. Copyright 2010, American Institute of Physics.

Fiber Reinforced Composites

MICROSTRUCTURE AND THERMODYNAMIC DESCRIPTIONS OF SIC-BASED CERAMIC FIBERS

Géraldine Puyoo, Georges Chollon, René Pailler, Francis Teyssandier
Laboratoire des composites thermostructuraux
3, allée de la Boétie, Pessac France
Pessac, France

ABSTRACT

SiC-based ceramic fibers are high performance materials designed to reinforce various kinds of matrices. Depending on the fiber generation, they are composed of almost pure SiC or can include a significant additional amount of carbon and oxygen. Though a large number of papers have been devoted to understanding the microstructure of SiC-based ceramic fibers, the nature of the phases that include C and O is still controversial.

This communication is intended to propose a microstructure description of these fibers according to their composition in the Si-C-O isothermal section of the phase diagram and the corresponding thermodynamic description.

INTRODUCTION

Silicon carbide Ceramic Matrix Composites (CMCs) have been initially developed to be used in severe environments such as, rockets, gas turbines and jet engines, for relatively short life time and high temperature applications (600°C-1200°C). The reinforcement made of fibers weaved in a multidirectional preform is densified by a silicon carbide matrix deposited from a gaseous precursor. Using these composite materials instead of superalloys allows both an increase of the engine operating temperature and a significant weight saving. Such high performance composite materials were rather aimed at being commercialized for military and space applications to withstand high temperature during short lifetime.

Today the challenge is to adapt SiC/SiC composites to be used in civil aircraft engines (i.e. at lower temperatures but also during much longer lifetime). Composite materials and the SiC fibers have been improved for 20 years in order to increase the creep, oxidation resistance and particularly lifetime in oxidizing atmospheres. Three classes of fibers have been successively developed from the first generation that included oxygen and carbon in excess, to the third generation, close to stoichiometric silicon carbide. Unfortunately, the latter fibers remain too costly for civil applications, and the improvement of first generation Si-O-C fibers is currently studied for such applications. The durability of SiC/SiC composites is ultimately related to the fibers lifetime, i.e. their sensitivity to oxidation or to sub-critical crack growth. This latter mechanism is dictated by the composition and the structure of the Si-O-C fibers. It is expected to be accelerated by the presence of free carbon. The percentage of oxygen and the SiC grain size may also have a significant influence on the fiber reactivity. For this reason further analyses have been carried out on a wide variety of 1st generation of silicon carbide fibers in order to provide a detailed description of their microstructure and properties.

The microstructure of oxycarbide fibers is globally known and is described as a continuum consisting in pure nanometric β-SiC crystals, embedded in a glassy silicon oxycarbide phase, together with free carbon nanodomains (basic structural units, BSUs) [1,2,3,4,5].

The intergranular phase SiO_xC_y, in the first generation of fibers, acts as a binder keeping the β-SiC nanocrystals and the amorphous phase in a metastable state. A slight crystallization is first observed above 1100°C. It corresponds to the SiO_xC_y amorphous phase transformation into β-SiC(s), SiO_2(s) and C(s). At higher temperature, the SiC crystals grow and the formation of gaseous species such as CO(g) and SiO(g) happened at around 1400°C. These transformations have detrimental effects on the mechanical properties of the fibers[6,7].

Many authors have tried to identify the microstructure of the various constituents of the Si-O-C fibers. In 1989, Laffon et al.[1] studied the Nicalon[TM] NG100 and NG200 fibers mainly by EXAFS and X-ray diffractometry and proposed a model of structure including β-SiC nanocrystals embedded in a continuum of mixed SiO_xC_y (x+y=4) tetrahedra. The free carbon is represented in the model as aggregates of few graphene layers saturated with hydrogen at the edges, i.e. without any chemical bond between the continuum and the carbon phase. The same year, Porte et al.[2] tried to explain how the oxygen is incorporated in the structure by XPS spectroscopy. They noticed that there is no significant change in the carbon content between surface and bulk. The composition of the Nicalon fiber as deduced from the XPS Si2p and C1s peaks analysis is: $SiC:SiO_xC_y:C:SiO_2 = 1:0.5:0.75\pm0.25:0.008$.

In 1993, Le Coustumer et al.[3] studied the Nicalon fiber NLM202. They concluded that the fiber composition is: 55wt.% of β-SiC crystals, 40wt.% of $SiO_{1.15}C_{0.85}$ and 5wt.% of free carbon. To quantify the free carbon phase, they assumed that it is present as perfect coronene molecules and calculated its amount from the structure of basic structural units (BSUs) and the proportion of hydrogen. From the percentage of free carbon with the atomic chemical composition and the hypothesis of having a minimum of SiC_xO_y phase, it is possible to find out a description of the fiber and determine the proportion of each phase. An average grain size of 1.6 nm was measured using (111)-SiC Dark Field imaging.

In 1995, Bodet et al.[7] calculated the phase proportions and composition of the Nicalon fiber NLP201 from the atomic chemical composition and the ratio SiC/SiC_xO_y as determined by XPS analysis (deconvolution of the Si2p peak). An average β-SiC grain size of 2.7 nm was measured on a (111)-SiC DF-TEM image.

More recently, a model for the nanodomains in polymer-derived Si-O-C ceramics was proposed by Saha et al.[9]. These various studies on Si-O-C materials derived from polymer precursors reveal interesting features of these unusual amorphous ceramics.

From a thermodynamic point of view, Si-O-C phases can only be simulated, at the moment, as the combination of silicon carbide, graphite and silica. Such a mixture does not account for the real structure of the fibers nor their actual reactivity towards gaseous species.

In 2007 and 2008, T. Varga[6] and R.M. Morcos[8] et al. have measured by high temperature calorimetry the enthalpy of formation of Si-O-C solid solutions of various compositions. These measurements revealed an unexpected stability of Si-O-C phases as compared to the stability of silicon carbide, graphite and silica.

In the present work, microstructural and chemical characterizations have been carried out on Nicalon[TM] and Tyranno Si-O-C fibers: Nicalon NLP101, NLM202, NLM207 and Tyranno ZMI, LOX-M, S, AM in order to improve the understanding of the particular fiber structure to propose a relevant thermodynamic description of the Si-O-C system. An original thermodynamic study of the SiO_xC_y phase at 298K is also proposed. The aim of this study is to simulate the reaction between the fibers (Nicalon[TM] fibers or Tyranno fibers) and a gaseous phase.

EXPERIMENTAL

Percentage of phases in first generation fibers

The β-SiC phase proportion in the Si-O-C fibers was estimated by X-ray diffraction (D8 X-ray diffractometer, Bruker). The amount of β-SiC crystals was quantitatively evaluated by use of crystalline silicon as an internal standard. For that purpose, 20 wt.% of pure silicon was ground together with 80 wt.% of a Si-O-C fiber during 5 min at 25 Hz in a steel pot[9]. The area under the Si peaks located at $2\theta=28.44°$ and $2\theta=47.21°$ and the β-SiC peaks at $2\theta=35.5°$ and $2\theta=41.35°$, were measured to estimate the proportion of phases according to the following equation:

$$\frac{[SiC]}{[Si]} = K \frac{\sum SiC\ peak\ areas}{\sum Si\ peak\ areas} \tag{1}$$

A theoretical diffraction pattern corresponding to a mixture including 20wt.% of silicon and 80wt.% of silicon carbide was simulated. The application Topas P3 was used to fit the peaks and to calculate accurately a theoretical value of the constant K. The as determined K value was confirmed experimentally by the analysis of a diffraction pattern of a mixture composed of 20 wt.% of pure silicon and 80wt.% of SiC micrometric crystals (β-SiC from Alfa Aesar 99.8%). Additional experiments were carried out in order to assess the SiC quantification process. Different well defined proportions of micrometric β-SiC powder and amorphous silicon oxycarbide (from poly(phenyl-methylsilsesquioxane), 90% phenyl, 10% methyl after pyrolysis at 1000°C) containing a light amount of free carbon were mixed together with 20wt.% of pure silicon. The reproducibility error of this process is 1% and the maximum error in the determination of SiC content is 4% as a result of the sample preparation.

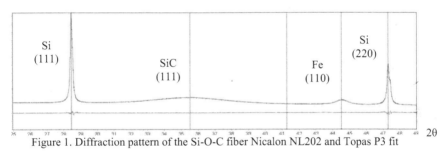

Figure 1. Diffraction pattern of the Si-O-C fiber Nicalon NL202 and Topas P3 fit

The presence of iron is due to the grinding process in steel pots. It is not affecting the quantification. Indeed the iron peak is neither overlaying the silicon nor silicon carbide peaks. The β-SiC peak ($2\theta = 35.6°$) is broad due to the nanometric size of the crystallites. The measurement of the peak width allows an estimation of the average β-SiC crystal size using the Scherrer formula (2).

$$H = \frac{k\lambda}{\tau\cos\theta} \qquad (2)$$

Where H is the peak width, k a form factor equal to 0.9, λ=0.15406 nm the wavelength of the X-ray beam and τ the mean diameter of the β-SiC crystals.

The free carbon content in each fiber was estimated from valence of silicon calculations. All the silicon atoms are included in a continuum consisting of an amorphous SiO_xC_y phase and β-SiC nanocrystals. The valence number for the silicon atom in the continuum is 4 and can be written as $v_{Si} = 4n_{Si}$ with n_{si} the atomic percentage of silicon in a fiber. The free carbon amount can be deduced by assuming that all the oxygen atoms $v_O = 2n_O$, are bonded to silicon. $v_{C-SiOC} = 4n_{Si} - 2n_O$ chemical bonds with silicon are available for carbon atoms in the SiO_xC_y continuum. All the remaining carbon atoms can be attributed to the free carbon phase.

$$n_{C-free} = n_{C-total} - n_{C-SiOC} \qquad (3.1)$$
$$n_{C-free} = n_{C-total} - n_{Si} + \frac{n_O}{2} \qquad (3.2)$$

These valence conditions lead to a composition of the oxycarbide phase that can be written $SiC_xO_{2(1-x)}$ (0<x<1).

RESULTS AND DISCUSSION

The quantitative elemental analysis (EA) of the fibers was performed by the analysis center of CNRS in Solaize, France. The weight concentrations of the elements in the fibers and thus the proportion of the different phases constituting the fibers, as deduced from the above assumptions, are presented in table I and II.

Table I. Elemental analyses of the Si-O-C fibers

Wt.%	O	Si	C	N	Ti	Al	Zr	H
Nicalon 101	14,32	54,72	30,10	<0,1				<0,30
Nicalon 202	12,90	55,76	29,20	0,14				0,31
Nicalon 207	11,95	55,94	29,46	0,1				<0,30
Tyranno LOXM	8,88	52,89	34,05	<0,10	2,25			<0,30
Tyranno S	16,50	47,39	30,52	<0,10	1,91			0,34
Tyranno ZMI	8,55	53,06	34,82	<0,10			1,07	0,33
Tyranno AM	12,27	49,94	34,12	<0,10		0,63		<0,30

The fibers are composed of three phases: free carbon, nanometric β-SiC crystals and an amorphous phase SiO_xC_y. The weight concentration of β-SiC phase in the fibers was measured by X-ray diffraction (figure 2) and the weight fraction of the free carbon phase can be inferred directly from the overall composition by means of the following equation:

$$wt.\%Cfree = \frac{at.\%Cfree * M(C) * 100}{at.\%Ctotal * M(C) + at.\%O * M(O) + at.\%Si * M(Si)} \qquad (4)$$

Table II. Quantification of weight percents of β-SiC and of free carbon

	wt.% β-SiC	wt.% Cfree	wt.% SiOxCy	β-SiC grain size
Nicalon 101	42,34	12,18	45,48	1,71
Nicalon 202	53,77	10,42	35,81	1,81
Nicalon 207	62,90	10,30	26,80	1,87
Tyranno LOXM	71,69	15,41	12,90	1,87
Tyranno ZMI	74,76	15,91	9,34	2,18
Tyranno AM	51,51	18,03	30,46	2,16
Tyranno S	35,12	17,42	47,46	1,20

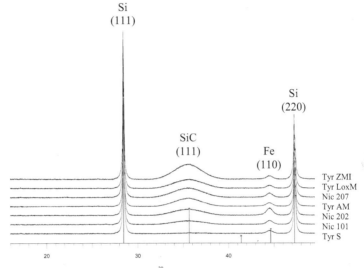

Figure 2. Diffraction spectra of the Si-O-C fibers with 20wt.% of Si

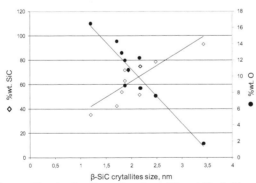

Figure 3. The β-SiC crystallite size is a linear function of the wt.% of silicon carbide and oxygen in the Si-O-C fibers. The β-SiC wt.% of Tyranno ZE and ZM fibers were also determined.

The size and the proportion of β-SiC crystallites is clearly correlated to the oxygen content: the higher the oxygen content, the smaller the β-SiC crystallites, in terms of both the weight fraction and the mean grain size.

The free carbon phase in the Si-O-C fibers can be easily evidenced as basic structural units (BSU) (thin graphene layers stacks in a turbostratic form) by TEM. Moreover, Raman and XPS analyses confirm the presence of sp^2 carbon.

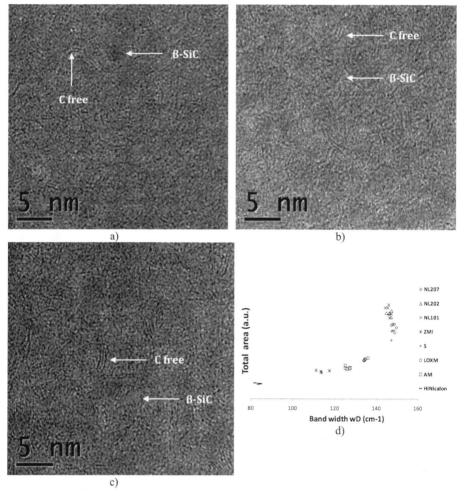

Figure 4 a) TEM image of the Nicalon NL207, b) Tyranno LOXM, c) Tyranno ZMI fiber and d) Raman data, area as a function of the D band width.

Raman analyses clearly demonstrate the sp^2 structure of the carbon in the fibers. Moreover, the TEM images also highlight the presence of graphene layers in all the fibers. The Raman spectroscopy reveals not only a G band characteristic of graphite but also a D band with a width depending on the disorder into the carbon structure.

Further analyses are currently carried out on the fibers to complete the knowledge on the SiC$_x$O$_y$ phase. Quantitative phase measurements may also be obtained by solid NMR, from the analysis of the integration of the ^{29}Si peaks. The comparison of the results is needed to finalize the thermodynamic description of these fibers.

THERMODYNAMICS

Introduction

Morcos[8] and Varga[6] have determined that polymer derived ceramics (PDC) have negative enthalpies of formation as compared to the stable crystalline constituents, i.e. silicon carbide, graphite and crystallized silica. As the amorphous state has the highest entropy, this suggests that PDCs made by a controlled pyrolysis of the crosslinked polysiloxanes could be thermodynamically stable. It is experimentally observed that incorporation of carbon into silica glass improves high temperature mechanical properties and increases the devitrification resistance of the amorphous material. Actually, the incorporation of carbidic carbon (sp^3) into the structure of amorphous silica leads to the strengthening of the network by increasing the bond density.

The structure of these materials could also explain the unexpected stability of these materials up to 1500°C[6]. A structural model proposed by Saha and al.[10] describes the structure of these materials as a network of graphene sheets sharing nanometric domains of silica. The graphene network, limiting the long-range diffusion of silica, improves the resistance to crystallization of Si-O-C materials as compared to silica.

In a work published in 2003, P. Kroll[11] proposed an atomistic model of stoichiometric amorphous silicon oxycarbide glasses (Si-O-C) based on a random network. The generic composition of this phase is the result of a rule of mixture applied to silicon carbide and silica: SiC$_x$O$_{2(1-x)}$ = xSiC + (1 − x)SiO$_2$. The variation of the structure of the material at the atomic scale is calculated by ab initio methods when incorporating tetrahedral carbon, sp^3-C into an ordered amorphous silica network. A first critical composition is observed (Si$_{0.38}$O$_{0.50}$C$_{0.12}$) for the maximum carbon content that the amorphous Si-O-C structure can incorporate without network structure disruption. Beyond this composition, the strain generates the reorganization of the network and causes partial modification of the carbon structure from sp^3 to sp^2 state. Beyond composition Si$_{0.36}$C$_{0.08}$O$_{0.56}$, a one dimensional −Si−C− chain is formed in the network. This one dimensional chain is extended to a three-dimensional Si–C substructure beyond composition Si$_{0.33}$C$_{0.13}$O$_{0.50}$.

The same year, Otta and Pezzotti[12] prepared an almost pure oxycarbide glass from siloxane precursors with a composition very close to the theoretical SiC$_x$O$_{2(1-x)}$ stoichiometry, i.e. almost no free carbon. This ceramic was prepared from vinylmethylsiloxane cyclics (VHM-005) and polymethylhydrosiloxane (HMS-991). The final material was obtained after two pyrolysis steps at 813K and 1173K. The last preparation step, which differs from common Si-O-C glasses synthesis methods, consisted in hot pressing cycle at 2073K (5min, 30MPa). NMR analyses showed that the glass was mainly a pure silicon oxycarbide phase with a chemical composition of $Si_{0.42}O_{0.29}C_{0.29}$. X-ray analyses detected only 5 vol.% of crystallized silicon carbide and Raman analyses revealed a free carbon amount of ≈ 7 wt.%. This phase is not only located almost on the quasi binary axis SiC-SiO$_2$, but it also seems to be at the limit between a binary ($SiC_xO_{2(1-x)}$ and C) and a ternary domain ($SiC_xO_{2(1-x)}$, C, β-SiC).

Based on these studies, we propose an isothermal section of the ternary Si-O-C phase diagram that incorporates Si-O-C glasses as a stable amorphous phase. In the diagram presented in figure 5, the following main domains are identified:

- A quasibinary SiC-SiO$_2$ amorphous phase of generic composition $SiC_xO_{2(1-x)}$ corresponding the random structure proposed by Kroll[11]. This phase extends from amorphous SiO$_2$ up to a limit corresponding to the formation of crystalline SiC. According to the results of Otta and Pezzotti[12] this composition is identified as: $Si_{0.42}O_{0.29}C_{0.29}$. Beyond this limit, a two-phased domain $Si_{0.42}O_{0.29}C_{0.29}$ + SiC is observed.
- In the O rich region, the $SiC_xO_{2(1-x)}$ amorphous phase is in equilibrium with two two-phased domains:
$SiC_xO_{2(1-x)}$ + C and $SiC_xO_{2(1-x)}$ + Si
- In the Si rich region of the isothermal section of the phase diagram, the two-phased $Si_{0.42}O_{0.29}C_{0.29}$ + SiC domain is in equilibrium with two three-phased domains: $Si_{0.42}O_{0.29}C_{0.29}$ + SiC + C and $Si_{0.42}O_{0.29}C_{0.29}$ + SiC + Si.

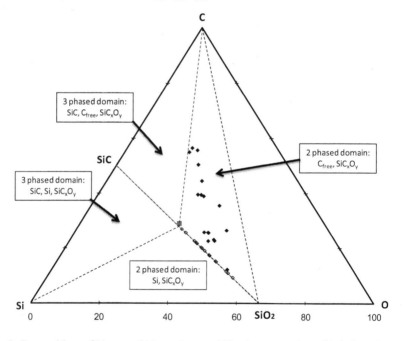

Figure 5. Compositions of Varga and Morcos' oxycarbide glasses together with their projections on the quasi-binary axis SiC-SiO$_2$.
■ represents the composition of Otta and Pezzotti[12]

The compositions of $Si_xO_yC_z$ materials published in the literature are in accordance with such a diagram (figure 6). Most of these materials were prepared from polymer pyrolysis or sol-gel process. The materials located in the two-phased domain ($SiC_xO_{2(1-x)}$ + C) of the proposed phase diagram do not include SiC and belong to the blackglass family. The three points located in the ternary domain (SiC, C, $Si_{0.42}O_{0.29}C_{0.29}$) are compositions prepared by Kaneko[13] et al. by oxygen-controlled pyrolysis of polycarbosilane (PCS). The PCS precursor is also used for the synthesis of Nicalon fibers. The pyrolysis of this polymer yields to the formation of a three phases material, β-SiC crystals, SiO_xC_y and free carbon. The HR-TEM images revealed that the ceramics having the highest oxygen content tended to remain in the amorphous state, whereas β-SiC crystals are observed in the ceramics presenting the lowest amount of oxygen.

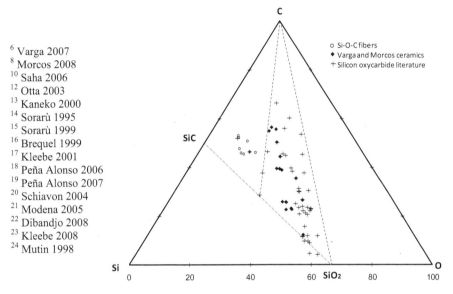

[6] Varga 2007
[8] Morcos 2008
[10] Saha 2006
[12] Otta 2003
[13] Kaneko 2000
[14] Sorarù 1995
[15] Sorarù 1999
[16] Brequel 1999
[17] Kleebe 2001
[18] Peña Alonso 2006
[19] Peña Alonso 2007
[20] Schiavon 2004
[21] Modena 2005
[22] Dibandjo 2008
[23] Kleebe 2008
[24] Mutin 1998

Figure 6. Compositions of oxycarbide materials from literature plotted on the proposed isothermal section of the ternary Si-O-C diagram.

Calculations

The proposed isothermal section of the Si-O-C phase diagram can be modeled by use of a Gibbs free energy of formation of the $SiC_xO_{2(1-x)}$ amorphous phase. Correspondence between phase diagram and diagram of enthalpy is summarized in figure 7.

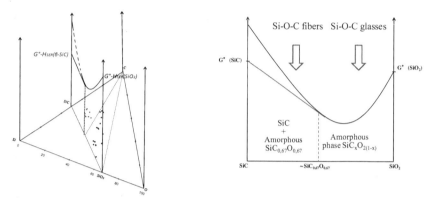

Figure 7 a) Theoretical enthalpy diagram, b) SiC-SiO$_2$ quasi binary section of the 3D Gibbs energy diagram.

The Gibbs energy of formation of the amorphous phase $SiC_xO_{2(1-x)}$ is modeled as a substitutional solid solution. It is calculated as the sum of a mechanical mixture of the pure components (reference state), an ideal entropy of mixing and an excess quantity corresponding to the enthalpy of mixing. The reference state was chosen as the Gibbs energy of a mechanical mixture of Si, O and C in the liquid state which is the most representative of the amorphous state. Thermodynamic data used in our calculations come from the Scientific Group Thermodata Europe (SGTE) data bank. The lattice stabilities were taken from the Dinsdale[25] compilation. The Gibbs energy of mixing is written:

$$G_m = \sum_i x_i \, {}^\circ G_i - T S_m^{id} + {}^{XS} G_m$$

with $S_m^{id} = -R \Sigma x_i \ln x_i$ (5)

The excess Gibbs energy corresponds to the enthalpy of formation of the solid solution $SiC_xO_{2(1-x)}$. Data related to this phase were deduced from the measurement of enthalpies of formation of $Si_xO_yC_z$ polymer-like amorphous ceramic materials, carried out by Varga[6] and Morcos[8]. These authors performed calorimetric measurements for a large range of compositions of amorphous silicon oxycarbide ceramics. By use of thermochemical cycles they deduced the enthalpy of formation of the $Si_xO_yC_z$ materials without any assumption concerning the phase composition. In the present paper, we used Varga[6] and Morcos[8] calorimetric measurements, but we assumed that their materials were composed of a solid solution of generic composition $SiC_xO_{2(1-x)}$ and free carbon supposed to have a graphite-like structure. Such a structural hypothesis is confirmed by P. Kroll[26] using ab initio molecular dynamic simulations at elevated temperatures. Starting from the elemental composition as measured by Varga[6] and Morcos[8] (figure 5) we calculated the composition of the solid solution $SiC_xO_{2(1-x)}$ and the proportion of free carbon assuming that each sample of global composition $Si_xO_yC_z$ is in fact a mixture: $(1 - a) SiC_xO_{2(1-x)} + a\ C_{free}$.

The following modified thermochemical cycle was used to determine the enthalpies of formation of the pure oxycarbide phases from the elements at 298.15K.

$$(1 - a)Si_xO_yC_z (\text{pure amorphous phase, 298.15K}) + (1 - a)\left(x + z - \tfrac{y}{2}\right)O_2(g, 1078.15K) \rightarrow$$
$$(1 - a)xSiO_2(cr, 1078.15K) + (1 - a)zCO_2(g, 1078.15K) \qquad (6.1)$$

$$aC_{free}\ (\text{graphite, 298.15K}) + aO_2(g, 1078.15K) \rightarrow aCO_2(g, 1078.15K) \qquad (6.2)$$

The as determined enthalpies of formation of the pure oxycarbide phase corresponding to the various samples studied by Varga[6] and Morcos[8] are plotted in figure 8. The enthalpy of formation of SiO_2 quartz is also plotted in the graph. This value is a good approximation of the value corresponding to the SiO_2 glass state as the Gibbs energy of amorphous SiO_2 estimated by Golczewski[27] et al.. It is very close at room temperature to the value corresponding to the crystallized state. All these data were fitted as a regular solution with interaction parameters between Si/O and Si/C. (figure 8).

Enthalpy of formation (J/mol)

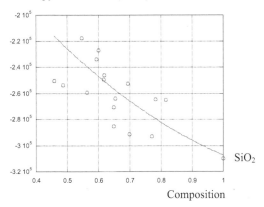

Figure 8. Enthalpies of formation from the elements at 298K of the pure oxycarbide phase corresponding to the various samples studied by Varga and Morcos as a function of composition of the solid solution.
Composition corresponds to x in the formula $xSi_{0,33}O_{0,67} + (1-x)Si_{0,5}C_{0,5}$

The fitted expression was further used to determine the Gibbs energy of the solid solution as a function of composition at 298K by use of equation 5. The corresponding curve plotted in figure 9 together with the Gibbs energy of $Si_{0.5}C_{0.5}$ reveals two domains in accordance with the proposed isothermal section of the ternary Si-O-C phase diagram (figure 5).

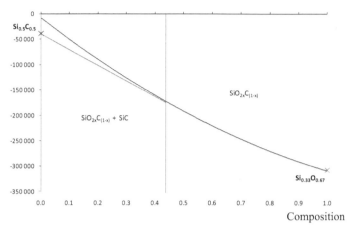

Figure 9. Gibbs energy of $SiO_{2x}C_{(1-x)}$ phase at 298K.
Composition corresponds to x in the formula $xSi_{0.33}O_{0.67} + (1-x)Si_{0.5}C_{0.5}$

The calculation is currently completed to provide a description of the solid solution as a function of both composition and temperature. The complete model will allow us to simulate reactivity of such a material with gas phase or other phases.

Conclusion
Microstructural and chemical characterizations have been carried out on Nicalon[TM] and Tyranno Si-O-C fibers: Nicalon NLP101, NLM202, NLM207 and Tyranno ZMI, LOX-M, S, AM in order to improve the understanding of the particular fiber structure and to propose a relevant thermodynamic description of the Si-O-C system. DRX analyses were carried out to find out the percentage of β-SiC crystallites in the fibers. Raman and HR-TEM techniques were used to investigate in detail their microstructure. The description of the microstructure of the fibers is nevertheless still uncomplete and further analyses are currently performed.

An original thermodynamic study of the $SiO_{2(1-x)}C_x$ phase belonging to the blackglass family is derived from the results of the literature. A description of the isothermal section of the Si-O-C phase diagram including the amorphous $SiO_{2(1-x)}C_x$ phase is proposed. It is in accordance with most of the experimental results and ab initio simulation.

REFERENCES
[1]C. Laffon, A.M. Flank, P. Lagarde, M. Laridjani, "Study of Nicalon-based ceramic fibers and powders by EXAFS spectrometry, X-ray diffractometry and some additional methods," *J. of Mater. Sci.*, **24**, 1503-1512 (1989).
[2]L. Porte, A. Sartre, "Evidence for a silicon oxycarbide phase in the Nicalon silicon carbide fibre," *J. of Mater. Sci.*, **24**, 271-275 (1989).

[3]P. Le Coustumer, M. Monthioux, A. Oberlin, "Understanding Nicalon Fiber," *J. of the Eu. Ceram. Soc.*, **11**, 95-103 (1993).

[4]N. Hochet, M. H. Berger, A. R. Bunsell, "Microstructural evolution of the latest generation of small-diameter SiC-based fibers tested at high temperatures," *J. of Microscopy*, **185**, 243-258 (1997).

[5]A.R. Bunsell, "A review of the development of three generations of small diameter silicon carbide fibres," *J.Mater.Sci* **41**, 823-839 (2006).

[6]T. Varga, A. Navrotsky, J. L. Moats, R. M. Morcos, "Thermodynamically Stable SixOyCz Polymer-Like Amorphous Ceramics," *J. Am. Ceram. Soc.*, **90 [10]**, 3213-3219 (2007).

[7]R. Bodet, N. Jia, R.E. Tressler, "Microstructural Instability and the Resultant Strength of Si-C-O (Nicalon) and Si-N-C-O (HPZ) Fibers," *J. Am. Ceram. Soc.*, **16**, 653-664 (1996).

[8]R. M. Morcos, A. Navrotsky, T. Varga, "Energetics of SixOyCz Polymer-Derived Ceramics Prepared Under Varying Conditions," *J. Am. Ceram. Soc.*, **91 [9]**, 2969-2974 (2008).

[9]A. Saha, R. Raj, "Crystallization Maps for SiCO Amorphous Ceramics," *J. Am. Ceram. Soc.*, **90 [2]**, 578-583 (2007).

[10]A. Saha, R. Raj, "A model for the nanodomains in polymer-derived SiOC," *J. Am. Ceram. Soc.*, **89 [7]**, 2188-2195 (2006).

[11]P. Kroll, "Modelling and simulation of amorphous silicon oxycarbide", *J. Mater. Chem.*, **13**, 1657-1668 (2003).

[12]K. Ota, G. Pezzotti, "Internal friction analysis of structural relaxation in Si-O-C glass," *J. of Non-Crystalline Solids*, **318**, 248-253 (2003).

[13]K. Kaneko, "HRTEM and ELNES analysis of polycarbosilane-derived Si-O-C bulk ceramics," *J. of Non-Crystalline Solids*, **270**, 181-190 (2000).

[14]G. D. Sorarù, G. D'Andrea, R. Campostrini, "Structural Characterization and High-Temperature Behavior of Silicon Oxycarbide Glasses Prepared from Sol-Gel Precursors Containing Si-H Bonds," *J. Am. Ceram. Soc.*, **78 [2l]**, 379-87 (1995).

[15]G. D. Sorarù, D. Suttor, "High Temperature Stability of Sol-Gel-Derived SiOC Glasses," *Journal of Sol-Gel Science and Technology* **14**, 69–74 (1999).

[16]H. Brequel, J. Parmentier, G.D. Sorarù, L. Schiffini, S. Enzo, "Study of the phase separation in amorphous silicon oxycarbide glasses under heat treatment," *NanoStructured Materials*, **11 [6]**, 721-731 (1999).

[17]H.J. Kleebe, and C. Turquat, "Phase Separation in an SiCO Glass Studied by Transmission Electron Microscopy and Electron Energy-Loss Spectroscopy," *J. Am. Ceram. Soc.*, **84 [5]**, 1073–80 (2001).

[18]R. Peña-Alonso, G. D. Sorarù, R. Raj, "Preparation of Ultrathin-Walled Carbon-Based Nanoporous Structures by Etching Pseudo-Amorphous Silicon Oxycarbide Ceramics", *J. Am. Ceram. Soc.*, **89 [8]**, 2473–2480 (2006).

[19]R. Pea-Alonso, G. Mariotto, C. Gervais, F. Babonneau, G. D. Sorarù, "New Insights on the High-Temperature Nanostructure Evolution of SiOC and B-Doped SiBOC Polymer-Derived Glasses," *Chem. Mater.*, **19 [23]**, 5694-5702 (2007).

[20]M. A. Schiavon, C. Gervais and F. Babonneau, G. D. Sorarù, "Crystallization Behavior of Novel Silicon Boron Oxycarbide Glasses", *J. Am. Ceram. Soc.*, **87 [2]**, 203–208 (2004).

[21]S. Modena, G. D. Sorarù, Y Blum, R. Raj, "Passive Oxidation of an Effluent System: The Case of Polymer-Derived SiCO", *J. Am. Ceram. Soc.*, **88 [2]**, 339–345 (2005).

[22]P. Dibandjo, S. Dirè, F. Babonneau and G. D. Sorarù, "New insights into the nanostructure of high-C SiOC glasses obtained via polymer pyrolysis", *Eur. J. Glass Sci. Technol.*, **49** (4), 175–178 (2008).

[23]H.-J. Kleebe, Y. D. Blum, "SiOC ceramic with high excess free carbon," *Journal of the European Ceramic Society*, **28,** 1037–1042 (2008).

[24]P. H. Mutin, "Control of the Composition and Structure of Silicon Oxycarbide and Oxynitride Glasses Derived from Polysiloxane Precursors", *Journal of Sol-Gel Science and Technology* **14**, 27–38 (1999).

[25]A. Dinsdale, " SGTE data for pure elements", *CALPHAD*, **15,** 317-425, (1991)

[26]P. Kroll, "Modeling the free carbon phase in amorphous silicon oxycarbide", *J. of Non-Crystalline Solids.*, **351**, 1121-1126 (2005).

[27]J.A. Golczewski, H.J. Seifert, and F. Aldinger, "A thermodynamic model of amorphous silicates", *Calphad*, **22 [3]**, 381-396 (1998).

STATIC FATIGUE OF MULTIFILAMENT TOWS AT HIGH TEMPERATURES ABOVE 900°C

A. Laforêt and J. Lamon
CNRS/University of Bordeaux
Laboratoire des Composites Thermostructuraux
3 Allée de La Boétie
33600 Pessac, France
lamon@lcts.u-bordeaux1.fr

ABSTRACT

Previous work has shown that SiC-based fibers are sensitive to delayed failure at intermediate temperatures $\leq$ 800°C. This behaviour has been attributed to slow crack growth driven by the oxidation of carbon at grain boundaries. The paper examines the static fatigue behaviour of SiC-based Hi-Nicalon fibers at higher temperatures ($\geq$ 900°C), when oxidation is enhanced. The influence of the oxide layer which grows at the surface of fibers is investigated. Fiber tows were subjected to static fatigue tests at 900°C and 1000°C. The slow crack growth based model which has been established earlier was revisited in order to introduce the contribution of the oxide coating. The model was found to describe satisfactorily the static fatigue behaviour of fiber tows at these temperatures.

INTRODUCTION

Ceramic matrix composites (CMCs) are very attractive for structural applications at high temperatures. Nowadays, they are essentially used in space and defence applications. They are now considered for introduction in aircraft engines components. The control and prediction of CMC lifetime thus becomes a crucial issue.

Delayed failure of SiC/SiC composites, and SiC fibers has been observed under stresses far smaller than the stress-to-failure, essentially at intermediate temperatures $\leq$ 800°C [1, 2]. The stress-rupture behavior of Nicalon/SiC at 950°C was also reported [3].

Previous work has also shown that the delayed failure of SiC-based fibers and tows was satisfactorily predicted by a slow crack growth based model, in the previously mentioned range of intermediate temperatures (< 800°C) [4]. The present paper focuses on higher temperatures (900°C and 1000°C), when additional effects related to oxidation rate and temperature are expected. The static fatigue behaviour of tows of SiC fibers (Hi-Nicalon, Nippon Carbon Co., Japan) has been investigated. The experimental data were analyzed using the subcritical crack growth model that has been proposed earlier for single fibers [4]. The model has been revisited so as to account for oxidation related features. The model is based on the following crack velocity equation [4]:

$$V = A_1 K_I^n \qquad (1)$$

Where V is crack velocity, A_1 and n are constant and K_I is the stress intensity factor.

Under a constant stress, the lifetime of a single filament is the time required for the critical flaw to grow from initial size c_j to critical length a_c:

$$t= \int_{c_j}^{a_c} \frac{da}{V} \qquad (2)$$

Introducing fracture mechanics-based crack size-stress relationships and integrating equation (2) gave:

$$t = \frac{2K_{IC}^{2-n}}{\sigma^n A_1 (n-2) Y^2} \left[\sigma_f^{n-2} - \sigma^{n-2} \right] \qquad (3)$$

With $A_1 = A_{10} \exp \left[-\frac{E_a}{RT} \right]$, K_{IC} is the critical stress intensity factor, σ is the applied stress, Y is a crack shape factor, E_a is the activation energy relative to the chemical reaction at crack tip, T is temperature, R= 8.314 J K^{-1} mol^{-1}, σ_f is fiber strength in the absence of environmental effect.

The statistical distribution of fiber strengths is described satisfactorily by the conventional Weibull equation:

$$P = 1 - \exp \left[-\frac{V_f}{V_o} \left(\frac{\sigma_f}{\sigma_o} \right)^m \right] \qquad (4)$$

Where P is failure probability, V_f is the fiber volume, V_o is the reference volume, σ_o and m are statistical parameters.

EXPERIMENTAL PROCEDURE
        Hi-Nicalon tows contain 500 filaments of 15 µm average diameter each. During the static fatigue tests, the gauge length (25 mm) was located in the furnace hot-zone at a uniform temperature (hot grip technique). The gauge length is defined as the distance between the grips. A silica tube protected the tow against possible pollution from furnace elements. It also allowed environment control through a constant gas flow (N$_2$/O$_2$). The test specimens were heated up to the test temperature before loading (heating rate ~ 20°C/min). Then a dead-weight-load was hung slowly (this operation took < 10 s). Lifetime was captured automatically by a computer when specimen failed.
        Tow ends were affixed within alumina tubular grips using alumina-based cement. Much care was taken during test specimen preparation and handling. A device to ensure alignment of tows and of filaments within the tows was used. For this purpose, a low load (12 g) was applied on tows.
        Stress on multifilament tows was derived from the applied load, using the following equation which takes into account the individual fibre failures induced by the applied load:

$$\sigma = \frac{F g \rho l N_0}{m_t \left[N_0 - N(F) \right]} \qquad (5)$$

Where F is the applied load, g is gravity constant, ρ is tow density, l is the gauge length, and m_t is tow mass. N_o is the initial number of unbroken fibres and N(F) the number of fibres broken under F.

A few static fatigue tests were carried out using the cold grip technique, on much longer tows (gauge length = 300 mm). Specimen temperature varied from the test temperature in the middle of tow located in the furnace, to room temperature in the grips. Tow deformation was measured using a LVDT extensometer.

Monotonic tensile tests were conducted at room temperature in order to determine the residual stress-strain behaviour after short static fatigue tests. A standardized testing procedure was applied for the tensile tests [5]. The tows were loaded at a constant deformation rate (5 mm/min). The static fatigue test conditions were as follows: hot grip technique, gauge length =25mm, 20 minutes for heating up and for cooling down, 15 minutes at the test temperature, test temperatures = 500, 800, 900 and 1000°C, applied stress = 1.7 MPa.

The fractured specimens were examined using scanning electron microscopy.

RESULTS AND DISCUSSION
Static fatigue under a constant temperature (hot grip technique)

Figure 1 shows plots of the rupture times that were measured under various constant loads. The stress-rupture time dependence is well fitted by the slow crack growth law that has been established at lower temperatures (500 and 800°C) [4,6]. According to equations in the introduction, the law reduces to:

$$t\sigma^n = A_0 \exp \left[\frac{E_a}{RT} \right] \qquad (6)$$

With $A_0 = 5.62 \ 10^{17} \ \text{s}^{-1} \ \text{MPa}^{-n}$, n= 8.4 and $E_a = 181.6 \ \text{kJ mol}^{-1}$ [6].

Equation (6) describes quite well the average stress-rupture time behaviour at 900°C and 1000°C. At 1000°C, the following features can be noticed:
- the scatter in rupture times at a given stress is much larger when comparing to 900°C,
- the stress at instantaneous fracture decreases tremendously as temperature increases.
Furthermore, the comparison of data obtained at 900 and 1000°C clearly indicates that a certain amount of specimens exhibited longer lifetimes at 1000°C.

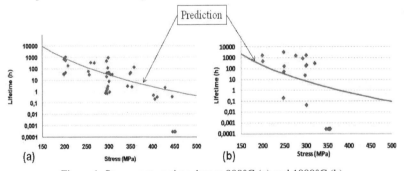

Figure 1. Stress-rupture time data at 900°C (a) and 1000°C (b).

Figure 2 shows the temperature dependence of tow instantaneous failure. It appears that this trend is at variance with that one observed on single filaments, which suggests the contribution of an extraneous effect involving the fiber framework. This weakening of tows can be logically attributed to fiber interactions [7]. Under the current environment, it seems reasonable to relate the fiber interactions to oxidation, since it is well known that an oxide layer forms on the fibers, and that the growth rate increases with temperature [8,9].

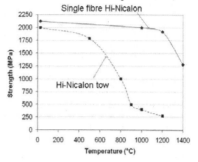

Figure 2. Temperature dependence of strength for Hi-Nicalon tows and single filaments.

Static fatigue under a temperature gradient (cold grip technique)

Figure 3 shows the location of fracture. It can be noticed that the tow failed from outside the furnace, where the temperature was around 600°C, whereas the portion subjected to 1000°C survived. This result demonstrates that lifetime may be longer at 1000°C. This is consistent with the trend indicated by the stress-rupture data at 1000°C versus those at 900°C (figure 1). Furthermore, it is worth pointing out that the part of fibers that was exposed to 1000°C was coated by a thick layer of oxide. From Figure 3, it can be also noticed that tow deformations remained quite constant during the tests: their amplitude did not exceed 20 μm (corresponding average strain 0.0016%) after 20 days under 200 MPa. This result demonstrates that creep did not occur at the temperatures ≤ 1000°C, under a stress of 200 MPa. This is consistent with previous results obtained on single SiC fibers for which creep was observed only at temperatures from 1200°C. It also suggests that the weakening of tows does not result from a degradation of the whole fiber, since this would cause a strain increase, or a stiffness decrease. It is in agreement with a local effect of damage such as slow growth of pre-existing flaws.

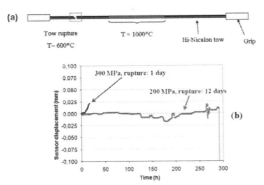

Figure 3. Static fatigue using the cold grip technique: (a) schematic diagram showing the specimen and failure location, (b) deformations of tows

Residual stress-strain behaviour

Figure 4 shows the stress-strain behaviour of tows that experienced short static fatigue at temperatures ranging from 500 to 1000°C. It appears that the stress-strain behaviour is highly dependent on the test temperature.

After fatigue at 500°C, the stress-strain behaviour is identical to that reference one obtained on as-received tows. The rupture stress is unchanged (1800 MPa, 130 N) and the stress-strain curves exhibit the typical downward curvature reflecting individual fiber breaks prior to ultimate failure [7].

After fatigue at higher temperatures, the stress-strain behaviour was tremendously affected.

- The failure stress decreased to 500 MPa (35 N). The strength decrease was commensurate with the test temperature,
- The curve does not exhibit a continuous downward curvature, but instead load drops and linear segments, suggesting that groups of fibers failed.

The groups of fibers which failed were estimated from load drops [7] using the following equation:

$$\frac{N_2}{N_1} = \frac{F_2}{F_1} \tag{7}$$

Where N_1 and N_2 are numbers of surviving fibers, F_1 and F_2 are corresponding forces at a constant strain.

Figure 4 shows that failures of groups of fibers produce the features displayed by the experimental curves.

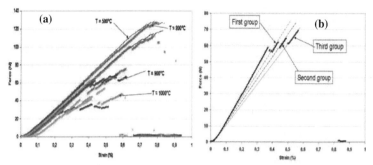

Figure 4. Force strain curves under monotonic tensile loading at room temperature on tows that experienced static fatigue at high temperatures during 15 minutes: (a) experimental results, (b) predictions

Fracture surfaces

Figures 5 and 6 show the typical fracture patterns with successive crack fronts that have been detected on Hi-Nicalon tows after static fatigue at 900°C and 1000°C. This demonstrates the presence of slow crack growth. It supports results reported above (figure 1). Figures 5 and 6 also show the presence of an oxide layer on the fibers. It is obvious from the micrographs that the layer thickness is much more significant on those tows which experienced fatigue at 1000°C. It can also be noticed from figure 6 that some fibers were stuck by oxide and from figure 7 that groups of fibers failed simultaneously.

Figure 8 provides evidence of groups of fibers which failed together during residual tests. It is worth pointing out that these tows had been subjected to short static fatigue (15 minutes).

Thus, microscopy supports the analysis of experimental results.

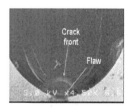

Figure 5. SEM micrograph showing the fracture surface of a Hi-Nicalon fiber after static fatigue on a tow at 900°C, under a stress of 250 MPa during 14 days.

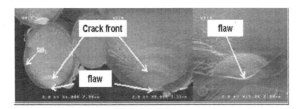

Figure 6. SEM micrograph showing fracture surfaces of fibers in a Hi-Nicalon tow after static fatigue at 1000°C, under a stress of 250 MPa during 13 days.

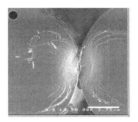

Figure 7. Micrograph showing a group of fiber failures after static fatigue at 1000°C, under a stress of 250 MPa during 13 days.

Figure 8. SEM micrograph showing fracture surfaces of fibers after tensile test at room temperature on a tow which had been subjected to static fatigue at 1000°C for 15 minutes.

PREDICTION OF LIFETIME

The model of static fatigue of fibers has been revisited in order to take into account the effects of the oxide coating at temperatures $\geq$ 900°C, i.e. lifetime and variability increases.

The lifetime increase can be logically attributed to protection by the oxide layer which caused reduction in the oxidation rate at crack tip. It was treated by introducing crack slowing down, through constant A_1 which determines constant $A = A_0 \exp [\frac{E_a}{RT}]$, as demonstrated in a previous paper[6]. For this purpose, a time dependent retardation function $f(t_0)$ was used [11], so that A takes the following form:

$$A = A_0 \exp [\frac{E_a}{RT} + f(t_0)] \qquad (8)$$

$f(t_0)$ operates when the oxide coating reaches a critical thickness at time t_0:

$$f(t_0) = f.H (t-t_0) \qquad (9)$$

With H = 0 when $t < t_0$, and H = 1 when $t \geq t_0$; f is a constant (f was set to 3); $t_0 = \frac{d_{oxide}^2}{kT}$

d_{oxide} is the critical thickness of the oxide layer:

$$k = 1.07 \ 10^{-11} \exp\left[\frac{20326}{T}\right] \qquad \text{when } T < 1000°C \ ^{12}$$

$$k = 2.5 \ 10^{-19} \exp\left[\frac{91000}{RT}\right] \ P_{02}^{0.89} \qquad \text{when } T = 1000°C \ ^{12}$$

It was considered that, at critical time t_0, grain boundaries were completely filled by oxide:

$$d_{oxide} = d_f + \frac{1}{2} \ d_{gb} \qquad\qquad (10)$$

Where d_f is the thickness which has been consumed by oxide production at fiber surface, and d_{gb} is the width of grain boundaries. d_{gb} was measured by image analysis of SEM micrographs: d_{gb} = 40 nm. d_f was estimated from the stoichiometric equation of SiC fiber oxidation. d_{oxide} was estimated to be 50 nm, from equation (10). Table 1 summarizes the values of k and t_0.

Figure 9 shows the crack velocity decrease predicted using the retardation function. Figure 10 shows that the upper rupture times were satisfactorily described by the model. Fiber characteristics introduced into the equations are summarized in table 2.

Table 1. Values of constant k and critical time t_0.

	k (s.m^{-2})	to (s)
500°C	4,08.10^{-24}	612927268
800°C	6,4.10^{-21}	393152
900°C	3,2.10^{-20}	78200
1000°C	3,1.10^{-19}	8057

Table 2. Fiber characteristics that were used for lifetime predictions.

	900°C	1000°C
A_0 (s^{-1} MPa^{-n})	5,62 E+17	5,62 E+17
E_a (J/mol)	181600	181600
R (J/mol/K)	8,314	8,314
Y	1,13	1,13
K_{IC} (MPa. $\sqrt{m}$)	1,25	1,25
$\sigma_{r\,min}$ (MPa)	350	350
$\sigma_{r\,max}$ (MPa)	1790	1790
$\sigma_{r\,moy}$ (MPa)	1100	1100
V (m^3)	3,56 E-12	3,56 E-12
A_1 (MPa^{-n} s^{-1} m$^{1-n/2}$)	1,54 E-08	6,66 E-08
f	3	3
k (m^2/s)	3,10 E-19	3,20 E-20
n	8,4	8,4
Critical silica layer (m)	5,00 E-08	5,00 E-08

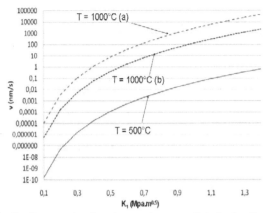

Figure 9. Crack velocity diagram at various temperatures, predicted using the slow crack growth equation $V = A_1 K_I^n$ for Hi-Nicalon fibers (a) at 1000°C when $t < t_0$ and (b) at 1000°C when $t \geq t_0$.

It was considered that variability is related to interactions caused by the oxide fiber coating. As shown by figure 2, the strength of tows is comprised between a temperature dependent minimum value and a maximum value corresponding to reference tow strength at room temperature:

$$\sigma_{r\,min} (T) \leq \sigma_r \leq \sigma_{r\,max} (=1800 \text{ MPa}) \tag{11}$$

Therefore, lower and upper bounds for the stress-rupture time data were obtained by inserting $\sigma_f = \sigma_{min}$ (T) and $\sigma_f = \sigma_{max}$ (=1800 MPa) into equation (3) (table 2). Figure 11 shows that the scatter in lifetimes was satisfactorily predicted for the above data.

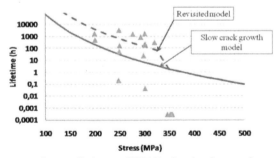

Figure 10. Stress-rupture time predictions at 1000°C using the slow crack growth model and the model revisited.

CONCLUSIONS

Experimental results, SEM fractography and the slow crack growth based model lead to the same conclusions. At 900°C and 1000°C, the delayed failure of tows is caused by a slow crack phenomenon, as observed earlier at lower temperatures. As temperature increases, the fibers are coated by a thicker oxide layer, which influences the slow crack growth phenomenon. The crack velocity slows down, and fiber interactions are enhanced, which leads to longer lifetimes under low stresses and smaller stresses at instantaneous failure. The oxide coating associated effects were introduced into the slow crack growth based model. The lifetimes that were calculated were found in excellent agreement with experimental results. In particular, variability associated to fiber interactions was satisfactorily described. Finally, it is worth pointing out that the tests using the cold grip technique showed that deformations remained constant at temperatures ≤ 1000°C. This demonstrates that creep did not occur at these temperatures and that the phenomena involved do not affect the whole fiber, which is consistent with slow crack growth.

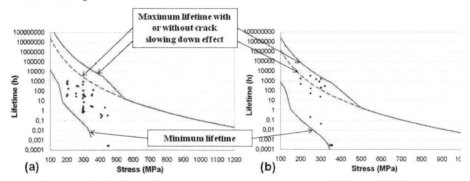

Figure 11. Predictions of scatter in stress-rupture time data (a) 900°C, (b) 1000°C.

ACKNOWLEDGEMENTS
The support of CNRS, SNECMA PROPULSION SOLIDE and DGA is highly appreciated. The work was carried out within the frame of a CPR program entitled "Modelling, prediction and validation of CMC lifetime" and involving the following laboratories: INSA Lyon, LMT Cachan, PROMES Perpignan, CEAT Toulouse.
The authors are indebted to Mrs. J. Forget for help in preparation of manuscript.

REFERENCES
[1]Bertrand S., Pailler R., and Lamon J., "Influence of strong fiber/coating interfaces on the mechanical behaviour and lifetime of Hi-Nicalon/(PyC/SiC)$_n$/SiC minicomposites", *J. Am. Ceram. Soc.*, 84, 4, 787-94 (2001).
[2]Forio P., Lavaire F and Lamon J., "Delayed failure at intermediate temperatures (600°-700°C) in air in silicon carbide multifilament tows", *J. Am. Ceram. Soc.*,87, 5, 888-893 (2004).
[3]Lara-Curzio E., "Stress-rupture of Nicalon/SiC continuous fiber ceramic composites in air at 950°C", *J. Am. Ceram. Soc.*, 80, 12, 3268-72 (1997).

[4]Gauthier W., Lamon J., "Delayed failure of silicon carbide fibers in static fatigue at intermediate temperatures (500-800°C) in air", *Ceram. Engineering and Science Proceedings*, Volume 28, Issue 2, edited by Edgar Lara-Curzio, John Wiley & Sons, Inc., USA, 423-431 (2008).

[5]European Standard, Advanced technical ceramics – Ceramic Composites – Method of test for reinforcement – Part 5: Determination of distribution of tensile strength and of tensile strain-to-failure of filaments within a multifilament tow at ambient temperature, ENV 1007-5 (1998).

[6]Gauthier W., Lamon J., "Delayed failure of Hi-Nicalon and Hi-Nicalon S multifilament tows and single filaments at intermediate temperatures (500-800°C)" *J. Am. Ceram. Soc.*, in press (2009).

[7]Calard V., Lamon J., "Failure of fiber bundles", *Composites Science and Technology*, 64, 701-710 (2004).

[8]Gogotsi Y., Yoshimura M., "Oxidation and properties degradation of SiC fibers below 850°C", *J. Materials Science Letters*, 13, 680-683 (2004).

[9]Shimoo T., Kakehi Y., Kakimoto K. and Kamura K.O., *J. Japan Inst. Metals*, 56, 175 (1992).

[10]Lara-Curzio E., "Oxidation induced stress-rupture of fiber bundles", *J. Engineering Materials and Technology*, 120, 105-109 (1998).

[11]S.M. Wiederhorn, "Influence of water vapor on crack propagation in Soda-lime glass", *Journal of the American Ceramic Society*, 50, 407–414 (1967).

[12]Avril L., Rebillat F., Legallet S., Louchet C., Guette A., "Quantitative approach to oxidation of Hi-Nicalon SiC fibers", presented during "Journées du Groupe Français des Céramiques", Paris (2005).

3D MULTISCALE MODELING OF THE MECHANICAL BEHAVIOR OF WOVEN COMPOSITE MATERIALS

G. Couégnat, E. Martin, J. Lamon
Laboratoire des Composites Thermostructuraux
Université Bordeaux 1 / CNRS
3, Allée de la Boétie, F-33600 Pessac, France

ABSTRACT

The present paper proposes a multiscale model of the mechanical behavior of woven composite materials. The DMD model is based on a physical description of the reinforcement geometry of the material and the damage mechanisms. The model is applied to a woven-ceramic-matrix composite material and its predictive capabilities are investigated.

INTRODUCTION

There is a growing interest in the use of textile architectures for advanced composite applications. They offer a number of attractive properties compared to their non-woven counterparts, such as improved inter-laminar properties and impact resistance. However, their complex and multiscale architecture (Fig. 1) makes the analysis and the modeling of their behavior quite challenging. Most of the available models for woven composites are based on continuum damage mechanics. They can provide a good approximation of the overall response of the material but do not take into account their specificities. And despite many attempts to develop effective models for woven composites, there is currently no approach that can accurately capture all of the important aspects of fabric deformation and effectively predict both the macroscopic non-linear mechanical response, as well as the response of the constituents at the micro and mesostructural level.

Recently, multiscale damage models based on micromechanical considerations have been developed for composite laminates[1,2]. In these models, the effective macroscopic non-linear behavior of the material is derived from the numerical homogenization of representative unit cells. The extension of such modeling approaches from unidirectional laminates to complex multilayer woven materials requires the development of (i) an appropriate multiscale modeling framework and (ii) specialized numerical tools.

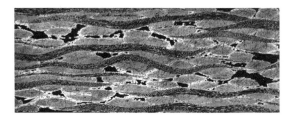 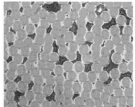

Figure 1: Typical microstructure of a woven composite material: scale of the textile reinforcement – mesocale (*left*) and scale of a yarn – microscale (*right*).

The purpose of the present paper is to introduce a multiscale modeling approach based on a physical description of the geometry of the woven reinforcement, the properties of the constituents and their damage mechanics for the derivation of the effective macroscopic constitutive behavior of woven composite materials.

THE DISCRETE MICRO DAMAGE (DMD) FRAMEWORK

Damage modeling

The DMD damage model is written at the macroscopic scale in the framework of the continuum damage mechanics. A major difference with the other classic macroscopic damage models emerges from the nature of the damage variables. Here, the internal state variables are chosen to directly measure the state of the microstructural damage instead of just measuring a loss of elastic properties. In the DMD approach, the damage variables are defined as crack densities and interfacial debonded areas. Thus, the effective elastic stiffness tensor $\tilde{\mathbf{C}}$ is directly defined from a set of micromechanical damage variables $(d_1,...,d_n)$:

$$\tilde{\mathbf{C}} = f(d_1,...,d_n) \tag{1}$$

The effective elastic strain energy Φ can then be written as follows:

$$2\Phi = (e - e^0 - e^{th}):\tilde{\mathbf{C}}:(e - e^0 - e^{th}) + (e - e^0 - e^{th}):\tilde{\mathbf{C}}^0 : e^0 + e^0 : \tilde{\mathbf{C}}^0 : (e - e^0 - e^{th}) \tag{2}$$

where $\tilde{\mathbf{C}}$ and $\tilde{\mathbf{C}}^0$ are respectively the effective elastic stiffness tensor and the initial elastic stiffness tensor, e is the total strain, e^{th} the thermal strain and e^0 is the residual inelastic strain.

The effective elastic stiffness tensor $\tilde{\mathbf{C}}$ is related to the damage variables d_i through a damage effect tensor $\mathbf{D}$:

$$\tilde{\mathbf{C}} = (\mathbf{C}^0 - \mathbf{D}(d_1,...,d_n)) \tag{3}$$

The damage effect tensor $\mathbf{D}$ is defined as:

$$\mathbf{D} = D_{ij} = h_{ij}(\eta_1 d_1,...,\eta_n d_n) \cdot C_{ij}^0 \tag{4}$$

where h_{ij} represents the relative variation of the (i,j)-component of the stiffness tensor for a damage state $(d_1,...,d_n)$ and η_k is a damage deactivation index allowing the partial or complete recovery of the initial elastic stiffness under a reverse loading.

This formulation allows one to obtain an effective macroscopic behavior law through the derivation of the effective strain energy that is defined from a micromechanical description of the damage state. The scale transition is performed through the evaluation of the damage effect tensor $\mathbf{D}$. Hence, the main step of the derivation of the behavior law consists in the estimation of the damage effect tensor.

Numerical evaluation of the damage effect tensor

The damage effect tensor is evaluated numerically using a finite element (FE) homogenization procedure[3]. Specific damage configurations are introduced in a discrete manner into finite element periodic cells that are representative of the microstructure of the material, i.e. cracks and debonding surfaces are generated in the finite element meshes of these cells. The effect of the damage on the effective elastic properties h_{ij} $(d_1,...,d_n)$ is evaluated using a standard finite element homogenization procedure. Finally, the computed values of h_{ij} are interpolated over the whole damage space $(d_1,...,d_n)$

and the evolution of the macroscopic damage effect tensor $\mathbf{D}$ is fitted as a function of the microscopic damage variables $(d_1,....,d_n)$. The response surfaces are fitted with n-dimensional polynomial functions using a least square fitting technique. This allows a more efficient and numerically stable way of evaluating the values of the response surfaces and their derivatives than performing a nearest-neighbor interpolation. Practically, with fourth-order polynomial surfaces, the maximal error between the computed values and the response surface is less than 2 percent.

The numerical procedure associated with the DMD model is based on the homogenization of damaged unit cells and thus requires generating representative cells of the microstructure of the material at different scales. One must be able to determine geometrical and morphological properties of the microstructure of the material, and then have to create finite element meshes of representative unit cells.

CONSTRUCTION OF REPRESENTATIVE UNIT CELLS

Construction of representative unit cells at the microscale
The morphology of the microstructure at the microscale is governed by the pseudo-random arrangement of the fibers (Fig 2). Statistical descriptors like covariance[4] could here be used to quantify the heterogeneity of the microstructure.

Representative unit cell could then be constructed using this statistical information. Given an initial random distribution of N fibers inside a unit cell, the fiber positions are optimized until the covariance of the unit cell replicates the one from the complete microstructure. At the end of this iterative optimization process, the unit cell has a covariance similar to the microstructure one and can be considered as statistically representative of the whole microstructure. Once the position of the fibers is determined, it is straightforward to add the interphase and the matrix layers around the fibers and to generate a FE mesh of the representative unit cell (Fig 2).

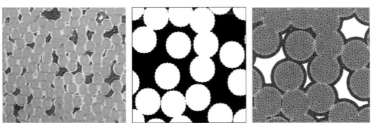

Figure 2: Typical microstructure of a yarn (*left*), covariance-optimal fiber positions (*center*), and finite element mesh of a representative cell (*right*)

The geometrical and mechanical representativity of these covariance-optimized unit cells has been investigated for a growing number of fibers. It has been shown that unit cells containing only 20 to 30 fibers are sufficient to obtain a good overall estimation of the effective elastic properties of a yarn that are made of several hundreds of fibres[5].

Construction of representative unit cells at the mesoscale
The construction of unit cells at the mesoscale is a two-step procedure. First, the geometry of the woven reinforcement is modeled from the yarns interleaving sequence and their geometrical properties. The total bending energy of the textile reinforcement is minimized to calculate its internal

geometry in a relaxed state. This geometrical model is then transformed into a volumetric FE mesh of the reinforcement.

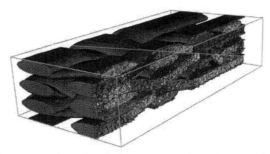

Figure 3: Representative unit cell of a woven ceramic matrix composite material.

The second step consists in generating a three-dimensional model of the matrix at the mesoscale. In the case of ceramic matrix composite, the matrix is made of a thin layer deposited almost uniformly around the yarns. Thus, the volumetric FE meshes of the yarns are first dilated by the thickness of the layer of matrix. Then the inner part corresponding to the yarns initial volume is removed, and the new entity is remeshed accordingly. All the Boolean operations (union, difference) between the volumes are performed directly on the FE meshes[6]. The FE mesh of the representative unit cell is finally obtained by merging the yarns meshes and the matrix mesh (Fig. 3).

APPLICATION OF THE DMD MODEL TO A WOVEN CERAMIC-MATRIX COMPOSITE

Identification of the damage effect tensor

The DMD approach has been applied to a multilayer woven ceramic-matrix composite (CMC). The damage mechanisms that have been observed experimentally for CMC materials are (i) intra-yarns matrix microcracking and (ii) fibre breaking at the microscale, and (iii) inter-yarns matrix cracking and (iv) transverse yarns cracking at the mesoscale (Fig. 4). It is worth noting that each elementary damage mechanism is accompanied by debonding at the fibre/matrix and yarn/matrix interfaces.

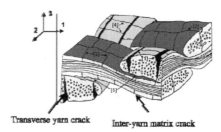

Transverse yarn crack   Inter-yarn matrix crack

Figure 4: Illustration of the elementary damage mechanisms at the mesoscale.

A damage variable is introduced for each elementary damage mechanism. The damage variables are directly defined in terms of crack density and debonded area, eventually normalized with respect to the maximal observed values. It is important to remark that for this specific CMC material the damage is always orientated by the directions of the reinforcement, whatever the direction of the loading is, which dramatically reduces the number of possible damage configurations.

As discussed earlier, the damage effect tensor is identified using representative unit cells at the micro and mesoscale in which damage has been introduced (Fig 5).

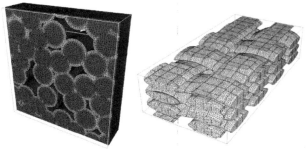

Figure 5: Finite element meshes of representative unit cells at the microscale (*left*) and at the mesoscale (*right*). Note that some cracks and interfacial debonding have been introduced in the mesh of the mesoscopic unit cell.

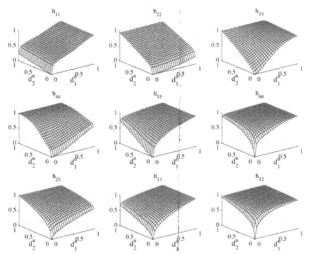

Figure 6: Evolution of the damage effect tensor components h_{ij} as a function of two micro damage variables $\left(d_1^n, d_2^n\right)$ corresponding to inter-yarns matrix cracking, respectively in weft and warp direction.

The damaged yarns are supposed to be homogeneous at the mesoscale. Their effective properties are computed from the homogenization of representative unit cells at the microscale. These properties are then used for the computation of the effective properties of the material at the mesoscale. The damage effects h_{ij} are obtained as functions of the elementary damage mechanisms variables and the damage effect tensor **D** can then be directly computed from these response surfaces for any damage configuration. For example, the effect of the inter-yarns matrix cracking in weft and warp direction is illustrated in Fig 6. It is worth noting that the effect of matrix cracking is almost symmetric with respect to reinforcement directions (h_{11} and h_{22}). The matrix cracking also leads to an important increase of the out-of-plane compliance of the material (h_{33}) due to the yarn/matrix debonding accompanying the matrix fragmentation process. Finally, one can see the strong coupling between the matrix cracking and its effect on the in-plane shear stiffness of the material (h_{12}).

Identification of the damage kinematics

As mentioned previously, for this specific CMC material, the damage is always orientated by the directions of the reinforcement, even for off-axis loading. Thus, only tensile tests in the axis of the textile reinforcement are necessary to identify the damage kinetics. Moreover, as the damage effect seems to be symmetric with respect to the reinforcement direction, the identification of the damage kinematics could be performed in either weft or warp direction. Finally, the damage kinematics, i.e. the evolution of the damage variables versus the applied load, has been identified on a monotonic tensile test in the weft direction (Fig. 7). The optimal set of parameters is determined by minimizing the deviation between the experimental and simulated strain-stress curve using an evolutionary algorithm[7]. The physical admissibility of the identified kinematics has been validated thanks to *in-situ* experimental micrographic observations of the crack patterns during tensile tests on polished specimens.

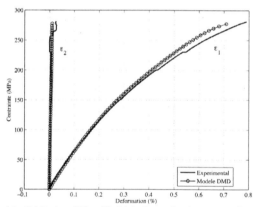

Figure 7: Experimental (*solid line*) and identified (*circle markers*) stress-strain curves for a monotonic tensile test in the direction of the reinforcement axis.

Validation of the DMD model

The predictive character of the DMD model has been investigated by simulating off-axis tensile tests, which from the material point of view are a combination of tensile and shear loading. The stress-strain curves of simulated 45° and 22.5° off-axis tests are depicted in Fig. 8.

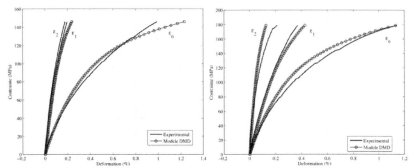

Figure 8: Experimental (*solid line*) and simulated (*circle markers*) stress-strain curves for off-axis monotonic tensile tests at 45° (*left*) and 22.5° (*right*).

The results show a good agreement between the stress-strain curves simulated with the DMD model and the experimental results. One has to remind here that the identification of the parameters of the damage kinematics is performed on only on-axis tensile test. Hence, the numerical multiscale identification of the damage effects accurately predicts the tensile-shear couplings and their effect on the effective behavior of the material. These couplings are directly determined by the previously computed damage effect response surfaces, without the need for introducing any extra material parameters.

Application of the DMD model to a structural test

The DMD has been implemented into the general finite element code ZeBuLoN[8] and applied to small industrial structural tests, namely a tensile test on a shifted double-notched specimen, a torsion test on a holed plate and a flexure test on a L-angle corner.

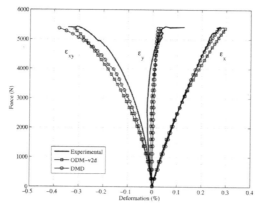

Figure 9: Experimental (*solid line*) and simulated (*reference macroscopic model with square markers and DMD model with circle markers*) load-strain curves for a tensile test on a CMC shifted double-notched specimen.

The DMD simulation results have been compared to those obtained with a reference macroscopic model[9] for the same structural tests. In all the investigated cases, the global response of the structure simulated with the DMD model is close to the experimental one, and at least as accurate as the results obtained with the reference macroscopic model (see for instance, Fig. 9)

Moreover, as the damage variables of the DMD model directly represent the micromechanical damage, one can immediately evaluate the local evolution of the crack densities, which is not possible with a purely macroscopic model. As an example, the evolution of the inter-yarns matrix crack density for the shifted double-notched specimen tensile test is depicted in Fig. 10. The crack density is found to be maximal in the centre zone between the notches, which corresponds to the fracture zone of the specimen (Fig. 11). This micromechanics-based character increases the confidence of the results obtained with the DMD model since the simulations could be compared with the experiments, not only in terms of global structural response, but also in terms of local damage-driven microstructural evolution. This could also be of great interest for further extensions of the DMD model to the simulation of thermo-mechanical fatigue loadings under oxidizing atmosphere, which requires an accurate local description of the evolution of the microstructure of the material.

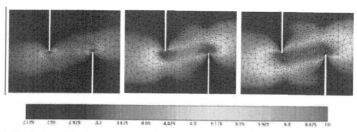

Figure 10: Evolution of the inter-yarns matrix cracks density (mm^{-1}) for the shifted double-notched specimen tensile test. The corresponding applied loads are F=2270N (*left*), F=4140N (*centre*), and F=5320N (*right*).

Figure 11: Fracture pattern of the shifted double-notched specimen.

CONCLUSION

A multiscale model of the mechanical behavior of woven composite materials has been proposed. The DMD model is based on a physical description and modeling of the material microstructure and the elementary damage mechanisms. The damage variables introduced in the DMD model directly describes the local damage state of the material, as they are defined as crack densities and debonded areas. The effective macroscopic mechanical behavior is obtained through a numerical homogenization procedure that requires the construction of representative unit cells. Specific computational tools have been developed in order to model the geometry of those unit cells and create finite element meshes of them.

The DMD model has been identified for a woven ceramic-matrix composite. It has been validated on off-axis tensile tests and its predictive capabilities have been evidenced. Although the identification of the damage kinematics requires only one on-axis tensile test, the DMD model is capable to correctly predict the tensile/shear couplings and simulate the behavior of the materials for off-axis tests. The DMD model has also been applied to simple structural tests and has been proven to be as accurate as a reference macroscopic model. As a multiscale model, the DMD model also allows to describe the local damage-driven microstructural evolution.

ACKNOWLEDGMENTS

This work was carried out under the AMERICO project (Multiscale Analysis: Innovating Research for Composites) directed by ONERA and funded by DGA/STTC (French Ministry of Defense) that is gratefully acknowledged. The authors would also thank Snecma Propulsion Solid (SAFRAN group) for their collaboration.

REFERENCES

[1] C. Huchette. *Sur la complémentarité des approches expérimentales et numériques pour la modélisation des mécanismes d'endommagement des composites stratifies* (in French). PhD thesis, Université de Paris 6, 2005.

[2] P. Ladeveze, G. Lubineau, D. Marsal, Towards a bridge between the micro- and mesomechanics of delamination for laminated composites, *Composites Science and Technology*, 66:698-712, 2006.

[3] A. Caiazzo and F. Costanzo. On the constitutive relations of materials with evolving microstructure due to microcracking. *International Journal of Solids and Structures*, 37:3375–3398, 2000.

[4] D. Jeulin, M. Ostoja-Starzewski, *Mechanics of Random and Multiscale Microstructures*, Springer, 2001.

[5] G. Couégnat. *Approche multiéchelle du comportement mécanique de matériaux composites à renfort tissé* (in French). Ph.D. thesis, Université Bordeaux 1, 2008.

[6] S. H. Lo and W. X. Wang. Finite element mesh generation over intersecting curved surface by tracing neighbours. *Finite Elements in Analysis and Design*, 41:351–370, 2005

[7] G. Couégnat. *Identification numérique et expérimentale d'un modèle de comportement appliqué à un matériau composite à matrice céramique* (in French). Master thesis, Ecole des Mines d'Albi, 2004.

[8] http://www.nwnumerics.com/Zebulon/

[9] L. Marcin, N. Carrère, and J. Maire. A macroscopic visco-elastic-damage model for three-dimensional woven fabric composites. In *Proceeding of ECCM13 - 13th European Conference on Composite Materials*, Stockholm, Sweden, 2008.

MODE I INTERLAMINAR FRACTURE TOUGHNESS TESTING OF A CERAMIC MATRIX COMPOSITE

Ojard, G.[1], Barnett. T.[2], Dahlen, M.[2], Santhosh, U.[3], Ahmad, J.[3], and Miller, R.[1]

[1]Pratt & Whitney
400 Main Street
East Hartford, CT 06108

[2]Southern Research Institute
757 Tom Martin Drive
Birmingham, AL 35211

[2]Research Applications, Inc.
11772 Sorrento Valley Rd,
San Diego, CA

ABSTRACT

As ceramic matrix composites are being targeted for aerospace applications, key material properties need to be understood. This is especially true since the current 2-D architecture-based materials are fabricated by stacking plys that are prime paths for delaminations. Hence, an effort to understand the current test method for delamination toughness was initiated. Specifically, interest was in using a notch for the test and not a starter crack. A model ceramic matrix composite of a SiC fiber with a polymer infiltration pyrolysis derived ceramic matrix was used in this investigation. This material was machined as a double cantilever beam specimen. Testing was done under displacement control. This effort will discuss the testing and analysis of the double cantilever beam specimen.

INTRODUCTION

As interest is expressed in ever increasing higher temperature capable materials, Ceramic Matrix Composites (CMCs) become more attractive than monolithic ceramics since they exhibit the capability to handle damage (increased strain capability) [1]. This is shown by the interest in CMCs for gas turbine engines for turbine and combustor applications [2,3]. Applications are also being pursued in the energy field where improved efficiencies are being pursued such as the use of candle filters [4]. These applications take advantage of the CMC material capability to handle elevated temperatures with limited cooling as well as the improved thermal shock resistance [4, 5]. In the area of propulsion, there is the added benefit of using a lower density material which leads to increased energy efficiencies [5,6].

Even with the potential that CMCs possess, there is a need to fully test and understand the material capabilities. This extends to the delamination capability of the material. Currently, most efforts are supplemented with sub-element testing and increasingly complex rig tests to show that delaminations are not an issue for a given application. Additionally, efforts are undertaken to consider 3D weaves or interlocked architectures to limit delamination growth in ceramic matrix composites [7]. By changing the architecture, the technology readiness level of the material is lowered and additional testing will be required. These approaches can incur significant testing cost. A lot of this testing or consideration of alternate weaves can be reduced or eliminated by determining the material resistance of crack propagation between plies. This can be done by conducting interlaminar fracture toughness testing of the composite focused on propagating cracks within the matrix of the CMC.

With this in mind, a series of limited tests were undertaken to look into the delamination toughness capability of a ceramic matrix composite [8]. The focus of this effort was to start the testing using a notch as a more reproducible approach and see if the test generated practical results. (The use of a notch and not a starter crack out of a notch would reduce material usage as well.) Additionally, the test results will be compared against results documented by others on other material systems to see if there are some universal lessons learned out of this test method.

PROCEDURE

Material Description
Due to the experimental nature of the testing and the desire to explore variations of 2D weaves and thicknesses, a polymer infiltration pyrolisis (PIP) system was chosen as the model material for this effort. This type of experimental system allows similar fabrication regardless of the weave and thickness. Two different weaves were tested: cross-ply (CP) and quasi. The cross-ply material was either an 8 ply ($[0/90]_{2s}$) or 18 ply layup. The quasi material was either an 8 ply ($[0/90/45/-45]_s$) or 16 ply layup. The room temperature average properties for these two layups are shown in Table I. The test results indicate that the in-plane elastic modulus is relatively independent of ply lay-up (cross-ply or quasi).

Table I. Room Temperature Tensile Properties (In-Plane Properties)

Lay-Up	Modulus (E11)	Ultimate Tensile Strength	Strain-to-failure
	GPa (StDev)	MPa (StDev)	mm/mm (StDev)
Cross Ply	111 (7)	246 (30)	0.004 (0.001)
Quasi	107 (12)	199 (16)	0.0036 (0.001)

Specimen
To machine the double cantilever beam specimens, the specimens were initially machined into rectangles that were 101.6 mm in length and were 12.7 mm in width. Specimen size was based on availability of material. The specimen thickness was not modified during the machining and was left in the as-received state. To fabricate the notch, diamond sawing was done to a depth of 25.4 mm. This is shown schematically in Figure 1. An actual machined notch is shown in Figure 2. The notch was centered in the thickness and the average thickness of the notch was 0.66 mm.

Testing
The schematic for this test is shown in Figure 3. This testing configuration is taken from ASTM D5528 [8]. Loading was done using loading blocks which where adhesively attached to the specimen. The piano hinge approach was not used. The crack opening displacement was measured using extensometer arms as shown in Figure 4. All testing was done under displacement control and the data was recorded electronically.

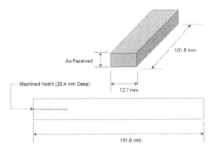

Figure 1. Specimen schematic

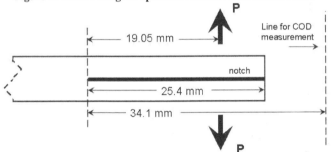

Figure 2. Macro image of specimen focused on centered notch

Figure 3. Test schematic showing loading scheme
(loading done with loading blocks)

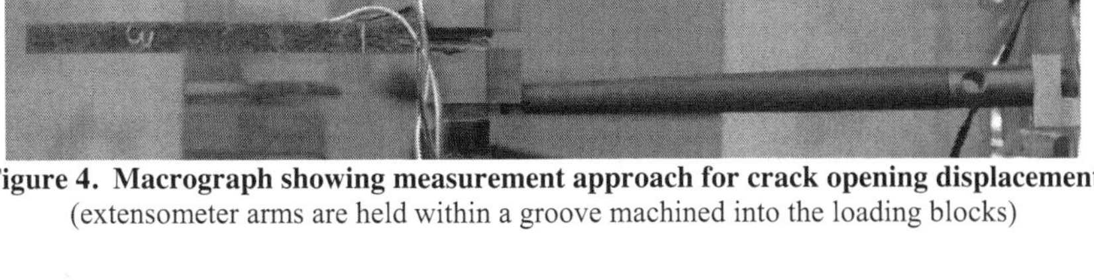

Figure 4. Macrograph showing measurement approach for crack opening displacement
(extensometer arms are held within a groove machined into the loading blocks)

During the testing, detailed images were taken periodically during the loading process on one side of the specimen to document the progress of cracks emanating from the machined notch. These photos were carefully reviewed to allow determination of the crack length generated. This would allow generation of the strain energy release rate (G) for different crack lengths.

The data generated from the testing effort was calculated per the procedures discussed in ASTM D-5528 [8]. The data analysis presented in this paper is from the Modified Beam Theory (MBT) Method [8]. The other approaches for determining the Interlaminar Energy Release Rate (G_I) will not be presented.

RESULTS

8 Ply Material Specimens

The cross-ply and quasi material specimens that were 8 plies thick had an average thickness of 3 mm. With the notch width of 0.66 mm, the resulting arms for the specimen had a thickness of 1.2 mm. This made the arms very slender and during setup for testing it was clearly noted that slight loads resulted in large crack opening displacements. This can clearly be seen in Figure 5 (specimen photo under load) and Figure 6 (load versus crack opening) for the different 8 ply material tested (cross-ply and quasi). Data analysis proved difficult as the data was not consistent with other data sets (to be discussed next for thicker specimens) and it was clear that large displacement corrections were required (per ASTM procedure [8]). These corrections for both lay-ups were large and were multiplicative. It was clear that the 8 ply data could not be confidently presented as the corrections applied did not bring the data into the data range seen for the higher ply count specimens. No additional analysis was done on the 8 ply tests performed but future numerical analysis may be able to correct the data (for cross ply and quasi layups).

16 and 18 Ply Material Specimens

As noted previously, the quasi material was also made as a 16 ply layup with an average thickness of 5.96 mm while the cross-ply material was made as an 18 ply layup with an average thickness of 6.86 mm. With the notch width of 0.66 mm, this meant that the average arms for the DCB test were 2.6 mm in thickness or greater (double the 1.2 mm for the 8 ply material). This increased thickness generated less deflection compared to the 8 ply material as shown by the crack opening displacement (COD) data as shown in Figure 7 as well as a representative image form testing as shown in Figure 8. (COD shown in Figure 7 is half that shown in Figure 6.)

Figures 7 and 8 clearly show that there is not as much deflection as reported for the 8 ply material. The crack opening displacement for the thicker material is half that of the 8 ply material. Therefore, there was no need for corrections to the collected data.

Figure 5. Image from 8 ply quasi specimen in Figure 6 showing large deflection
(Overall (and initial) specimen thickness is 1.2 mm)

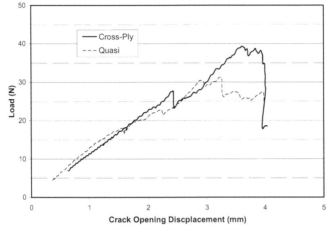

Figure 6. Load versus crack opening displacement for an 8 ply quasi specimen
(raw data shown – not corrected for offset of extensometer arms)

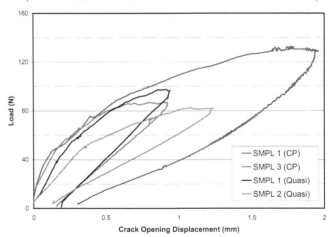

Figure 7. Load versus crack opening displacement for 16 and 18 ply specimens
(Quasi at 16 ply and Cross-ply at 18 ply (raw data shown))

Figure 8. Test image during testing from 18 ply specimen
(specimen thickness of 6.86 mm)

During the testing, photos were taken at increasing load levels to monitor crack formation and length (See Figure 9). Figure 9 shows that the porosity required careful review of the specimens to be sure that the crack was being tracked. The results of this for the specimens are shown in Table II. This allowed the determination of the Interlaminar Energy Release Rate (G_I) for various delamination lengths. Here, the crack opening displacement was corrected. As shown in Figures 3 and 4, the COD was measured offset from the load line and by using the method of similar triangles; the COD was corrected to the value shown in Table II. (The data is not corrected for the graphical images such as Figures 6 and 7.)

In reviewing the calculation approaches listed in the ASTM standard [8] and considering the slight difference between the methods discussed, it was decided to use the correction to the Modified Beam Theory. For this analysis, the Interlaminar Energy Release Rate (G_I) is given by:

$$G_I = 3P\delta \: / \: 2b(a + |\Delta|)$$

where P is the load, δ is the load point deflection (or opening displacement), b is the specimen width, a is the delamination length and Δ is a delamination length correction. This correction is determined experimentally by generating a least squares plot of the cube root of the compliance versus delamination length [8]. (The compliance is determined by the ratio of the load point displacement to the applied load (δ/P).) The delamination behaves as if it is a longer delamination by the Δ length. (Δ is determined by finding x-intercept of the least squares fit line.) Figure 10 shows that ~8.8 mm needs to be added to the delamination length for the G_I calculation shown above (previous equation).

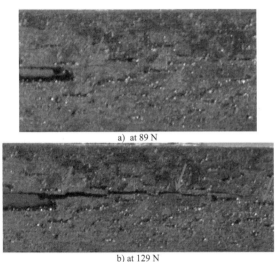

a) at 89 N

b) at 129 N

Figure 9. Images from SMPL 1 (CP) at different load levels

Table II. Measurements Taken of Specimens during Testing

Specimen Identification	Load (N)	Crack Length* (mm)	COD** (mm)
CP SMPL 1	89	22	0.35
CP SMPL 1	111	26	0.66
CP SMPL 1	129	33	1.13
CP SMPL 2	49	20	0.11
CP SMPL 2	80	22	0.26
CP SMPL 2	102	25	0.46
CP SMPL 3	71	20	0.23
CP SMPL 3	80	22	0.30
CP SMPL 3	87	27	0.45
CP SMPL 4	49	20	0.14
CP SMPL 4	62	22	0.29
CP SMPL 4	71	27	0.48
Q SMPL 1	58	20	0.17
Q SMPL 1	93	27	0.51
Q SMPL 2	73	22	0.47
Q SMPL 2	82	26	0.68

* = machined notch length and subsequent crack emanating from the notch (from the load line)
** = COD is corrected for extensometer arm being offset from load line

Additionally, this approach allows a check on the approach by determining the modulus of the material [8]. The modulus can be determined by:

$$E = 64((a + |\Delta|)^3 \, P \, / \, \delta bh^3$$

where h is the specimen thickness. Consistent with this approach, the modulus was found to be independent of the delamination length [8]. For this effort, the modulus for the material was determined to be 130 GPa (+/- 30 GPa). This value is higher than the data shown in Table I but with the wide scatter seen, it is within the range reported earlier in Table I. (Table I is also based on a larger data set and 130 GPa is on the high end of the data but outside the standard deviation.)

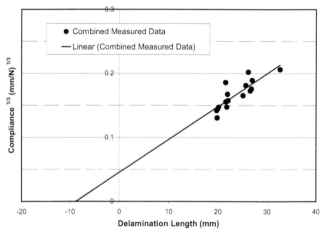

Figure 10. Determination of Delta (Δ) for the Specimens Tested

The resultant interlaminar energy release rate versus crack length was determined for all the specimens tested for varying crack lengths. The resultant data is shown in Figure 11. As the crack length increases, there is a greater range of scatter seen in the data.

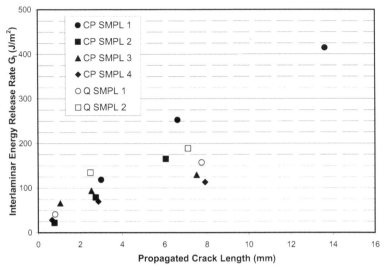

Figure 11. Interlaminar Energy Release Rate versus crack length for the specimens tested
(16 ply quasi material and 18 ply cross-ply material)

There was an outlier in the data that was not included in the analysis shown in Figure 11. One cross-ply specimen (SMPL-CP-5) had a significant crack propagate at a low load. This is shown in Figure 12. The load to propagate this crack was 39 N and this is low compared to the loads shown in Figure 7.

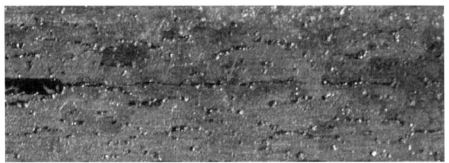

Figure 12. Image from SMPL 5 (CP) at 39 N

All of the above work looked at the Interlaminar Energy Release Rate. What is of greater interest is the Interlaminar Fracture Toughness (G_{IC}). To generate this, the load versus crack opening displacement were linearly fit from a load of 20-35 N and then that line was extrapolated along the curve and where the load deviated from the linear line was recorded and then used to generate the G_{IC} values. This data was analyzed using the standard Modified Beam Theory (there was no delta applied to the delamination (crack) length). The equation for the Modified Beam theory is as follows:

$$G_I = 3P\delta / 2ba.$$

The results of this testing are shown in Table III. The data for the two layups overlaps, even for this limited number of quasi specimens. Therefore, differences in the Interlaminar Fracture Toughness (G_{IC}) are inconclusive between layups.

Table III. G_{IC} Values –Modified Beam Theory (Standard)

Specimen	Layup	G_{IC} (J/m^2)	Average (J/m^2)	St Dev. (J/m^2)
SMPL-CP-1	Cross Ply	7	18	8
SMPL-CP-2	Cross Ply	18		
SMPL-CP-3	Cross Ply	24		
SMPL-CP-4	Cross Ply	23		
SMPL-Q-1	Quasi	21	38	24
SMPL-Q-2	Quasi	55		

DISCUSSION

8 Ply Material

The testing of the 8 ply material showed very large deflections, requiring significant data corrections. In similar testing, the use of such corrections by other investigators was not reported [9,10]. This is a combination of two factors: (1) the use of cut notches versus pre-generated cracks made the arms of the double cantilever beam relatively thin; and (2) the lower modulus of the material system used in this study versus the modulus values reported by other investigators [9]. This may indicate a lower limit of modulus when testing thin (e.g., 8-ply) specimens. The testing comparison here indicated that for thin material, sharp pre-cracks are needed in order to keep the arms of the double cantilever beam as thick as possible. Such limitations to a double-cantilever test raise concerns for test reliability. If pre-generation of cracks are required for thinner specimens, for example, this could lead to highly variable testing conditions and the corresponding need for more material. As noted above for the outlying specimen, a crack jump of 20 mm or greater can occur, leading to ineffectual consumption of material when performing characterization efforts on high cost material.

16 and 18 Ply Material

The testing on the thick specimens generated G_I (Interlaminar Energy Release Rate) values from 20 to 420 J/m^2 (See Figure 12). These values are within the range documented by other investigators but on the low end for the multiple SiC/SiC CMC systems tested where G_I was found to range from 200 to 600 J/m^2[9]. The work reported by others was on CMC systems that were either fabricated by Chemical Vapor Infiltration (CVI), Hot Press (HP) or Melt Infiltrated (MI). These systems all create matrices that do not have the remnant PIP cracks present from repeated infiltrations cycles. While porosity may be a concern in the comparison here, it should be noted that CVI systems may have a matrix that has double the porosity of a PIP system

Interlaminar Fracture Toughness – G_{IC}

The initiation values were determined for the series of tests run. The results are reported in Table III. For this analysis, the standard Modified Beam Theory was used based on the limited damage and crack rotation expected at the start of the test. The values reported in Table III are low but they are consistent with an extrapolation of the data shown in Figure 12 when looking at a zero propagated crack length. (The value is also consistent with the MBT analysis of the large crack extension seen in the outlying specimen discussed above (See Figure 12).)

Additionally, some early testing data generated by the authors was reviewed to determine the interlaminar fracture toughness to see how it compared against the data shown in Table III. At the start of testing, a series of tests were run without documenting the crack growth during the test. These tests were run as a check of the testing approach since this material had an accidental intermediate temperature oxidation exposure done to it that showed debited properties from expectation in standard tensile testing. Both 18 ply cross ply and 16 ply quasi material was tested. The G_{IC} values from the cross ply testing was determined to be 28 and 34 J/m^2 (there were two tests) while the G_{IC} value from the quasi testing was determined to be 42 J/m^2 (there was only one test). These values are in line with the results reported previously in Table III. This indicates that the embrittlement seen from the intermediate temperature oxidation was limited to the interface coating and did not affect the matrix of the material. Additionally, the shape of the curve seen during testing was consistent with the unaffected material that was presented in Figure 7.

CONCLUSION

A series of double cantilever beam tests were done on a Polymer Infiltration Pyrolisis system with varying thickness and ply count. It was clear that a machined notch, engineered for test repeatability, did not yield acceptable crack propagation in 8 ply test specimens. Testing of thicker specimens, of varying crack length and specimen layup, had G_I values from 20 to 420 J/m^2. The range of values arrived at for the model material system are on the lower end of documented results for a range of Ceramic Matrix Composites fabricated by chemical vapor infiltration or melt infiltration means [9]. This is due to the heavily cracked nature of a Polymer Infiltration Pyrolisis fabricated CMC with the repeated thermal cycles that the material is subjected to during manufacturing. This is also shown in the average G_{IC} value of 45 J/m^2 that was determined (Average of all values reported in Table III).

This effort did show that a notched approach could be considered for such testing but not for thin materials (as noted above). The appropriate thickness will vary from CMC to CMC based on material properties and manufacturing methods. For the series of tests conducted here, the corrected Modified Beam Theory approach was used. Additionally, the method presented a way to check the approach by allowing the modulus of the material to be calculated. This is a clear benefit on such a challenging test.

FUTURE WORK

Additional data collected during the testing has not been analyzed at this time. The strain gauges placed on the specimen are still available for data analysis and will be considered for future work. Additionally, crack propagation and micromechanical modeling and analysis is planned and will be presented in the future. A numerical approach may be pursued to help determine the deviation from linearity that was used in the G_{IC} evaluation.

ACKNOWLEDGMENTS

This work was funded under AF Contract FA8650-07-C-5219 (Phase II SBIR Contract) awarded to Research Applications, Inc. Pratt & Whitney was a subcontractor to RAI.

REFERENCES

1. Brewer, D., 1999, "HSR/EPM Combustion Materials Development Program", Materials Science & Engineering, 261(1-2), pp. 284-291.
2. Brewer, D., Ojard, G. and Gibler, M., "Ceramic Matrix Composite Combustor Liner Rig Test:, ASME Turbo Expo 2000, Munich, Germany, May 8-11, 2000, ASME Paper 2000-GT-670
3. Verrilli, M. and Ojard, G., "Evaluation of Post-Exposure Properties of SiC/SiC Combustor Liners testing in the RQL Sector Rig", Ceramic Engineering and Science Proceedings, Volume 23, Issue 3, 2002, p. 551-562.
4. K.K. Chawla (1998), *Composite Materials: Science and Engineering, 2nd Ed.*, Springer, New York
5. Bouillon, E.P., Lamouroux, F., Baroumes, L., Cavalier, J.C., Spriet, P.C. and Habarou, G., 2002, "An Improved Long Life Duration CMC for Jet Aircraft Engine Applications", ASME paper No. GT-2002-30625.
6. M.F. Ashby and D.R.H. Jones (2001), **Engineering Materials 1: An Introduction to their Properties and Applications, 2nd Ed.**, Butterworth Heinemann, Oxford
7. Yun, H. M., DiCarlo, J. A., and Fox, D. S., "Issues on Fabrication and Evaluation of SiC/SiC Tubes with Various Fiber Architectures", High Temperature Ceramic Matrix Composites 5, pp. 537-542. 2005
8. ASTM D-5528-01, "Standard Test Method for Mode I Interlaminar Fracture Toughness of Unidirectional Fiber Reinforced Polymer Matrix Composites", ASTM International, 100 Barr Harbor Drive, West Conshohocken, PA.
9. Choi, S.R., and Kowalik, R.W., "Interlaminar Crack Growth Resistances of Various Ceramic Matrix Composites in Mode I and Mode II Loading", Journal of Engineering for Gas Turbine Power, May 2008, Vol. 130
10. Hojo, M., Yamao, T., Tanaka, M., Ochiai, S., Iwashita, N., and Sawada, Y., "Effect of Interface Control on Mode I Interlaminar Fracture Toughness of Woven C/C Composite Laminated", JSME International Journal, Series A, Vol. 44, No. 4, pp. 573-581, 2001

COMPARATIVE STUDY OF TENSILE PROPERTIES OF UNI-DIRECTIONAL SINGLE-TOW SIC-MATRIX COMPOSITES REINFORCED WITH VARIOUS NEAR-STOICHIOMETRIC SIC FIBERS

Kazumi Ozawa, Yutai Katoh, Edgar Lara-Curzio, Lance L. Snead
Materials Science and Technology Division, Oak Ridge National Laboratory
P.O. Box 2008, Oak Ridge, TN 37831, USA

Takashi Nozawa
Fusion Research and Development Directorate, Japan Atomic Energy Agency
2-4 Shirakata Shirane, Tokai, Ibaraki 319-1195, Japan

ABSTRACT

Tensile properties of unidirectional single tow SiC/SiC mini-composites reinforced with four different commercial and experimental near-stoichiometric SiC fibers were evaluated. The composites reinforced with Hi-Nicalon™ Type-S (HNLS), Tyranno™-SA3, experimental Sylramic™ and Sylramic-iBN fibers exhibited ultimate tensile stresses equivalent to ~73%, ~53%, ~69%, and ~81% of the single fiber strength at 25 mm, respectively. The ultimate tensile stress appeared to increase with the estimated interfacial sliding stress with the exception of the HNLS composite. The HNLS composite seems to have achieved the high tensile strength due to very low interfacial sliding stress arising from the larger radial tensile residual stress and its smooth fiber surface. In contrast, the fracture behavior of the other composites may have been strongly affected by clamping stress produced by the relatively rough fiber surfaces.

INTRODUCTION

Silicon carbide (SiC) -based ceramics and composites (SiC/SiC) are considered for applications in various components of fusion and advanced fission reactor systems and fuel assemblies, due primarily to their superior irradiation performance and thermo-physical, -chemical, and -mechanical properties[1,2]. Continuous SiC-fiber, SiC-matrix composites are of particular importance for applications which require reliability, toughness, and near-net shape fabrication.

One focus of irradiation studies of SiC/SiC composites in the past few decades has been determination of the fundamental response of the materials produced with various constituent options and processing routes. Through this effort, composites containing near-stoichiometric SiC fibers, such as Tyranno™-SA3 (SA3) and Hi-Nicalon™ Type-S (HNLS), and high crystallinity SiC matrices produced by chemical vapor infiltration (CVI) or nano-infiltration and transient eutectic-phase (NITE) processes are found to possess acceptable irradiation stability. However, a design scheme for the optimum interphase for radiation service remains unresolved due primarily to the lack of understanding of the irradiation effects on the fiber/matrix interfacial properties.

The mini-composites, or composites uni-directionally reinforced with a single fiber tow, is an appropriate approach for evaluating the interfacial properties[3]. The general advantages of the mini-composite approach are as follows: 1) due to its simple uni-directional fiber architecture, the in-situ fiber strength and interfacial properties can be determined in relatively simple ways[4,5], and 2) a large number of specimens can be fabricated at a low cost in relatively a short time. Moreover, the very small volume of the mini-composite samples is attractive for irradiation studies due to the very high per-unit-volume cost of irradiation and the radiological concerns associated with the evaluation of irradiated material.

In the present work, a procedure for evaluating mechanical properties of the SiC/SiC mini-composite samples was established at Oak Ridge National Laboratory, with a specific consideration of minimizing the personal exposure during testing of the radiological samples. In the course of the test procedure development, CVI SiC matrix mini-composite samples reinforced with four different near-stoichiometric SiC fibers were evaluated in the unirradiated condition.

EXPERIMENTAL
Materials
        Table I lists the information describing the mini-composites and optical micrographs of polished cross sections for each mini-composite are shown in Fig. 1. In brief, four types of unidirectional single-tow mini-composites were prepared in this study. Tyranno-SA3, Hi-Nicalon Type-S, experimental Sylramic™ and experimental Sylramic™-iBN SiC fibers were used as reinforcements. The Sylramic-iBN, made from Sylramic, is the fiber with the excess boron used as a sintering aid removed from the fiber bulk. This boron formed a thin in-situ BN coating on the fiber surface, in order to improve thermo-structural and -chemical properties[6]. The unidirectional mini-composites were fabricated by the CVI process at Hyper-Therm High-Temperature Composites, Inc. (Huntington Beach, California). During the CVI process, the fiber bundle was held so that the cross sectional area of a mini-composite was minimized and the shape of the cross section was close to a perfect circle, as shown in Fig. 1. The fiber/matrix (F/M) interphase was single layer pyrolytic carbon (PyC) with a nominal thickness of 150 nm. The PyC interphase coating was not applied to the experimental Syl-iBN composite. Hereafter, the IDs indicated in Table I (SA3 for Tyranno-SA3 (UD)/150nm-PyC/CVI, HNLS for Hi-Nicalon Type-S (UD)/150nm-PyC/CVI, Syl-PyC for experimental Sylramic (UD)/150nm-PyC/CVI, and Syl-iBN for experimental Sylramic-iBN (UD)//CVI mini-composite) are used in this paper.

Tensile Tests
        The test apparatus is shown in Fig. 2. The tensile tests were conducted using an electromechanical testing machine (Insight 10, MTS Systems Co., Eden Prairie, Minnesota) with a load capacity of 10 kN. Strain was measured by a pair of linear variable differential transducers, LVDTs (Lucas Schaevitz GCA-121-125, Hampton, Virginia). We adopted an alignment system similar to the one used in single fiber tensile testing standardized in ASTM C 1557 in order to assure specimen alignment.
        Figure 3 is a drawing of a specimen and aluminum tabs. The pair of aluminum tabs was fastened to both sides of a mini-composite sample using an epoxy adhesive dispersed by a syringe. In order to assure specimen alignment, the specimen to be glued was fixed in a V-notched fixture with flathead screws, as shown in Fig. 4. A polyethylene spacer sheet and/or the combination of a silicon release spray and a silicone remover were used to remove the specimen from the fixture after epoxy curing. This fixture including the specimen was cured at 110°C for more than 3 hours in an air furnace in order to develop the adhesive's maximum bonding strength. The gauge length of the specimens is 18mm, defined by the distance between the inner ends of the aluminum tabs. Figure 5 shows the gripping fixture and the guide rail provided for alignment. The specimen was clamped into the simple V-notched fixtures with flathead screws. The guide rail was used to confirm specimen alignment and to avoid damaging the specimen during handling.
        The tests were conducted at ambient conditions under crosshead displacement control at 0.1 mm/min. Unloading/reloading cyclic tensile tests were conducted to evaluate interfacial properties. Previous to the tensile tests, compliance of the grip assembly was determined using a set of tungsten wires with varied gauge lengths. As a result, the system compliance of 1.32×10^2 μm/N was obtained as shown in Fig. 6. Tensile strain is determined by the following equation, $\varepsilon = (\Delta L - C_S P)/L_0$, where ε is the tensile strain, ΔL the average cross-head displacement recorded by LVDTs, C_S the system compliance, P the applied load, and L_0 the gauge length. The fiber volume fraction (f) of the mini-composites was determined by a combination of optical microscopy and image analysis. It is

noted that some errors may have been introduced due to surface damage during polishing. The mini-composite tensile stress (σ) is defined by $\sigma = P/(\pi r^2 N/f)$, with r the average fiber radius, and N the number of filament in a tow. The proportional limit load or proportional limit stress (PLL or PLS) was defined as the stress at 0.5% deviation from the extrapolated fit of the slope by the least squares method for the initial loading.

After the tensile tests, the fracture surfaces of the specimens were examined using a scanning electron microscope (SEM, Topcon SM-510, Paramus, New Jersey). Fiber pull-out length and matrix crack spacing were also measured, in order to estimate the interfacial sliding stress. This study included only tensile tests on unirradiated composites, but identical procedures will be used in the post irradiation experiments.

RESULTS AND DISCUSSION
Tensile Load-Strain Responses
       All four representative load-tensile strain curves are shown in Fig. 7, and the results of the mechanical properties of the mini-composites are listed in Table II. As shown in Fig. 7 (a), the HNLS mini-composites exhibited a typical pseudo-ductile fracture behavior; there was an initial steep linear region in the load-strain curve, with a second, nearly linear region at higher strains, repeated during tensile loading with multiple unloading-reloading sequences. The initial linear portion corresponds to the linear elastic deformation of the mini-composite, whereas the second linear portion corresponds to a process of progressive development and opening of multiple matrix micro-cracks. The SA3 mini-composites exhibited more brittle failure, with the narrower hysteresis loop width ($\sim$0.02%) and smaller elongation after the PL (an average of 0.174% for the fracture strain), compared to the HNLS mini-composites. The Syl-PyC and Syl-iBN composite exhibited rather brittle failure. The overall behavior of these mini-composites is similar to the SA3 mini-composite, implying fiber properties comparable to Tyranno-SA3 at room temperature. Of importance is the fiber strength (the RS values in Table II.). In the overall composite strength, 69% of Sylramic and 81% of Sylramic-iBN fiber strength was achieved in the mini-composite system; this compared to the HNLS (77%) and the SA3 (53%). This result indicates that the experimental fibers are capable of bearing loads similar to the two near-stoichiometric SiC fibers which are currently commercially available.

SEM observation
       Figure 8 shows SEM images of the fracture surface of each mini-composite. Table III lists the result of the average fiber pull-out length determination. Shorter pull-out lengths ($\sim$100 μm) were measured in SA3 and Syl-PyC compared with the HNLS ($\sim$300 μm), while very short pull-out length ($\sim$15 μm) was obtained in Syl-iBN (HNLS > SA3 $\sim$ Syl-PyC > Syl-iBN). In order to measure the matrix crack spacing, the fractured specimen was mounted on a glass platen with an epoxy adhesive and polished longitudinally, but unfortunately typical matrix crack spacing could not be detected for all specimens using an optical microscope. It is believed that such typical transverse matrix cracks were not seen along the whole specimen due to the premature failure before the matrix loading reached saturation in these composite systems. Therefore, the matrix crack spacing in a fiber bundle or the distance of crack deflection near a fracture surface were used as the matrix crack spacing in this study. Examples of these determinations are shown in Fig. 9.

Interfacial Sliding Stress
       The interfacial sliding stress (τ) was estimated from the hysteresis loop width, the matrix crack spacing, and the fiber pull-out length. First, in the hysteresis loop analysis by Lamon et al.[4] and Vagaggini et al.[5], the interfacial sliding stress (τ) can be obtained from the following equation[4]:

$$\tau = \frac{b_2(1-a_1f)^2}{2f^2E_m}\left(\frac{r}{d}\right)\left(\frac{\sigma_p^2}{d\varepsilon}\right)\left(\frac{\sigma}{\sigma_p}-\frac{\sigma_{min}}{\sigma_p}\right)\left(1-\frac{\sigma}{\sigma_p}\right) \tag{1}$$

where a_1 and b_2 are the Hutchinson-Jensen parameters[7], E_m the matrix modulus, d the average matrix crack spacing, $d\varepsilon$ the hysteresis loop width, σ_p the peak stress of the hysteresis loop, σ the stress where $d\varepsilon$ is measured, and σ_{min} the minimum stress of the hysteresis loop. τ was determined from the hysteresis loop immediately prior to failure, assuming the same matrix crack spacing at the failure of the composite systems.

τ is also estimated using the matrix crack spacing, expressed as[8]:

$$\tau = \frac{\sigma_s r E_m(1-f)}{2dEf} \tag{2}$$

where σ_s is the applied stress at matrix cracking saturation. The observed result of the matrix crack spacing suggests that the mini-composites fractured before reaching matrix crack saturation, therefore the ultimate tensile stress was taken as σ_S for convenience.

The third method for estimating τ is to use the fiber pull-out length. Curtin derived the following equation to estimate the interfacial sliding stress[9]:

$$\tau = \frac{r\lambda(m)\sigma_c}{4l_p} \tag{3}$$

where l_p is the average of the fiber pull-out length, σ_c and m the in-situ fiber fracture strength and the Weibull modulus, respectively, and $\lambda(m)$ is the known function of Weibull modulus as $\lambda(m) \approx 0.716 + 1.36/m^{0.6}$ for $m \geq 1$.[10] In this paper the fiber strength at fracture is used for the in-situ fiber fracture strength. There is limited data about the Weibull modulus of in-situ SiC fiber strength, but $m=3-9$ is assumed in Eq. (3), according to the references about in-situ strength of similar SiC fibers[10,11].

All results for the interfacial sliding stress are shown in Table III. A relatively low value of τ is obtained for the HNLS composite and very high τ values were derived for the three other composites. Because of the relatively brittle failure experienced by these composites, it is likely that the τ values were not correctly determined. However, it is noted that the ranking of the τ values seems to be qualitatively reasonable as they are inversely correlated with the fiber pull-out length (HNLS < SA3 ~ Syl-PyC < Syl-iBN).

The three analytical methods gave very major differences in τ values. During the unloading-reloading processes in the tensile testing, as seen in Fig. 7, the stable interfacial sliding was not achieved with an exception for the HNLS composite. In such a case, it is known that the hysteresis loop analysis gives an unrealistically large sliding stress value. In the matrix crack spacing analysis, the ultimate tensile strength was used to estimate the sliding stress instead of the stress for matrix crack saturation. Because the matrix crack saturation was obviously not achieved at the failure of the non-HNLS composites, this may have caused a slight underestimation of the sliding stress values. For the method based on the fiber pull-out length, the estimated sliding stress values may suffer substantial errors due to uncertainty in the pull-out length determination. Thus, the τ values in Table III are considered to have only qualitative significance, although those determined from the matrix crack spacing seem to be most realistic.

The interfacial sliding stress (τ) is essentially the frictional stress at the sliding interface, thus should macroscopically be a product of the clamping stress and the interfacial frictional coefficient, which is strongly related with the fiber surface roughness. The experimental observations with regard to τ values and the fiber surface roughness are summarized in Table IV. The relative τ is clearly dependent on the micro-topological features of the fiber surface.

The clamping stress, or the compressive radial stress at the interface, is generally induced both by the mismatch in coefficient of thermal expansion (CTE) between the fiber and the matrix (possibly involving the interphase), the physical roughness of the interface, and the compliance and strength of the interphase. We have considered the clamping stress in each mini-composite system qualitatively. The clamping stress (σ_{CL}) is expressed by the following equation[12],

$$\sigma_{CL} \sim \sigma_{CTE} + \sigma_R$$
$$= \sigma_{CTE} + \frac{E_m E_f}{E_f(1+v_m) + E_m(1-v_f)}\left(-\frac{A}{r}\right) \tag{4}$$

where σ_{CTE} and σ_R are the residual stress induced by CTE mismatch and the contribution of fiber roughness, the v_m, v_f are the Poisson's ratios of the matrix and fiber, E_f the Young's modulus of a fiber, and A the characteristic amplitude of roughness along the debonded interface. Figure 10 shows the calculated results for the residual stress of each composite system, and Table V gives the calculated result for the clamping stress in each composite system. The four phase model[13] was used to estimate σ_{CTE}. For calculation, v_m, v_f and A were assumed to be 0.2, 0.2 and $2R_q$, where R_q is the root mean square fiber surface roughness. The thickness of the BN layer on the Sylramic-iBN fiber was assumed to be 200 nm[6]. As shown in Table V, the contribution of fiber surface roughness is much larger than the CTE mismatch. Additionally, the clamping stress is strongly affected by the fiber surface roughness. This trend of the calculated clamping stress qualitatively corresponds to the misfit stress as described below. Therefore, it is concluded that the relationship of the tensile properties of the mini-composites studied is mainly attributed to the difference in the fiber surface roughness.

Misfit Stress (σ^T)

Misfit stress of each mini-composite was also measured using regression analysis of the tensile reloading segments, derived by Steen et al.[14] (Table II). It is noted that negative misfit stress ($\sigma^T < 0$) denotes compressive axial residual stress in the matrix and tensile residual stress in the radial direction. As shown in Table II, the larger σ^T values were obtained for the SA3, Syl-PyC, and Syl-iBN composites compared to the HNLS composites (HNLS < SA3 < Syl-PyC < Syl-iBN ~ 0).

Ultimate Tensile Strength (UTS)

The tensile properties of a unidirectional composite system are strongly affected by the Weibull modulus (m), the characteristic strength (σ_0) of the fibers, and the interfacial sliding stress (τ). Under assumption of the global load sharing (GLS) theory of Curtin[15], the fiber-averaged UTS ($\sigma_{U,f}$) in a unidirectional composite system is expressed as follows:

$$\sigma_{U,f}^{GLS} = \left(\frac{\sigma_0^m \tau L_0}{r}\right)^{\frac{1}{m+1}}\left[\frac{2(m+1)}{(m+2)m}\right]^{\frac{1}{m+1}}\left(\frac{m+1}{m+2}\right) \tag{5}$$

Figure 11 compares the fiber-averaged UTS results accompanied by the $\sigma_{U,f}^{GLS}$ calculated according to this theory. The interfacial sliding stress from the matrix crack spacing was used for τ. The Weibull modulus was assumed to be ~6.2 for HNLS[16,17], 8.2 for SA3[17], and 4.21 for the two experimental Sylramic fibers[18]. The fiber-averaged UTS of HNLS appeared to agree with the value calculated from the GLS theory. In contrast, this theory largely overestimates the fiber-averaged UTS for the three other composites. It is likely that, in these composites, the global load sharing is not achieved before fracture but the fracture is governed by the local load sharing (LLS), due to their relatively high interfacial sliding stress (Hence the use of Sylramic-iBN fiber without any interfacial coating in unidirectional architecture is generally not recommended due to the high fiber surface roughness (Ref. [6] and Table IV)). Based on one of the simplest LLS models, Zweben derived the following relationship for the fiber-averaged UTS of uni-directional composites[19]:

$$\sigma_{U,f}^{LLS} = \sigma_0 \left(\frac{L_0(m-1)}{LNm} \right)^{\frac{1}{m+1}}$$

(6)

where L is also the gauge length of the mini-composite. In this model, the early fiber failures tend to initiate catastrophic failure of the composite. Hence it is noted that this can be regarded as a lower bound on the composite strength. Figure 11 also shows the $\sigma_{U,f}^{LLS}$ calculated using Eq. (6). These values underestimate the experimental data. However, the difference between the fiber-averaged UTS and the $\sigma_{U,f}^{LLS}$ may be reasonable since the fracture mode is different. The calculated $\sigma_{U,f}^{LLS}$ are the values under the assumption of a catastrophic (brittle) fracture. However, the three composites actually fractured with pseudo ductile behavior as shown in Fig. 8. The part of the specimen where this pseudo ductile fracture has occurred should contribute to the increase of the composite strength.

CONCLUSION

Four types of mini-composites (with variations in fibers and interphase but with the same SiC matrices) were evaluated for tensile and fiber/matrix interfacial properties. The composites reinforced with HNLS, SA3, experimental Sylramic™ and Sylramic-iBN fibers exhibited ultimate tensile stresses equivalent to ~77%, ~53%, ~69%, and ~81% of the single fiber strength at 25 mm, respectively. The ultimate tensile stress appeared to increase with the estimated interfacial sliding stress, with the exception of the HNLS composite. The sliding stress increased with increasing the fiber surface roughness. The HNLS composite exhibited the high tensile strength, likely because the very low interfacial sliding stress enabled global load sharing. The SA3 and Sylramic composites exhibited premature failure, attributed to the very high interfacial sliding stresses due primarily to the physical roughness of the fiber surfaces. More compliant interphases would optimize these fiber composites with unidirectional architecture.

ACKNOWLEDGEMENT

The authors would like to thank to Drs. F.W. Wiffen and W.J. Kim for reviewing the manuscript. The experimental Sylramic and Sylramic-iBN fibers used in this study were supplied by COI Ceramics, Inc. (San Diego, California). This research was sponsored by the Office of Fusion Energy Sciences, US Department of Energy under contract DE-AC05-00OR22725 with UT-Battelle, LLC.

REFERENCES
[1]T. Nozawa, T. Hinoki, A. Hasegawa, A. Kohyama, Y. Katoh, L. L. Snead, C. H. Henager Jr, and J. B. J. Hegeman, Recent advances and issues in development of silicon carbide composites for fusion applications, *J. Nucl. Mater.*, **386-388**, 622-27 (2009).
[2]Y. Katoh, L. L. Snead, C. H. Henager Jr, A. Hasegawa, A. Kohyama, B. Riccardi, and H. Hegeman,

Current status and critical issues for development of SiC composites for fusion applications, *J. Nucl. Mater.*, **367-370**, 659-71 (2007).

[3]R. Naslain, J. Lamon, R. Pailler, X. Bourrat, A. Guette, and F. Langlais, Micro/minicomposites: a useful approach to the design and development of non-oxide CMCs, *Composites Part A: Applied Science and Manufacturing*, **30**, 537-47 (1999).

[4]J. Lamon, F. Rebillat, and A. G. Evans, Microcomposite Test Procedure for Evaluating the Interface Properties of Ceramic Matrix Composites, *J. Am. Ceram. Soc.*, **78**, 401-05 (1995).

[5]E. Vagaggini, J.-M. Domergue, and A. G. Evans, Relationships between Hysteresis Measurements and the Constituent Properties of Ceramic Matrix Composites: I, Theory, *J. Am. Ceram. Soc.*, **78**, 2709-20 (1995).

[6]H. M. Yun, J. Z. Gyekenyesi, Y. L. Chen, D. R. Wheeler, and J. A. DiCarlo, Tensile Behavior of SiC/SiC Composites Reinforced by Treated Sylramic SiC Fibers, *Ceram. Eng. Sci. Proc.*, **22**, 521-31 (2001).

[7]J. W. Hutchinson and H. M. Jensen, Models of fiber debonding and pullout in brittle composites with friction, *Mech. Mater.*, **9**, 139-63 (1990).

[8]F. W. Zok and S. M. Spearing, Matrix crack spacing in brittle matrix composites, *Acta Metallurgica et Materialia*, **40**, 2033-43 (1992).

[9]W. A. Curtin, In Situ Fiber Strengths in Ceramic-Matrix Composites from Fracture Mirrors, *J. Am. Ceram. Soc.*, **77**, 1075-78 (1994).

[10]I. J. Davies, T. Ishikawa, M. Shibuya, and T. Hirokawa, Fibre strength parameters measured in situ for ceramic-matrix composites tested at elevated temperature in vacuum and in air, *Compos. Sci. Technol.*, **59**, 801-11 (1999).

[11]S. Guo and Y. Kagawa, Tensile fracture behavior of continuous SiC fiber-reinforced SiC matrix composites at elevated temperatures and correlation to in situ constituent properties, *J. Eur. Ceram. Soc.*, **22**, 2349-56 (2002).

[12]R. J. Kerans and T. A. Parthasarathy, Theoretical Analysis of the Fiber Pullout and Pushout Tests, *J. Am. Ceram. Soc.*, **74**, 1585-96 (1991).

[13]Y. Mikata and M. Taya, Stress Field in a Coated Continuous Fiber Composite Subjected to Thermo-Mechanical Loadings, *J. Compos. Mater.*, **19**, 554-78 (1985).

[14]M. Steen and J. L. Valles, Unloading-reloading sequences and the analysis of mechanical test results for continuous fiber ceramic composites, *ASTM Special Technical Publication*, **1309**, 49-65 (1997).

[15]W. A. Curtin, B. K. Ahn, and N. Takeda, Modeling brittle and tough stress-strain behavior in unidirectional ceramic matrix composites, *Acta Materialia*, **46**, 3409-20 (1998).

[16]T. Nozawa, T. Hinoki, L. L. Snead, Y. Katoh, and A. Kohyama, Neutron irradiation effects on high-crystallinity and near-stoichiometry SiC fibers and their composites, *J. Nucl. Mater.*, **329-333**, 544-48 (2004).

[17]C. Sauder, A. Brusson, and J. Lamon, Mechanical properties of Hi-NicalonS and SA3 fiber reinforced SiC/SiC minicomposites, *Ceram. Eng. Sci. Proc.*, **29**, 91-100 (2008).

[18]M. C. Osborne, C. R. Hubbard, L. L. Snead, and D. Steiner, Neutron irradiation effects on the density, tensile properties and microstructural changes in Hi-Nicalon(TM) and Sylramic(TM) SiC fibers, *J. Nucl. Mater.*, **253**, 67-77 (1998).

[19]C. Zweben, Tensile Failure of Fiber Composites, *AIAA J.*, **6**, 2325-31 (1968).

[20]T. Hinoki, Investigation of Mechanical Properties and Microstructure of SiC/SiC Composites for Nuclear Application, *Doctoral Thesis*, Kyoto University, (2001).

[21]C. Sauder, J. Lamon, and A. Brusson, Mechanical Behaviour and Structural Characterizations of Minicomposites SiC/SiC based on Hi-Nicalon S and SA3 Reinforcement, *Presented at the 33rd International Conference and Exposition on Advanced Ceramics and Composites*, (2009).

[22]J. J. Brennan, Interfacial characterization of a slurry-cast melt-infiltrated SiC/SiC ceramic-matrix composite, *Acta Materialia*, **48**, 4619-28 (2000).

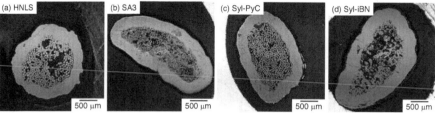

Figure 1. Optical micrographs of polished cross sections of (a) HNLS, (b) SA3, (c) Syl-PyC, and (d) Syl-iBN mini-composites.

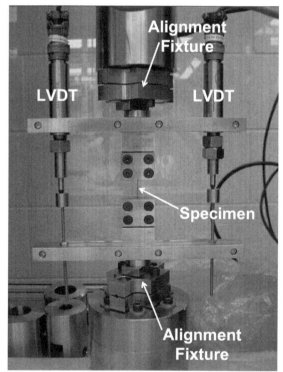

Figure 2. Mini-composite tensile test apparatus used in this study.

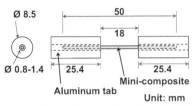

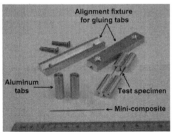

Figure 3. Dimension of a mini-composite and a pair of aluminum tabs.

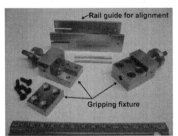

Figure 4. Mini-composite, aluminum tabs and alignment fixture for gluing tabs.

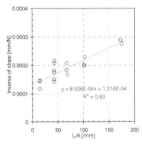

Figure 5. Gripping fixture and rail guide for specimen alignment.

Figure 6. Machine compliance of the mini-composite tensile system. l_0 and A denote the gauge length and the cross-sectional area of the specimen.

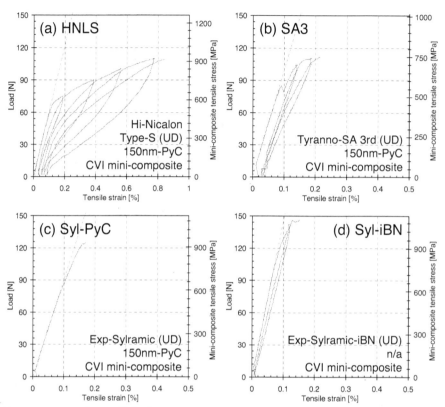

Figure 7. Representative tensile load-strain curves for (a) HNLS, (b) SA3, (c) Syl-PyC, and (d) Syl-iBN mini-composite. It is noted that the tensile strain axis is different for each mini-composite.

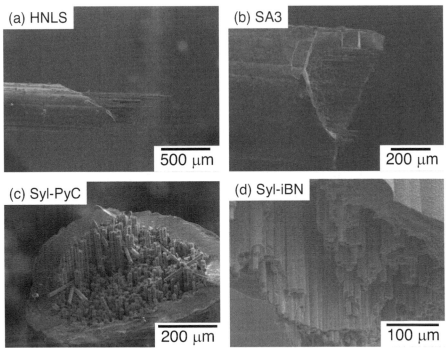

Figure 8.   SEM micrographs of the fracture surfaces of the (a) HNLS, (b) SA3, (c) Syl-PyC, and (d) Syl-iBN mini-composites.

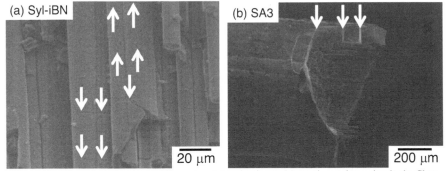

Figure 9.   Examples of matrix crack spacing determination.   (a) matrix crack spacing in the fiber bundle for Syl-iBN, and (b) distance between the crack deflection at outer SiC layer for SA3.

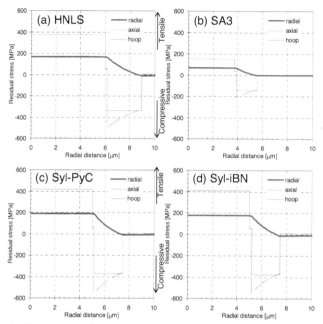

Figure 10. Residual stress analysis result for (a) HNLS, (b) SA3, (c) Syl-PyC, and (d) Syl-iBN composite, using the four phase model.

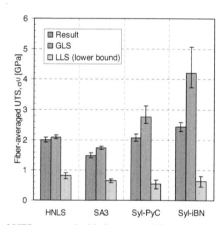

Figure 11. Fiber-averaged UTS compared with the two models. The error bars indicate the standard deviation for the result and ±1 change in the Weibull modulus for the GLS and the LLS.

Table I. Properties of mini-composites used in this study. Nominal mechanical properties of SiC fibers are also listed as supplied by their manufactures.

ID	Fiber	Int.	Matrix	r [μm]	N	f [%]	σ_f [GPa]	E_f [GPa]	ε_f [%]
HNLS	Hi-Nicalon Type-S	150 nm PyC	CVI	12	500	45-50	2.6	420	0.6
SA3	Tyranno-SA3	150 nm PyC	CVI	7.5	1600	42-47	2.8	409	0.7
Syl-PyC	Exp-Sylramic	150 nm PyC	CVI	10	800	43-50	3.0	380	0.8
Syl-iBN	Exp-Sylramic-iBN	iBN	n/a	CVI	10	800	44-47	3.0	380	0.8

Int. denotes interphase, r the fiber diameter, N the number of filaments in a tow, f the fiber volume fraction, σ_f the single fiber strength at 25 mm, E_f the fiber modulus, ε_f the fiber elongation, PyC pyrolytic carbon, CVI chemical vapor infiltration, and exp experimental.

Table II. Tensile test results for the mini-composites. Numbers in parenthesis show standard deviations.

ID	Fiber	Int.	Matrix	P_{fr} [N]	P_P [N]	ε_{fr} [%]	ε_P [%]	σ_U [MPa]	σ_P [MPa]	σ^I [MPa]	RS [%]	#
HNLS	Hi-Nicalon Type-S	150 nm PyC	CVI	113 (5)	67 (4)	0.877 (0.029)	0.152 (0.011)	923 (31)	552 (34)	-303 (75)	77 (3)	4
SA3	Tyranno-SA3	150 nm PyC	CVI	105 (6)	80 (8)	0.270 (0.072)	0.114 (0.041)	676 (52)	513 (51)	-110 (121)	53 (3)	4
Syl-PyC	Exp-Sylramic	150 nm PyC	CVI	130 (7)	91 (11)	0.201 (0.044)	0.121 (0.006)	958 (40)	675 (91)	-73 (-)	69 (4)	2
Syl-iBN	Exp-Sylramic-iBN	n/a	CVI	153 (10)	112 (18)	0.174 (0.033)	0.099 (0.012)	1115 (17)	819 (168)	0 (-)	81 (5)	2

P_{fr} denotes the load at fracture, P_P the load at proportional limit (PL), ε_{fr} the strain at fracture, ε_P the strain at PL, σ_U the ultimate tensile strength, σ_P the PL stress, σ^I the misfit stress, RS the relative strength of the mini-composite to the single fiber strength, and # the number of tests.

Table III. Results of the average fiber pull-out length, the average matrix crack spacing, and the interfacial sliding stress from the hysteresis loop analysis for the mini-composites.

	l_p [um]	d [μm]	τ from Eq. (1) [MPa]	τ from Eq. (2) [MPa]	τ from Eq. (3) [MPa]
HNLS	300	~400	5	8	11-14
SA3	100	~100	85	15	15-20
Syl-PyC	100	~200	170	16	28-37
Syl-iBN	15	~50	950	73	219-287

l_p, d, τ denote the average fiber pull-out length, the average matrix crack spacing, and the interfacial sliding stress, respectively.

Table IV. The relationship between microstructure, fiber roughness, and interfacial sliding stress of near-stoichiometric SiC fibers.

Fiber	Hi-Nicalon Type-S	Tyranno-SA3	Exp-Sylramic	Exp-Sylramic-iBN
Micro-structure	1 μm	1 μm	1 μm	1 μm
R_q [nm]	•0.87, Hinoki (2001) •2.33, Sauder et al. (2009) •<10, Yun et al. (2001)	•1.71, Hinoki (2001) •8.04, Sauder et al. (2009)	•~10, Yun et al. (2001) •17.561, Brennan (2000)	•~27, Yun et al. (2001)
τ [MPa]	•8, This work •14.46, Sauder et al. (2008)	•15, This work •223, Sauder et al. (2008)	•16, This work	•73, This work
Interphase	•150 nm PyC	•150 nm PyC	•150 nm PyC	•n/a (Sylramic-iBN)

R_q denotes the root mean square fiber surface roughness. References are from [6,20-22] for the fiber roughness, and from [17] for the interfacial sliding stress.

Table V. Calculated results of the radial residual stress by CTE mismatch (σ_{CTE}), the compressive stress caused by fiber surface roughness (σ_R), and the clamping stress (σ_{CL}). It is noted that negative value ($\sigma < 0$) denotes compressive stress.

	σ_{CTE} [MPa]	σ_R [MPa]	σ_{CL} [MPa]
HNLS	167	-172	-5
SA3	69	-939	-871
Syl-PyC	188	-1490	-1302
Syl-iBN	175	-2291	-2116

FOREIGN OBJECT DAMAGE IN AN N720/ALUMINA OXIDE/OXIDE CERAMIC MATRIX COMPOSITE

Sung R. Choi,[†] David C. Faucett, and Donald J. Alexander
Naval Air Systems Command, Patuxent River, MD 20670

ABSTRACT

Foreign-object-damage (FOD) phenomena of an N720/alumina oxide/oxide ceramic matrix composite (CMC) were characterized at ambient temperature using a velocity range of 180 to 340 m/s impacted by 1.59-mm diameter steel ball projectiles at a normal incidence angle. Two different types of target support were utilized: fully supported and partially supported. Surface damage of impact sites was typically in a form of craters and increased in size with increasing impact velocity. Subsurface damage beneath the craters was formed with a series of fiber breakage, collapse of inherent pores, compaction of the material, delamination of laminates, and formation of cone cracks. FOD was significantly greater in partial support than in full support, due to the presence of flexural tensile stresses on the backside of target specimens, verified from both impact morphology and post-impact strength. The material in partial support was on the verge of penetration by the projectile at a high impact velocity of 340 m/s.

INTRODUCTION

Because of their brittle nature, ceramics - either monolithic or ceramic matrix composites - are prone to localized surface/subsurface damage and/or cracking when subjected to impact by foreign objects. Foreign object damage (FOD) needs to be considered when those materials are designed for structural applications particularly in aeroengines such as airfoils and combustors. A significant amount of work on impact damage of brittle materials has been conducted both experimentally and analytically [1-14], including gas-turbine grade toughened silicon nitrides [15-17].

The previous work on FOD of monolithic silicon nitrides [15-17] was extended to gas-turbine grade, melt-infiltrated (MI) SiC/SiC [18] and N720™/aluminosilicate (N720/AS) CMCs [19]. Unlike the monolithic ceramics, the SiC/SiC and oxide/oxide CMCs exhibited no catastrophic failure up to 400 m/s, resulting in much increased FOD resistance particularly at higher impact velocities ≥300 m/s.

The current work, as an extension of the previous work, describes FOD behavior of an N720™ oxide fiber-reinforced alumina matrix (designated as 'N720/A') CMC. N720/alumina oxide/oxide targets in a flexure bar configuration were impacted at velocities from 180-340 m/s by 1.59-mm-diameter steel ball projectiles. Damage morphologies and post-impact strength of the composite were characterized. Additional information on Hertzian contact damage and its relation to FOD of the composite can be found in a separate paper in this volume [20].

[†] Corresponding author, sung.choi1@navy.mil

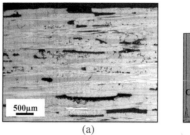
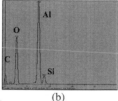

(a)   (b)

Figure 1. (a) Un-impacted Microstructure and (b) energy dispersive spectroscopy (EDS) of fiber interfaces of N720/alumino oxide/oxide CMC used in this work.

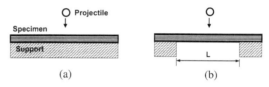

(a)   (b)

Figure 2. Two types of target specimen support used in FOD testing: (a) full support and (b) partial support (L=20 mm).

EXPERIMENTAL PROCEDURES

Material and Test Specimens
      The composite used in this work was a commercial, 2-D woven, N720™ fiber-reinforced alumina matrix oxide/oxide CMC, fabricate by ATK/COIC (San Diego, CA; Vintage 2008). N720™ oxide fibers, produced in tows by 3M Corp. (Minneapolis, MN), were woven into 2-D 8 harness-satin cloth. The cloth was cut into a proper size, slurry-infiltrated with the matrix, and 12 ply-stacked followed by consolidation and sintering. The fiber volume fraction of the composite panels was about 0.45. Typical microstructure of the composite is shown in Fig. 1. Significant porosity and microcracks in the matrix are characteristics which typify this class of oxide/oxide CMC and its means of damage tolerance. [21,22]. No interface fiber coating was applied (see Fig. 1). Porosity was about 25 %, bulk density was 2.74 g/cm^3, and elastic modulus was 81GPa by the impulse excitation of vibration technique [23]. Flexure bars measuring 12 mm in width, 50 mm in length, and about 3 mm in as-furnished thickness were machined from the composite panels for test specimens in FOD testing and post-impact strength and other related testing.

Foreign Object Damage Testing
      Foreign object damage testing was conducted using the experimental apparatus whose descriptions can be found elsewhere [15,16]. Briefly, hardened (HRC≥60) chrome steel-balls

 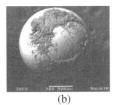 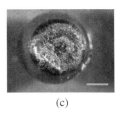

(a)　　　　　　　　　(b)　　　　　　　　　(c)

Figure 3. Steel ball projectiles retrieved upon impact in full support on: (a) N720/alumina oxide/oxide at 340 m/s (this study), (b) N720/aluminosilicate oxide/oxide at 400 m/s [19], and (c) MI SiC/SiC at 400 m/s [18].

with a diameter of 1.59 mm were inserted into a 300mm-long gun barrel. A helium-gas cylinder and relief valves were utilized to pressurize and regulate a reservoir to a specific level, depending on prescribed impact velocity. Upon reaching a specific level of pressure, a solenoid valve was instantaneously opened accelerating a steel-ball projectile through the gun barrel to impact a target specimen. The target specimens were fully or partially supported via rigid steel blocks, as shown in Fig. 2. Each target specimen was aligned such that the projectile impacted at the center of the specimen at a normal incidence angle. Four different impact velocities ranging from 180 to 340 m/s were employed with a total of four target specimens used at each impact velocity.

Post-Impact Strength Testing

As usual in any of previous FOD testing [15-19], post-impact strength was determined for target specimens impacted in order to assess more accurately the severity of impact damage. Strength testing was conducted using a four-point flexure steel fixture with 20mm inner and 40mm outer spans with an electromechanical test frame (Type 1122, Instron) at a crosshead speed of 0.5 mm/min. As-received strength of the composite was also determined with a total of six test specimens.

RESULTS AND DISCUSSION

It has been observed that the steel-ball projectiles impacting monolithic silicon nitrides (AS800 and SN282) [15-17] and MI SiC/SiC composite [18] were flattened or severely deformed or fragmented, depending on impact velocity. However, neither flattening nor noticeable deformation of the steel projectiles occurred in the N720/alumina composite even at the highest impact velocity of 340 m/s, as shown in Fig. 3. The similar trend was also observed previously from the N720/aluminosilicate composite [19]. This was attributed to the composites' 'soft' and porous nature, which may be also known from their relatively low elastic moduli: E = 67 and 81GPa for N720/AS [19] and N720/A, respectively; whereas, E = 220 GPa for MI SiC/SiC [18]. Included in the figure are the projectiles impacted at 400 m/s on N720/AS [19] and MI SiC/SiC [18] for comparison. Note that the target materials (fibers and matrices) of both N720/A and N720/AS CMCs were transferred to the projectiles upon impact, due to the soft and porous nature of both composites. Some limited material transfer also occurred in MI SiC/SiC.

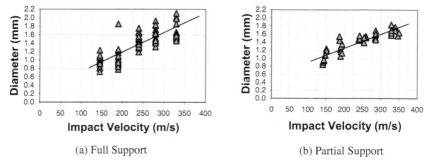

(a) Full Support   (b) Partial Support

Figure 4. Front impact-damage size as a function of impact velocity for N720/alumina oxide/oxide CMC impacted by 1.59-mm steel ball projectiles in full and partial supports.

Front impact damages of target specimens were in the form of craters, accompanying fiber/matrix breakage and material removal. Damage size (diameter) as a function of impact velocity is presented in Fig. 4. The damage size increased monotonically with increasing impact velocity and was seemingly independent of the type of target support within the statistical scatter. Figure 5 shows front impact damage with respect to impact velocity in both full and partial supports.

Backside damages in the target specimens were strongly dependent on the type of target support. In partial support, backside damage started even at a low impact velocity of 180 m/s and became significant at ≥280 m/s. On the contrary, no visible backside damage was observed in full support even at 340 m/s. This observation was in good agreement with both N720/AS [19] and MI SiC/SiC [18]CMCs. Figure 6 shows comparison of backside damage of N720/A target specimens with regard to impact velocity in both full and partial supports. The appreciable backside damage in partial support was due to the presence of backside flexural tensile stresses upon impact. The results of Fig. 6 indicate that the type of CMC component supports unto some substrates, if any, plays an important role to affect the overall structural integrity of related component when it comes to FOD.

Figure 7 shows the cross-sectional views of impact sites with respect to impact velocity in both full and partial supports. With increasing impact velocity, impact damage in both size (diameter) and depth increased. Some material removal from the craters was also evident. The material beneath the craters was subjected to some degree of compaction with its extent being dependent on impact velocity. This compaction can be understood based on the fact that the material is *soft* and *open* in its structure, as also observed in the N720/AS composite [19]. Also noted is the generation of cone cracks at higher impact velocities ≥280 m/s in both supports. Overall impact damage occurring at the highest impact velocity of 340 m/s in partial support is shown in Fig. 8. Fiber breakage, backside cracking, delamination, and cone cracking are all significantly induced damage features. The target, in fact, was on the brink of penetration by the projectile. Nevertheless, the target survived without any catastrophic failure, similar to the

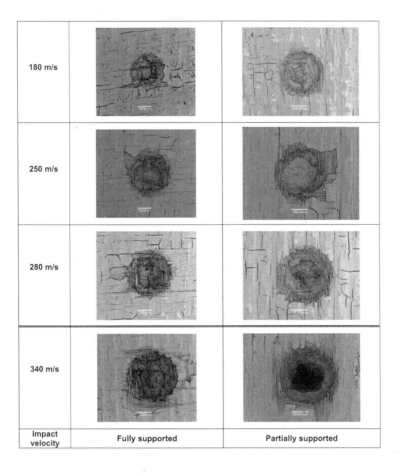

Figure 5. Images of front impact damage with respect to impact velocity in N720/alumina oxide/oxide CMC impacted by 1.59-mm steel ball projectiles in full and partial supports.

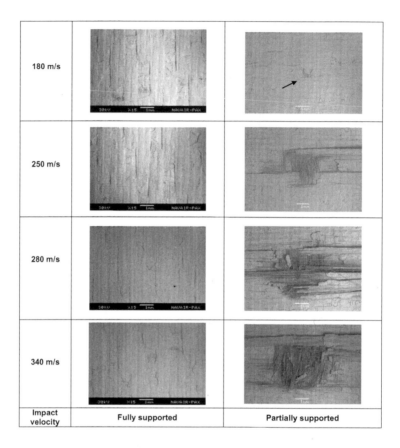

Figure 6. Images of backside impact damage with respect to impact velocity in N720/alumina oxide/oxide CMC impacted by 1.59-mm steel ball projectiles in full and partial supports. The arrow at 180 m/s in partial support indicates backside damage.

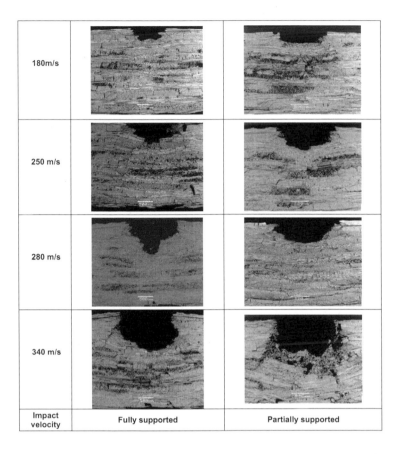

Figure 7. Images of cross-sectional views of impact sites with respect to impact velocity in N720/alumina oxide/oxide CMC impacted by 1.59-mm steel ball projectiles in full and partial supports.

Figure 8. Typical example of the cross-sectional view of a target specimen of N720/alumina oxide/oxide CMC impacted at 340 m/s by 1.59-mm steel ball projectiles in partial support.

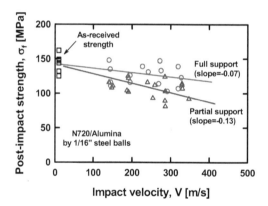

Figure 9. Post-impact strength as a function of impact velocity of N720/alumina oxide/oxide CMC impacted by 1.59 steel ball projectiles in both full and partial supports.

observations in MI SiC/SiC [18]and N720/AS [19] CMCs. This is in contrast with monolithic AS800 and SN282 silicon nitrides that exhibited catastrophic fracture upon impact at much lower velocities of 180-260 m/s [15]. In this respect, the CMCs revealed much more enhanced FOD resistance than monolithic counterparts.

The results of post-impact strength testing are depicted in Fig. 9, where post-impact strength was plotted as a function of impact velocity in both full and partial supports. As-received strength (= 145±13 MPa) was also plotted for comparison. Despite some inherent data scatter, a general trend is manifest such that post-impact strength decreased monotonically with increasing impact velocity and that for a given impact velocity greater strength degradation occurred in partial support than in full support. This phenomenon of post-impact strength degradation is all consistent with the results of impact morphologies as already seen from the frontal, backside, and cross-sectional impact damages. Without imposing any physical significance on the curve fitting, a linear regression analysis of post-impact strength (σ_f in MPa) versus impact velocity (V in m/s) was made and its result is shown as follows:

$$\sigma_f = -0.07V + 141$$
$$\sigma_f = -0.13V + 140$$

for full and partial supports, respectively. Their respective coefficient of correlation (r_{coef}) was $r_{coef} = 0.49$ and 0.774.

Finally, it should be stated that conventional optical and/or SEM examinations for the N720/A composite or any other fiber-reinforced CMCs are not sufficient in many cases to scrutinize the nature and aspect of impact damage. More appropriate nondestructive evaluation tools such as computed tomography and/or pulsed thermography techniques [24-26] are also needed.

CONCLUSIONS

1) The overall impact damage of the N720/alumina oxide/oxide composite was found to be greater in partial support than in full support.
2) In full support, frontal contact stresses played a major role in generating composite damage; whereas, in partial support, backside flexural tensile stresses were an additional source of damage generation.
3) Due to the composite's *soft* and *open* structure, compaction of the material beneath impact site occurred with its degree being dependent on impact velocity. Also, the formation of cone cracks took place making damage more significant at backside than at impact site.
4) Post-impact strength for a given impact velocity was greater in full support than in partial support, again an evidence of more enhanced damage occurring in partial support.

Acknowledgements

The authors acknowledge the support by the Office of Naval Research and Dr. David Shifler.

REFERENCES

1. Wiederhorn, S. M., and Lawn, B.R., 1977, "Strength Degradation of Glass Resulting from Impact with Spheres," J. Am. Ceram. Soc., **60**[9-10], pp. 451-458.
2. Wiederhorn, S. M., and Lawn B. T., 1979, "Strength Degradation of Glass Impact with Sharp Particles: I, Annealed Surfaces," J. Am. Ceram. Soc., **62**[1-2], pp. 66-70.

3. Ritter, J. E., Choi, S. R., Jakus, K, Whalen, P. J., and Rateick, R. G., 1991, "Effect of Microstructure on the Erosion and Impact Damage of Sintered Silicon Nitride," J. Mater. Sci., **26**, pp. 5543-5546.

4. Akimune, Y, Katano, Y, and Matoba, K, 1989, "Spherical-Impact Damage and Strength Degradation in Silicon Nitrides for Automobile Turbocharger Rotors," J. Am. Ceram. Soc., **72**[8], pp. 1422-1428.

5. Knight, C. G., Swain, M. V., and Chaudhri, M. M., 1977, "Impact of Small Steel Spheres on Glass Surfaces," J. Mater. Sci., **12**, pp.1573-1586.

6. Rajendran, A. M., and Kroupa, J. L., 1989, "Impact Design Model for Ceramic Materials," J. Appl. Phys, **66**[8], pp. 3560-3565.

7. Taylor, L. N., Chen, E. P., and Kuszmaul, J. S., 1986 "Microcrack-Induced Damage Accumulation in Brittle Rock under Dynamic Loading," Comp. Meth. Appl. Mech. Eng., **55**, pp. 301-320.

8. Mouginot, R., and Maugis, D., 1985, "Fracture Indentation beneath Flat and Spherical Punches," J. Mater. Sci., **20**, pp. 4354-4376.

9. Evans, A. G., and Wilshaw, T. R., 1977, "Dynamic Solid Particle Damage in Brittle Materials: An Appraisal," J. Mater. Sci., **12**, pp. 97-116.

10. Liaw, B. M., Kobayashi, A. S., and Emery, A. G., 1984, "Theoretical Model of Impact Damage in Structural Ceramics," J. Am. Ceram. Soc., **67**, pp. 544-548.

11. van Roode, M., et al., 2002, "Ceramic Gas Turbine Materials Impact Evaluation," ASME Paper No. GT2002-30505.

12. Richerson, D. W., and Johansen, K. M., 1982, "Ceramic Gas Turbine Engine Demonstration Program," Final Report, DARPA/Navy Contract N00024-76-C-5352, Garrett Report 21-4410.

13. Boyd, G. L., and Kreiner, D. M., 1987, "AGT101/ATTAP Ceramic Technology Development," Proceeding of the Twenty-Fifth Automotive Technology Development Contractors' Coordination Meeting, p.101.

14. van Roode, M., Brentnall, W. D., Smith, K. O., Edwards, B., McClain, J., and Price, J. R., 1997, "Ceramic Stationary Gas Turbine Development – Fourth Annual Summary," ASME Paper No. 97-GT-317.

15. (a) Choi, S. R., Pereira, J. M., Janosik, L. A., and Bhatt, R. T., 2002, "Foreign Object Damage of Two Gas-Turbine Grade Silicon Nitrides at Ambient Temperature," Ceram. Eng. Sci. Proc., **23**[3], pp. 193-202; (b) Choi, S. R., et al., 2004, "Foreign Object Damage in Flexure Bars of Two Gas-Turbine Grade Silicon Nitrides," Mater. Sci. Eng. **A 379**, pp. 411-419.

16. Choi, S. R., Pereira, J. M., Janosik, L. A., and Bhatt, R. T., 2003, "Foreign Object Damage of Two Gas-Turbine Grade Silicon Nitrides in a Thin Disk Configuration," ASME Paper No. GT2003-38544; (b) Choi, S. R., et al., 2004, "Foreign Object Damage in Disks of Gas-Turbine-Grade Silicon Nitrides by Steel Ball Projectiles at Ambient Temperature," J. Mater. Sci., **39**, pp. 6173-6182.

17. Choi, S. R., 2008, "Foreign Object Damage Behavior in a Silicon Nitride Ceramic by Spherical Projectiles of Steels and Brass," Mat. Sci. Eng. **A497**, pp. 160-167.

18. Choi, S. R., 2008, "Foreign Object Damage Phenomenon by Steel Ball Projectiles in a SiC/SiC Ceramic Matrix Composite at Ambient and Elevated Temperatures," J. Am. Ceram. Soc., **91**[9], pp. 2963-2968.

19. (a) Choi, S. R., Alexander, D. J., and Kowalik, R. W., 2009, "Foreign Object Damage in an Oxide/Oxide Composite at Ambient Temperature," J. Eng. Gas Turbines & Power, Transactions of the ASME, Vol. **131**, 021301. (b) Choi, S. R., Alexander, D. J., and Faucett, D. C., 2009, "Comparison in Foreign Object Damage between SiC/SiC and Oxide/Oxide Ceramic Matrix Composites," Ceram. Eng. Sci. Proc., **30**[2], pp. 177-188.
20. Faucett, D. C., Alexander, D. J., and Choi, S. R., 2010, "Static Contact Damage in N720/Alumina Oxide/Oxide Ceramic Matrix Composite with Reference to Foreign Object Damage," to be published in Ceram. Eng. Sci. Proc.
21. Mattoni, M. A., et al., 2005, "Effects of Combustor Rig Exposure on a Porous-Matrix Oxide Composite," J. App. Ceram. Tech., **2**[2], pp.133-140.
22. Simon, R. A., 2005, "Progress in Processing and Performance of Porous-Matrix Oxide/Oxide Composites," *ibid*, **2**[2], pp. 141-149.
23. ASTM C 1259, "Test Method for Dynamic Young's Modulus, Shear Modulus, and Poisson's Ratio for Advanced Ceramics by Impulse Excitation of Vibration," *Annual Book of ASTM Standards, Vol. 15.01*, ASTM, West Conshohocken, PA (2009).
24. Cosgriff, L. M., Bhatt, R., Choi, S. R., Fox, D. S., 2005, "Thermographic Characterization of Impact Damage in SiC/SiC Composite Materials," Proc. SPIE, Vol. 5767, pp. 363-372 in Nondestructive Evaluation & Health Monitoring of Aerospace Materials, Composites, and Civil Structure IV.
25. Bhatt, R. T., Choi, S. R., Cosgriff, L. M., Fox, D. S., Lee, K. N., 2008, "Impact Resistance of Environmental Barrier Coated SiC/SiC Composites," Mater. Sci. Eng. A 476, pp. 8-19.
26. Bhatt, R. T., Choi, S. R., Cosgriff, L. M., Fox, D. S., Lee, K. N., 2008, "Impact Resistance of Uncoated SiC/SiC Composites," Mater. Sci. Eng. A 476, pp. 20-28.

STATIC CONTACT DAMAGE IN AN N720/ALUMINA OXIDE/OXIDE CERAMIC MATRIX COMPOSITE WITH REFERENCE TO FOREIGN OBJECT DAMAGE

David C. Faucett, Donald J. Alexander, Sung R. Choi

Naval Air Systems Command, Patuxent River, MD 20670, USA

ABSTRACT

Static contact damage was characterized in an N720/alumina oxide/oxide ceramic matrix composite (CMC) by indenting CMC test coupons with 1.59 mm-diameter steel ball indenters. Responses of load-versus-displacement as well as damage morphologies were determined. The static Hertzian contact damage was compared with foreign object damage (FOD) imposed via ballistic impact by the same steel ball projectiles. Many features were in common in terms of the modes and aspects of deformation and damage in the vicinity of indents or impact sites, except for differences in damage severity. Prediction of quasi-static impact force was made based on a contact yield pressure analysis and was found to be in reasonable agreement with the experimental data.

INTRODUCTION

Either monolithic ceramics or ceramic matrix composites, because of their inherent brittle nature, are susceptible to localized damage and/or cracking when subjected to impact by foreign objects. Numerous cases of foreign object damage have been experienced in the hot sections of aeroengines in which combustion products, metallic particles, thermal-barrier coatings loosened, and/or small foreign objects ingested can cause damage to related hot components, resulting in serious structural and functional problems. As a consequence, foreign object damage (FOD) associated with particle impact needs to be addressed when advanced ceramics are designed for aeroengine hot-section components.

A series of work has been conducted to assess and characterize FOD phenomena of both monolithic silicon nitrides and CMCs including MI SiC/SiC and N720/aluminosilicate composites [1-6]. In general, ceramic targets were impacted by 1.59-mm (diameter) steel ball projectiles in a velocity range up to 400 m/s and their FOD aspects were characterized in terms of impact morphology and post-impact strength. Unlike monolithic ceramics, CMCs exhibited no catastrophic failure showing greater resistance to FOD over monolithic counterparts. For the N720/aluminosilcate CMC, some similarity in damage mode and features was also found in both static indent and ballistic impact, primarily due to the material's open and porous microstructure [7,8].

The work has been recently extended to a commercial, gas-turbine grade, N720/alumina oxide/oxide CMC to characterize its FOD behavior. Some results on FOD and static Hertzian contact damage on the composite were reported previously [9]. This paper presents additional results on static contact damage, compares them with FOD, and predicts quasi-static impact force based on a contact yield-pressure analysis using the static indent data. Detailed descriptions on FOD of the composite can be found in a companion paper in this volume [10].

EXPERIMENTAL PROCEDURES

Material and Test Specimens

The composite material used in this work has been described elsewhere [9-11]. Briefly, the N720/alumina oxide/oxide CMC was acquired from ATK/COIC (San Diego, CA; vintage 2008). N720™ oxide fibers, produced in tow form by 3M Corp. (Minneapolis, MN), were woven into 2-

D 8 harness-satin cloth. The cloth was cut into a proper size, slurry-infiltrated with the matrix (alumina), and 12 ply-stacked followed by consolidation and sintering. No interface fiber coating was employed. The fiber volume of the composite panels was about 45 %. Significant porosity and microcracks in the matrix were inherent characteristics to improve damage tolerance of this class of oxide/oxide CMCs [12-14]. Porosity was about 25 %, bulk density was 2.74 g/cm^3, flexure strength was 145±13 MPa, elastic modulus was 81GPa by the impulse excitation of vibration technique [15]. Flexure bars measuring 12 mm in width, 50 mm in length, and about 3 mm in as-furnished thickness were machined from the composite panels for test specimens.

Indentation Testing

Indentation testing for the composite was described elsewhere [7]. The testing was conducted by an electromechanical test frame (Type 1122, Instron, Canton, MA) using 1.59 mm-diameter, hardened (HRC≥60) chrome-steel ball indenters, the same as the ball projectiles that were utilized in the previous FOD work [1-6]. Five different indentation loads ranging from 490 N to 2450 N were used. A total of four indents for a given indentation load was utilized for each test specimen. Indentation load was applied onto the polished side of each specimen at a crosshead speed of 0.23 mm/min in displacement control. In-situ indentation displacement (depth) was determined using a linear variable displacement transducer (LVDT) (Type GT5000, RDP) to obtain load-versus-displacement curves. For some selected specimens, their cross-sectional views containing the indent sites were also examined to characterize their respective subsurface damage.

Foreign-Object-Damage Testing

FOD testing employed in this work has been described in the companion papers [9,10] and elsewhere [1-6]. Consequently, the description on FOD testing is omitted in this paper.

RESULTS AND DISCUSSION

1. Hertzian Contact Damage: Static Indentation

Load versus Displacement Curves: Typical loading/unloading curves of the N720/alumina composite obtained via *multiple* indentations up 2450 N are shown in Fig. 1. The curves were obtained through a series of multiple loading/unloading sequences by an increment of 500N. Regardless of the magnitude of indent load, significant inelasticity was evident in each of loading/unloading sequence. In other words, completely different behavior in loading and unloading was noted at each sequence, indicating that indentation gave rise to significant permanent deformation. No visible deformation in the steel ball indenters was observed, as shown in Fig. 2. Most of inelastic behavior of the indentation process was attributed exclusively to the significant permanent deformation of the composite. This appreciable deformation was caused by compaction or densification of the material underneath the ball indenters, evident of the material's *soft and open* microstructure. In the case of a dense, hard MI SiC/SiC CMC, most of indentation deformation took place in the ball indenters rather than in the hard composite. Note the difference in elastic modulus (E) between the two composites: E = 81 and 220 GPa for N720/alumina and MI SiC/SiC, respectively.

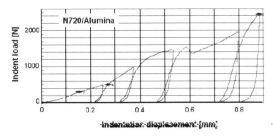

Figure 1. Typical indentation load vs. displacement curves of an N720/alumina oxide/oxide CMC, determined via multiple indents up to 2940 N by an 1.59 mm-diameter steel ball indenter.

Figure 2. A ball indenter used in indentation testing at P =2450 N, showing no visible deformation. Note some materials (fibers and matrices) were attached, transferred from the composite during indentation.

Also note that some materials of fibers and matrices were transferred from the composite to the ball indenter, as seen from Fig. 2, which is another indication of the material's soft and porous nature of microstructure. It is also noted from the result of Fig. 1 that the composite did not exhibit any hardening, typical of many dense ceramics, resulting in an almost similar slope in unloading and subsequent loading curves. Also noted from the figure are the peak loads that remained reasonably linear up to an applied load of 1960 N but with a load drop at around 1400 N. It was observed that there was no significant difference in the shape of curved between the multiple indents and the single indent provided the maximum indent load kept the same. However, it should be noted that even for a given test specimen the curves of all four indents were not identical, always having some variations from each other. This might be indicative of inhomogeneous and/or architectural nature of the composite. The composite's architectural nature, as to whether indentation was made either at matrix-rich region or at fiber-rich region or at in-between, would affect the results of indentation. A similar indent behavior was also observed in the N720/aluminosilicate oxide/oxide CMC [8,9] that exhibited the microstructure similar to N720/alumina.

Indentation Damage: Indent sites generated in the N720/alumina oxide/oxide specimens were typically in the form of craters, similar to the case of the N720/aluminisilicate [8,9]. Figure 3 presents typical indent morphology at an indentation load of 980 N. It is readily seen that the contact region (region 'C') was subjected to compaction under the ball indenter with many broken short fibers embedded in matrices, while the outside indentation boundaries were characterized by

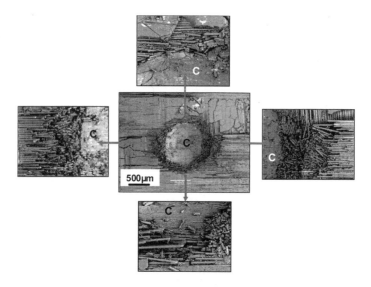

Figure 3. Examples of an indent site showing various fiber and matrix damage generated by indentation in a N720/alumina oxide/oxide CMC by 1.59-mm steel ball indenter at an indentation load of 980 N. The region 'C' indicates a central contact region.

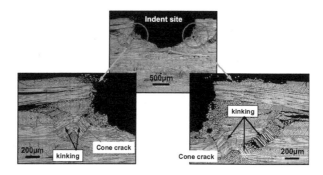

Figure 4. Cross-sectional view of an indent site of a N720/alumina oxide/oxide CMC indented by 1.59-mm steel ball indenter at an indentation load of 1470 N.

significant fibers/matrices breakage, due to compression or shearing or tension or all combined stresses during the indentation process. The overall indent sites tended to conform to the counter of spherical ball indenters.

Damage morphology can be seen more clearly from the cross-sectional views of indent sites, which is shown in Fig. 4 as an example. The figure presents various subsurface damage such as deformation and compaction of the material underneath the indenter, material removal, fibers/matrices breakage, kinking of fiber tows, cone cracking, etc. These features became more predominant with increasing indentation load. Of a special importance was a phenomenon of compaction of the material beneath the indenter. This compaction, accompanying significant reduction of pores in matrices, can be understood by considering the material's *soft, open, loosely-connected* microstructure. In some cases, as already seen in Fig. 3, a portion of the composite in contact with ball indenters was removed (plucked away) upon releasing indenters, due to appreciable contact stresses together with the soft, open nature of the composite.

Of another importance was cone cracking as well as the breakage and kinking of fiber tows, as seen in Fig. 4. The breakage of fiber tows was attributed to continuous stretching of the tows by the ball indenter during indentation; whereas, the kinking and cone cracking was due to the lateral displacement/deformation of the material as the indenter process continued. Similar features of compaction, fiber-tow breakage, and the occurrence of cone cracking were also observed in the N720/aluminosilicate composite [8,9]. Cone cracking has been well known as a common response of brittle materials when subjected to static Hertzian contact by hard probes such as balls, rollers, or spherical rods or punches [e.g., 16]. This is also true for the case of ballistic impact on monolithic ceramics or CMCs by spherical projectiles [1-6,17-21].

Figure 5 shows a summary of indent depth and diameter as a function of indentation load of the composite. Some variations in the data were inevitable but a general trend was self evident: both indent depth and diameter increase linearly with increasing indent load, a fact that the contact damage in magnitude was linearly dependent on indent load. Considering the indenter geometry and the depth of indent, a simple relationship can be obtained as

$$d = 2(zD - z^2)^{1/2} \tag{1}$$

where d is the diameter of indent site, z is the depth of indent site, and D is the diameter of an indenter. Plotting the data in Fig. 5 based on Eq. (1) yielded Fig. 6. As seen in the figure, the experimental data on diameter (d) and depth (z) were surprisingly in excellent agreement with the theoretical relationship of Eq. (1), despite the fact that demarcations of d and z were not always well defined because of the architectural complexity of the composite. This indicates that the composite was well conformed to the configuration of the spherical ball indenters, accompanying completely permanent (plastic) deformation of the material underneath the indenters. Some deviations from the relationship occurred at higher indentation loads, due to the material's significant transfer to the indenters and/or due to increased damage at indent boundaries via enhanced fibers/matrices cracking.

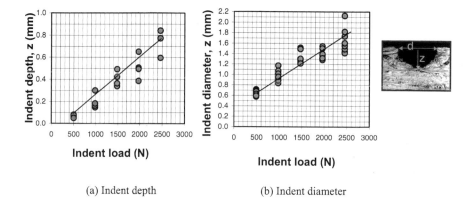

(a) Indent depth   (b) Indent diameter

Figure 5. Results of indent depth (z) and diameter (d) as a function of indent load determined for a N720/alumina oxide/oxide CMC indented by 1.59-mm steel ball indenters. The inset indicates z and d of an indent.

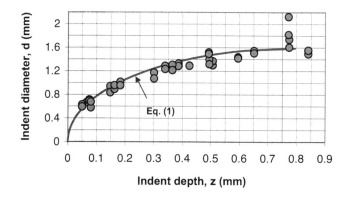

Figure 6. Comparison of the indent data on depth and diameter with the ideal curve as defined in Eq. (1) for a N720/alumina oxide/oxide CMC indented by 1.59-mm steel ball indenters.

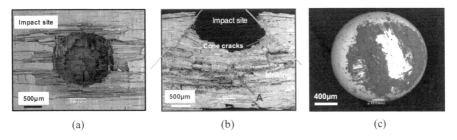

(a)   (b)   (c)

Figure 7. Examples of (a) impact site and (b) its cross-sectional view for a N720/alumina oxide/oxide CMC impacted at 340 m/s by 1.59 steel ball projectiles. The projectile retrieved right after impact at 280 m/s is shown in (c).

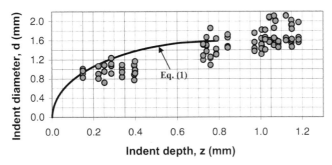

Figure 8. Comparison of impact damage data on depth and diameter with the theoretical curve (Eq. (1)) for a N720/alumina oxide/oxide CMC impacted by 1.59-mm steel ball projectiles at various impact velocities up to 340 m/s.

2. Foreign Object Damage

Detailed descriptions on FOD with respect to damage morphology and post-impact strength of the composite can be founded in a companion paper [10]. Only simple descriptions on impact damage will be presented here in this paper. The impact morphology was similar to the static contact damage with respect to the formation of craters, fibers/matrices breakage, deformation, compaction, kinking, and cone cracking, etc, as shown in Fig. 7. Also, the material transfer to the impacting projectiles was similar. However, due to the dynamic nature of impact, the damage severity was much greater and more violent in FOD than in static indent, also accompanying more material transfer from the target composite. The data on impact damage size (depth and diameter) was plotted in the same way as in Fig. 6 and are presented in Fig. 8. Comparing both the impact and the static data (Figs, 6 and 8), it is manifest that damage was much more random and significant in dynamic impact than in static indentation.

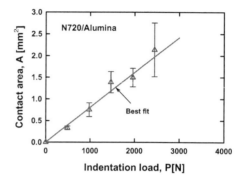

Figure 9. Contact area as a function of indentation load determined for an N720/alumina oxide/oxide CMC, indented by 1.59mm-diameter steel ball indenters. The line represents the best fit.

3. Prediction of Impact Forces

     Assuming that the impact event is quasi-static, a first-order approximation of impact force was made for the composite using a contact yield pressure analysis. The detailed analysis can be found in the previous work [3-(b)]. Figure 9 shows a relationship between contact indent area ($A=\pi d^2/4$) and indentation load (P), determined from the data in Fig. 5-b. As seen from the figure, an overall linearity between P and A was reasonably established. The linearity also implies that macroscopically the average 'contact yield pressure' ($=P/A$) would almost remain constant regardless of the magnitude of indentation load. The contact yield pressure, p_y, is defined as:

$$p_y = \frac{dP}{dA} = \frac{\Delta P}{\Delta A} \tag{2}$$

The value of p_y were estimated as $p_y = 1240\pm159$ MPa from the data in Fig. 9. Following the energy balance and the geometry of an impacting ball projectile, the resulting equation can be written as follows [3-(b)]:

$$m(1-e^2)V^2/2 = \pi p_y(Dz^2/2 - z^3/3) \tag{3}$$

where m is the mass of the projectile, e is the coefficient of restitution, defined as $e = -V_{bc}/V$ with V_{bc} being the bouncing-back velocity of the projectile and V being the impact velocity. The value of z in Eq. (3) can be solved as a function of impact velocity V for the given values of p_y, m, and e. Once z is solved, the impact force F can be calculated using the following equation [3-(b)]

$$F = [\pi(Dz - z^2)]p_y \tag{4}$$

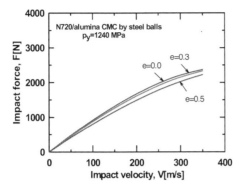

Figure 10. Predicted impact force as a function of impact velocity for different values of the coefficient of restitution (e) for an N720/alumina oxide/oxide composite impacted by 1.59mm-diameter steel ball projectiles.

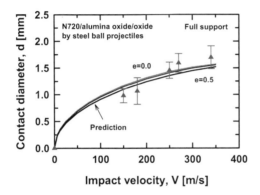

Figure 11. Comparison in impact damage size (d) as a function of impact velocity between the experimental data and the prediction for an N720/alumina oxide/oxide composite impacted by 1.59mm-diameter steel ball projectiles.

The resulting prediction of impact force within a geometrical limit $(D/z>2.0)$ is shown in Fig. 10 where three different values of $e=0.0$, 0.3, and 0.5 were employed for each composite. As seen in the figure, F increases almost linearly with increasing V. For a given V, F depends on the value of e. However, the difference in F among the values of e chosen was insignificant (i.e., <10% between $e=0$ and 0.5).

Since no impact force measurements were undertaken, the prediction of impact force can be no longer verifiable. However, the data on impact damage size (d) are available so that the impact force analysis can be indirectly check its validity by estimating the predicted damage size (d) as a function of impact velocity, based on Eq. (3) and (4). The result is shown in Fig. 11. The prediction, regardless of the value of e and some data scatter, is in good agreement with the experimental data. This verifies to some extent the validity of the analysis. However, it should be kept in mind that this type of approach to impact force prediction is only of first-order approximation because of many simplifying assumptions made and because of uncertainties associated with the violent and random nature of impact event as seen from Fig. 8. Pertinent test instrumentation/scheme should be sought to assess accurately dynamic impact event and related impact force.

CONCLUSIONS

1) The damage features of the N720/alumina oxide/oxide composite by 1.59 mm-diameter steel balls were similar in principle in both static Hertzian contact and dynamic impact. However, the overall severity of damage was much greater in impact than in static indentation.

2) Due to the composite's *soft* and *loosely-connected* structure, most of indent or impact damage occurred exclusively in the composite. Several different aspects of damage were involved including formation of craters, compaction of the material beneath indenters or impacting projectiles, fibers/matrices breakage, kinking, cone cracking, and material transfer to the indenters or projectiles, etc

3) A first-order, quasi-static approximation of impact force was made based on the contact yield pressure analysis, resulting in somewhat reasonable agreement with the experimental data. However, appropriate experimental techniques need to be exercised to verify its validity and to assess accurately time-varying impact events.

Acknowledgements
The authors acknowledge the supports of the Office of Naval Research and Dr. David Shifler.

REFERENCES
1. (a) S. R. Choi, J. M. Pereira, L. A. Janosik, and R. T. Bhatt, "Foreign Object Damage of Two Gas-Turbine Grade Silicon Nitrides at Ambient Temperature," *Ceram. Eng. Sci. Proc.*, **23**[3] 193-202 (2002); (b) *idem*, "Foreign Object Damage in Flexure Bars of Two Gas-Turbine Grade Silicon Nitrides," *Mater. Sci. Eng.*, **A 379**, 411-419 (2004).
2. (a) S. R. Choi, J. M. Pereira, L. A. Janosik, and R. T. Bhatt, "Foreign Object Damage of Two Gas-Turbine Grade Silicon Nitrides in a Thin Disk Configuration," ASME Paper No. GT2003-38544 (2003); (b) *idem*, "Foreign Object Damage in Disks of Gas-Turbine-Grade Silicon Nitrides by Steel Ball Projectiles at Ambient Temperature," *J. Mater. Sci.*, **39**, 6173-6182 (2004).
3. (a) S. R. Choi, J. M. Pereira, L. A. Janosik, and R. T. Bhatt, "Effect of Projectile Materials on Foreign Object Damage of a Gas-Turbine Grade Silicon Nitride," ASME Paper No. GT2005-68866 (2005); (b) S. R. Choi, "Foreign Object Damage Behavior in a Silicon Nitride Ceramic by Spherical Projectiles of Steels and Brass," *Mat. Sci. Eng.* **A497**, 160-167 (2008).

4. S. R. Choi, "Foreign Object Damage in Gas-Turbine Grade Silicon Nitrides by Silicon Nitride Ball Projectiles," ASME Paper No: GT 2009-59031, ASME Turbo Expo 2009, June 8-12, 2009, Orlando, FL.

5. idem, "Foreign Object Damage Phenomenon by Steel Ball Projectiles in a SiC/SiC Ceramic Matrix Composite at Ambient and Elevated Temperatures," J. Am. Ceram. Soc., 91[9] 2963-2968 (2008).

6. S. R. Choi, D. J. Alexander, and R. W. Kowalik, "Foreign Object Damage in an Oxide/Oxide Composite at Ambient Temperature," J. Eng. Gas Turbines & Power, Transactions of the ASME, Vol. 131, 021301 (2009).

7. S. R. Choi, D. J. Alexander, and D. C. Faucett "Comparison in Foreign Object Damage between SiC/SiC and Oxide/oxide Ceramic Matrix Composites," Ceram. Eng. Sci. Proc., 30[2] 177-188 (2009).

8. idem, "Foreign Object Damage vs. Static Indentation Damage in an Oxide/Oxide Ceramic Matrix Composite," in Processing & Properties of Advanced Ceramics & Composites, Ceramic Transactions, Vol. 203, pp.171-1800 (2009).

9. D. C. Faucett, D. J. Alexander, and S. R. Choi, "Static-Contact and Foreign-Object Damages in an Oxide/Oxide (N720/Alumina) Ceramic Matrix Composite: Comparison with N720/Aluminosilicate," presented at MS&T2009 Conference, October 25-29, Pittsburgh, PA; to be published in Ceramic Transactions (2010).

10. S. R. Choi, D. C. Faucett, and D. J. Alexander, "Foreign Object Damage in an N720/alumina Oxide/Oxide Ceramic Matrix Composite," presented at Daytona Beach Conference, January 24-28, 2010, Daytona Beach, FL; to be published in Ceram. Eng. Sci. Proc. (2010).

11. Ruggles-Wrenn, S. Mall, C. A. Eber, and L. B. Harlan, "Effects of Steam Environment on High Temperature Mechanical Behavior of Nextel720/Alumina (N720/A) Continuous Fiber Ceramic Composite," Composites A, 37[11] 2029-2040 (2006).

12. R.A. Simon, "Progress in Processing and Performance of Porous-Matrix Oxide/Oxide Composites," Int. J. Appl. Ceram. Technology, 2[2] 141-149 (2005).

13. F. Zok, "Developments in Oxide Fiber Composites," J. Am. Ceram. Soc., 89[11] 3309-3324 (2006).

14. M. A. Mattoni et al., "Effects of Combuster Rig Exposure on a Porous-Matrix Oxide Composite, Int. J. Appl. Ceram. Technology, 2[2] 133-140 (2005).

15. ASTM C 1259, "Test Method for Dynamic Young's Modulus, Shear Modulus, and Poisson's Ratio for Advanced Ceramics by Impulse Excitation of Vibration," Annual Book of ASTM Standards, Vol. 15.01, ASTM, West Conshohocken, PA (2009).

16. R. Mouginot and D. Maugis, Fracture Indentation beneath Flat and Spherical Punches, J. Mater. Sci., 20, 4354-4376 (1985).

17. Y. Akimune, Y. Katano, and K. Matoba, Spherical-Impact Damage and Strength Degradation in Silicon Nitrides for Automobile Turbocharger Rotors, J. Am. Ceram. Soc., 72[8], 1422-1428 (1989).

18. A. G. Evans, and T. R. Wilshaw, Dynamic Solid Particle Damage in Brittle Materials: An Appraisal, J. Mater. Sci., 12, 97-116 (1977).

19. A. D. Peralta and H. Yoshida, Ceramic Gas Turbine Component Development and Characterization, van Roode, M, Ferber, M. K., and Richerson, D. W., eds., Vol. 2, pp. 665-692, ASME, New York, NY (2003).

20. S. M. Wiederhorn and B. R. Lawn, Strength Degradation of Glass Resulting from Impact with Spheres, J. Am. Ceram. Soc., 60[9-10], 451-458 (1977).

21. G. Knight, M. V. Swain, and M. M. Chaudhri, Impact of Small Steel Spheres on Glass Surfaces, J. Mater. Sci., 12, 1573-1586 (1977).

EFFECTS OF ENVIRONMENT ON CREEP BEHAVIOR OF NEXTEL™ 720/ALUMINA-MULLITE CERAMIC COMPOSITE WITH ±45° FIBER ORIENTATION AT 1200 °C

M. Ozer and M. B. Ruggles-Wrenn*
Department of Aeronautics and Astronautics
Air Force Institute of Technology
Wright-Patterson Air Force Base, Ohio 45433-7765

ABSTRACT
    The tensile creep behavior of an oxide-oxide continuous fiber ceramic composite with ±45° fiber orientation was investigated at 1200 °C in laboratory air, in steam and in argon. The composite consists of a porous alumina-mullite matrix reinforced with laminated, woven mullite/alumina (Nextel™ 720) fibers, has no interface between the fiber and matrix, and relies on the porous matrix for flaw tolerance. The tensile stress-strain behavior was investigated and the tensile properties measured at 1200 °C. The elastic modulus was 38.9 GPa and the ultimate tensile strength (UTS) was 37 MPa. Tensile creep behavior was examined for creep stresses in the 13-32 MPa range. Primary and secondary creep regimes were observed in all tests. Tertiary creep was observed at 32 MPa in steam and at stress levels $\geq$ 26 MPa in argon environment. Creep run-out (set to 100 h) was achieved in all test environments for creep stress levels $\leq$ 20 MPa. At creep stresses > 20 MPa, creep performance was best in laboratory air and worst in steam. The presence of either steam or argon accelerated creep rates and significantly reduced creep life. Composite microstructure, as well as damage and failure mechanisms were investigated. Matrix degradation appears to be the cause of early failures in argon and in steam.

INTRODUCTION

    Advances in power generation systems for aircraft engines, land-based turbines, rockets, and, most recently, hypersonic missiles and flight vehicles have raised the demand for structural materials that have superior long-term mechanical properties and retained properties under high temperature, high pressure, and varying environmental factors, such as moisture [1]. Typical components include combustors, nozzles and thermal insulation. Ceramic-matrix composites (CMCs), capable of maintaining excellent strength and fracture toughness at high temperatures are prime candidate materials for such applications. Additionally, the lower densities of CMCs and their higher use temperatures, together with a reduced need for cooling air, allow for improved high-temperature performance when compared to conventional nickel-based superalloys [2]. Advanced reusable space launch vehicles will likely incorporate fiber-reinforced CMCs in critical propulsion components. In these applications, CMCs will be subjected to mechanical loading in complex environments. For example, a typical service environment for a reusable rocket engine turbopump rotor includes hydrogen, oxygen and steam, at pressures > 200 atm. Furthermore, concurrent efforts in optimization of the CMCs and in design of the combustion chamber are expected to accelerate the insertion of the CMCs into aerospace turbine engine applications, such as combustor walls [3-5]. Because these applications require exposure

---

* Corresponding author
The views expressed are those of the authors and do not reflect the official policy or position of the United States Air Force, Department of Defense or the U. S. Government.

to oxidizing environments, the thermodynamic stability and oxidation resistance of CMCs are vital issues.

The main advantage of CMCs over monolithic ceramics is their superior toughness, tolerance to the presence of cracks and defects, and non-catastrophic mode of failure. It is widely accepted that in order to avoid brittle fracture behavior in CMCs and improve the damage tolerance, a weak fiber/matrix interface is needed, which serves to deflect matrix cracks and to allow subsequent fiber pullout [6-9]. Historically, following the development of SiC fibers, fiber coatings such as C or BN have been employed to promote the desired composite behavior. However, the non-oxide fiber/non-oxide matrix composites generally show poor oxidation resistance [10, 11], particularly at intermediate temperatures (~800 °C). These systems are susceptible to embrittlement due to oxygen entering through the matrix cracks and then reacting with the interphase and the fibers [12-15]. The degradation, which involves oxidation of fibers and fiber coatings, is typically accelerated by the presence of moisture [16-22]. Using oxide fiber/ non-oxide matrix or non-oxide fiber/oxide matrix composites generally does not substantially improve the high-temperature oxidation resistance [23]. The need for environmentally stable composites motivated the development of CMCs based on environmentally stable oxide constituents [24-32].

More recently it has been demonstrated that similar crack-deflecting behavior can also be achieved by means of a finely distributed porosity in the matrix instead of a separate interface between matrix and fibers [33]. This microstructural design philosophy implicitly accepts the strong fiber/matrix interface. It builds on the experience with porous interlayers as crack deflection paths [34, 35] and extends the concept to utilize a porous matrix as a surrogate. The concept has been successfully demonstrated for oxide-oxide composites [24, 28, 32, 36-40]. Resulting oxide/oxide CMCs exhibit damage tolerance combined with inherent oxidation resistance. However, due to the strong bonding between the fiber and matrix, a minimum matrix porosity is needed for this concept to work [41]. An extensive review of the mechanisms and mechanical properties of porous-matrix CMCs is given in [1, 42].

In many potential applications oxide-oxide CMCs will be subject to multiaxial states of stress. The woven CMC materials developed for use in aerospace engine components are typically made from 0°/90° fiber architectures. However, the highest loads in structural components are not always applied along the direction of the reinforcing fibers. As a result, the components could experience stresses approaching the off-axis tensile and creep strengths. The objective of this effort is to investigate the off-axis tensile and creep behaviors of an oxide-oxide CMC consisting of a porous alumina-mullite matrix reinforced with the Nextel™ 720 fibers. A previous study examined high temperature mechanical behavior of this composite in the 0°/90° fiber orientation [43]. This study investigates tensile and creep behavior of the Nextel™ 720/alumina-mullite (N720/AM) composite in the ±45° orientation at 1200 °C in oxidizing (air and steam) and inert (argon) environments. Creep tests were conducted in air, steam and argon environments at stress levels ranging from 13 to 32 MPa. Results reveal that test environment has a noticeable effect on creep life. The composite microstructure, as well as damage and failure mechanisms are discussed.

MATERIAL AND EXPERIMENTAL ARRANGEMENTS

The material studied was Nextel™720/alumina-mullite (N720/AM), an oxide-oxide ceramic composite composed of Nextel™720 fibers and a porous matrix, which consists of mullite and alumina particles in a sol-gel derived alumina. There is approximately 12.5% (by volume) of mullite in the matrix composition. The composite, manufactured by COI Ceramics (San Diego, CA), was supplied in a form of 3.2 mm thick plate, comprised of 12 0°/90° woven layers, with a density of ~2.57 g/cm³ and a fiber volume of approximately 40.0%. Composite porosity was ~28.3%. The laminate was fabricated following the procedure described elsewhere [44]. No coating was applied to the fibers. The damage tolerance of the N720/AM composite is enabled by a porous matrix. The overall microstructure of the untested material is presented in Fig. 1, which shows 0° and 90° fiber tows as well as numerous matrix cracks. In the case of the as-processed material, most are shrinkage cracks formed during processing rather than matrix cracks generated during loading.

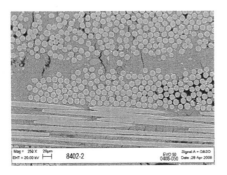

Fig. 1. Typical microstructure of the N720/AM ceramic composite.
Micrograph courtesy of A. Szweda, COI Ceramics Inc.

A servocontrolled MTS mechanical testing machine equipped with hydraulic water-cooled wedge grips, a compact two-zone resistance-heated furnace, and two temperature controllers was used in all tests. An MTS TestStar II digital controller was employed for input signal generation and data acquisition. Strain measurement was accomplished with an MTS high-temperature air-cooled uniaxial extensometer of 12.5-mm gage length. For elevated temperature testing, thermocouples were bonded to the specimens using alumina cement (Zircar) to calibrate the furnace on a periodic basis. The furnace controller (using a non-contacting thermocouple exposed to the ambient environment near the test specimen) was adjusted to determine the power setting needed to achieve the desired temperature of the test specimen. The determined power setting was then used in actual tests. Thermocouples were not bonded to the test specimens after the furnace was calibrated. Tests in steam environment employed an alumina susceptor (tube with end caps), which fits inside the furnace. The specimen gage section is located inside the susceptor, with the ends of the specimen passing through slots in the susceptor. Steam is introduced into the susceptor (through a feeding tube) in a continuous stream with a slightly positive pressure, expelling the dry air and creating a near 100% steam environment inside the susceptor. An alumina susceptor was also used in tests conducted in argon environment. In this case ultra high purity argon gas (99.999% pure) was introduced into the susceptor from a high-pressure cylinder creating an inert gas environment around the test section of the specimen. The power setting for testing in steam (argon) was determined by placing the specimen instrumented with thermocouples in steam (argon) environment and repeating the furnace calibration procedure. Fracture surfaces of failed specimens were examined using SEM (FEI Quanta 200 HV) as well as an optical microscope (Zeiss Discovery V12). The SEM specimens were carbon coated.

All tests were performed at 1200 °C. In all tests, a specimen was heated to test temperature in 25 min, and held at temperature for additional 15 min prior to testing. Dog bone shaped specimens of 152 mm total length with a 10-mm-wide gage section were used in all tests. Tensile tests were performed in stroke control with a constant displacement rate of 0.05 mm/s in laboratory air. Creep-rupture tests were conducted in load control in accordance with the procedure in ASTM standard C 1337 in laboratory air, steam and argon environments. In all creep tests the specimens were loaded to the creep stress level at the stress rate of 15 MPa/s. Creep run-out was defined as 100 h at a given creep stress. In each test, stress-strain data were recorded during the loading to the creep stress level and the actual creep period. Thus both total strain and creep strain could be calculated and examined. To determine the retained tensile strength and modulus, specimens that achieved run-out were subjected to a tensile test to failure at 1200 °C. It is worthy of note that in all tests reported below, the failure occurred within the gage section of the extensometer.

RESULTS AND DISCUSSION
Monotonic Tension

Tensile stress-strain behavior at 1200 °C is typified in Fig. 2. The stress-strain curves obtained for the 0°/90° fiber orientation are nearly linear to failure. Material exhibits typical fiber-dominated composite behavior. The average ultimate tensile strength (UTS) was 153 MPa, elastic modulus, 74.5 GPa, and failure strain, 0.34 % [43]. In the case of the ±45° orientation, the nonlinear stress-strain behavior sets in at fairly low stresses (~15 MPa). As the stress exceeds 32 MPa, appreciable inelastic strains develop rapidly. The specimen achieves a strain of 0.3% at the maximum load. Once the UTS is reached, the softening commences. These observations are consistent with the results reported earlier for the porous-matrix ceramic composites [45, 46]. The elastic modulus (38.9 GPa) and UTS (37 MPa) obtained for the ±45° orientation are considerably lower than the corresponding values for the 0°/90° specimens.

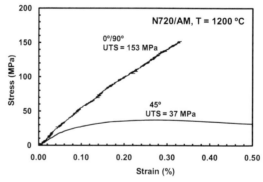

Fig. 2. Tensile stress-strain curves for N720/AM ceramic composite at 1200 °C. Data for the 0°/90° fiber orientation from Ruggles-Wrenn and Genelin [43] is also shown.

Creep-Rupture

Results of the creep-rupture tests for N720/AM composite with ±45° fiber orientation are summarized in Table I, where creep strain accumulation and rupture time are shown for each creep stress level and test environment. Creep curves obtained in air, steam and argon are shown in Figs. 3, 4 and 5, respectively. The time scale in Figs. 4(b) and 5(b) is reduced to clearly show the creep curves produced at 30 and 32 MPa.

Table I. Summary of creep-rupture results for the N720/AM ceramic composite with ±45° fiber orientation at 1200 °C in laboratory air, steam and argon environments.

Environment	Creep Stress (MPa)	Creep Strain (%)	Time to Rupture (s)
Air	13	0.81	360,000*
Air	30	4.23	360,000*
Air	32	4.43	360,000*
Steam	13	4.40	360,000*
Steam	20	12.48	360,000*
Steam	26	18.25	357,840
Steam	30	7.78	3,888
Steam	32	0.80	72
Argon	20	16.72	360000*
Argon	26	19.25	273,600
Argon	30	12.97	12,312
Argon	32	2.44	288

* Run-out

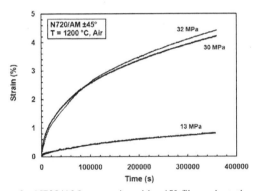

Fig. 3. Creep curves for N720/AM composite with ±45° fiber orientation at 1200 °C in air.

Creep curves produced in all tests conducted in air exhibit primary and secondary creep regimes, but no tertiary creep. Transition from primary to secondary creep occurs late in creep life, primary creep persists during the first 50-60 h of the creep test. In contrast, the creep curves obtained in steam and in argon show primary, secondary and tertiary creep. Tertiary creep is observed at 32 MPa in steam and at stresses > 20 MPa in argon. At 32 MPa in steam, transition from primary to secondary creep occurs after about 30 s of creep, and secondary creep transitions to tertiary creep during the last third of the creep life. At 26 MPa and at 30 MPa in argon, primary creep persists during the first half of the creep life, and secondary creep transitions to tertiary creep at the very end of the creep life. At 32 MPa in argon, transition from primary to secondary creep occurs almost immediately, and secondary creep transitions to tertiary creep during the second half of the creep life. Note that creep run-out of 100 h was achieved in all tests performed in air. In steam, creep run-out was achieved at stresses ≤ 26 MPa. In argon, creep run-out was achieved only at 20 MPa. While the test environment appears to have little influence on the appearance of the creep curves obtained at stresses ≤ 20 MPa, it has a noticeable effect on the strain accumulated during 100 h of creep. For a given creep stress, the largest creep strains were accumulated in argon, followed by those accumulated in steam and in air.

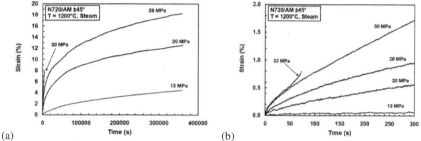

(a)   (b)

Fig. 4. Creep curves for N720/AM composite with ±45° fiber orientation at 1200 °C in steam: (a) time scale chosen to show creep strains accumulated at stresses ≤ 30 MPa and (b) time scale reduced to show the creep curve obtained at 32 MPa.

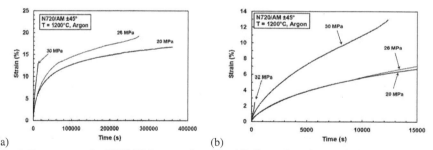

(a)   (b)

Fig. 5. Creep curves for N720/AM composite with ±45° fiber orientation at 1200 °C in argon: (a) time scale chosen to show creep strains accumulated at stresses ≤ 26 MPa and (b) time scale reduced to show the creep curve obtained at 30 MPa and at 32 MPa.

Minimum creep rate was reached in all tests. Creep rate as a function of applied stress is presented in Fig. 6, where results for N720/AM composite with 0°/90° fiber orientation from prior work [43] are included for comparison. It is seen that in air the secondary creep rate of the ±45° orientation can be nearly 10^3 times that of the 0°/90° orientation. In steam and in argon, the

±45° creep rate can be as high as 10^4 times the corresponding 0°/90° rate. This result is hardly surprising, considering that the creep rupture of the 0°/90° orientation is likely dominated by creep rupture of the Nextel 720 fibers. It is recognized that Nextel™720 fiber has the best creep performance of any commercially available polycrystalline oxide fiber. The superior high-temperature creep performance of the Nextel™720 fibers results from the high content of mullite, which has a much better creep resistance than alumina [47]. Conversely, the creep rupture of the ±45° orientation is largely dominated by an exceptionally weak porous alumina matrix.

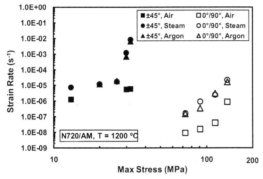

Fig. 6. Minimum creep rate as a function of applied stress for N720/AM ceramic composite with ±45° fiber orientation at 1200°C in laboratory air, steam and argon. Data for the 0°/90° fiber orientation from Ruggles-Wrenn and Genelin [43] are also shown.

For both fiber orientations, the minimum creep rates increase with increasing applied stress. In the case of the 0°/90° orientation, the secondary creep rate increases by two orders of magnitude as the creep stress increases from 73 to 136 MPa. For a given creep stress, creep rate in steam is approximately two orders of magnitude higher than that in air. Notably, creep rates obtained in argon are close to those produced in steam for a given applied stress. In the case of the ±45° fiber orientation, for stresses ≤ 26 MPa creep rate is relatively unaffected by environment. Creep rates obtained in all tests at stresses ≤ 26 MPa are less than 2 x 10^{-5} s^{-1}. As the creep stress increases to 32 MPa, the creep rate in air remains below 10^{-5} s^{-1}. However, creep rates in steam and in argon increase by ~ 3 orders of magnitude. Note that the creep rates obtained in steam remain close to those obtained in argon. The creep rates produced in steam and in argon are three orders of magnitude higher than the rates obtained in air at 30 MPa and at 32 MPa.

Stress-rupture behavior is summarized in Fig. 7, where results for N720/AM composite with 0°/90° fiber orientation from prior work [43] are included for comparison. As expected, creep life decreases with increasing applied stress for both fiber orientations. In the case of the 0°/90° orientation, the presence of steam dramatically reduced creep lifetimes. The reduction in creep life due to steam was ≥ 95% for applied stress levels ≥ 91 MPa, and 63% for the applied stress of 73 MPa. Because the creep performance of the 0°/90° orientation is dominated by the fibers, fiber degradation is a likely source of the composite degradation. Recent studies [48, 49] suggest that the loss of mullite from the fiber may be the mechanism behind the degraded creep performance in steam. Alternatively, poor creep resistance in steam may be due to an environmentally-assisted crack growth (or stress-corrosion) mechanism [43]. In this case, crack growth in the fiber is caused by a chemical interaction of water molecules with mechanically strained Si-O bonds at the crack tip with the rate of chemical reaction increasing exponentially with applied stress [50-58]. In the case of the ±45° orientation, environment has little effect on the creep lifetimes (up to 100 h) for applied stresses ≤ 20 MPa. For stresses ≥ 30 MPa, creep lifetimes can be reduced by nearly two orders of magnitude in the presence of steam and as much as an order of magnitude in the presence of argon.

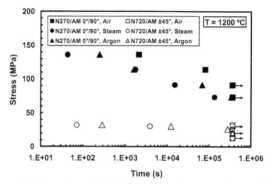

Fig. 7. Creep stress vs time to rupture for N720/AM ceramic composites at 1200°C in laboratory air, steam and argon. Data for 0°/90° fiber orientation from Ruggles-Wrenn and Genelin [43].

Retained strength and modulus of the specimens that achieved a run-out are summarized in Table II. Tensile stress-strain curves obtained for the specimens subjected to prior creep are presented in Fig. 8 together with the tensile stress-strain curve for the as-processed material. Prior creep in air, steam or argon appears to have increased tensile strength and stiffness. Furthermore, prior creep considerably reduced the composite's capacity for inelastic straining. The pre-crept specimens produced higher proportional limits and much lower failure strains than the as-processed material. This indicates that additional matrix sintering may be taking place which causes strengthening of the matrix.

Table II. Retained properties of the N720/AM specimens with ±45° fiber orientation subjected to prior creep at 1200 °C in laboratory air, steam and argon environments.

Environment	Creep Stress (MPa)	Retained Strength (MPa)	Retained Modulus (GPa)	Strain at Failure (%)
Air	13	44.3	44.6	0.14
Air	30	48.1	38.9	0.18
Air	32	44.1	38.8	0.16
Steam	13	48.8	40.1	0.15
Steam	20	54.0	40.0	0.17
Argon	20	54.5	39.6	0.24

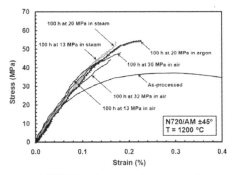

Fig. 8. Effects of prior creep at 1200°C in laboratory air, steam and argon environments on tensile stress-strain behavior of N720/AM ceramic composite with ±45° fiber orientation.

Composite Microstructure

Figures 9 and 10 show the fracture surfaces produced in tensile tests on specimens with ±45° fiber orientation subjected to prior creep at 1200 °C in air. It is seen that the fracture occurred along the plane at ~45° to the loading direction. The failure occurs primarily through the matrix, with only minimal fiber damage. A diffuse localized deformation band can be seen in all specimens in Fig. 9. Once the deformation is localized, considerable additional straining occurs within the localized band prior to specimen failure. Deformation within this band is accommodated by extensive fragmentation of the matrix and rotation of the fiber tows towards the loading direction. Pronounced through-thickness swelling is also observed. Note that the SEM micrographs in Fig. 10 show a predominantly fibrous fracture surface and reveal only small amounts of matrix particles bonded to the fiber surfaces.

Fig. 9. Fracture surfaces obtained in tensile tests on specimens subjected to prior creep at 1200 °C in air: (a) at 13 MPa, $t_f > 100$ h; (b) at 30 MPa, $t_f > 100$ h and (c) at 32 MPa, $t_f > 100$ h.

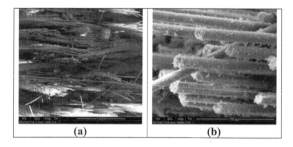

Fig. 10. SEM micrographs of the fracture surfaces obtained in tensile tests on specimens subjected to prior creep at 1200 °C in air: (a) at 13 MPa, t_f > 100 h; (b) at 30 MPa, t_f > 100 h and (c) at 32 MPa, t_f > 100 h.

Fracture surfaces of the N720/AM specimens with ±45° fiber orientation tested at 1200 °C in argon are shown in Figs. 11 and 12. The fracture surfaces in Fig. 11 are very similar to those produced in air (Fig. 9).

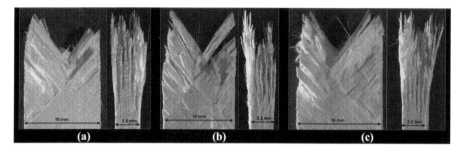

Fig. 11. Fracture surfaces obtained in tests conducted at 1200 °C in argon: (a) at 20 MPa, t_f > 100 h; (b) at 30 MPa, t_f = 3.42 h and (c) at 32 MPa, t_f = 288 s.

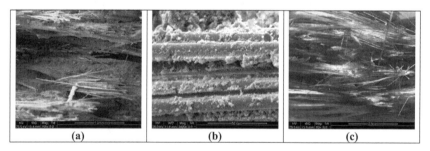

Fig. 12. SEM micrographs of the fracture surfaces obtained in tests conducted at 1200°C in argon: (a) at 20 MPa, t_f > 100 h; (b) at 20 MPa, t_f > 100 h and (c) at 32 MPa, t_f = 288 s.

Likewise, the fracture surface in Fig. 12(c) obtained after 288 s of creep at 32 MPa in argon is fibrous and similar to that obtained at 30 MPa in air (Fig. 10(a)). However, the fracture surface in Fig. 12(a), produced in tensile test on a specimen subjected to 100 h of prior creep at 20 MPa in argon, shows regions of fiber/matrix bonding as well as fibrous regions. Furthermore, a somewhat greater amount of matrix remains bonded to the fibers as seen in Fig. 12(b). Apparently the longer duration of the 20 MPa creep test in argon had a noticeable effect on the fracture surface appearance. It is likely that additional fiber/matrix bonding occurred during the 100 h of creep at 20 MPa in argon.

Fracture surfaces produced in steam are presented in Figs. 13 and 14. In air the effects of test duration on failure mechanisms were negligible. In argon, the test duration had a noticeable effect on the fracture surface appearance. In steam, the effects of test duration (i. e. exposure to steam environment under load) on the fracture surface appearance and on failure mechanisms are amplified considerably. Specimen subjected to 100 h of prior creep in steam and failed in a subsequent tensile test (Fig. 13(a)) fractured along the plane nearly orthogonal to the loading direction. Fracture surface in Fig. 13(a) suggests that the specimen pre-crept in steam failed catastrophically at the maximum load, with the majority of the fibers breaking in the process. Such failure process is typical for dense-matrix CMCs. As seen in Fig. 14 (a), the fracture surface is dominated by planar regions of coordinated fiber failure.

Fig. 13. Fracture surfaces obtained in tests conducted at 1200°C in steam: (a) at 20 MPa, $t_f > 100$ h; and (b) at 32 MPa, $t_f = 72$ s.

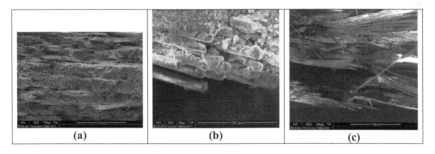

Fig. 14. SEM micrographs of the fracture surfaces obtained in tests conducted at 1200°C in steam: (a) at 20 MPa, $t_f > 100$ h; (b) at 20 MPa, $t_f > 100$ h and (c) at 32 MPa, $t_f = 72$ s.

Fig. 14(b) shows large amounts of matrix material bonded to the fiber surfaces. By contrast, the specimen that failed after only 72 s at 32 MPa in steam (Fig. 13(b)) exhibits the V-shaped fracture typically seen in the porous matrix composites with ±45° orientation. The SEM micrograph in Fig. 14(b) shows a predominantly fibrous fracture surface. In this case the failure occurred primarily through matrix damage and interplay delamination with little fiber fracture.

Recent studies [59-61] investigated effects of thermal aging on the physical and mechanical properties of composites consisting of Nextel™ 720 fibers and a porous matrix of mullite and alumina. For a composite with a pure alumina matrix, a porosity reduction of ~6% was observed after a 10-min exposure at 1200 °C [60, 61]. For a composite with a mullite/alumina matrix, strengthening of the matrix and the fiber-matrix interfaces was observed following aging at 1200 °C [59]. Additional sintering of the matrix during the aging treatments was considered to be associated predominantly with Al_2O_3. In the present study, it is likely that additional sintering of the matrix occurred during 100-h creep tests, especially those conducted in steam and in argon. The resultant strengthening of the matrix is also manifested in the retained properties of the composite. Results in Table II show that after prior creep, the modulus and the tensile strength increase, and the inelastic straining capabilities of the composite decrease. The increases in modulus and tensile strength are particularly pronounced in the case of specimens pre-crept in steam and argon. The matrix strengthening is also manifested in the change in the failure mechanism. The failure of the composite pre-crept in steam is dominated by fiber fracture, while the as-processed material fails predominantly through matrix damage and interplay delamination. The specimens pre-crept in air also exhibit an increase in the modulus and strength after 100 h of creep. However, in this case the fracture surface does not suggest a change in the failure mechanism. It is possible that the sintering of the matrix is accelerated in the presence of steam.

CONCLUDING REMARKS

The tensile stress-strain behavior of the N720/AM composite with ±45° fiber orientation was investigated and the tensile properties measured at 1200 °C. The stress-strain behavior departs from linearity at a low stress of ~15 MPa. Once the UTS = 37 MPa is reached, the softening commences. Considerable inelastic strains develop at stresses $\approx$ 32 MPa.

The creep-rupture behavior of the N720/AM composite with ±45° fiber orientation was characterized for stress levels ranging from 13 to 32 MPa at 1200°C in laboratory air, steam and argon environments. In air, the material exhibits primary and secondary creep regimes, but no tertiary creep. In steam, tertiary creep was observed at 32 MPa. In argon, tertiary creep was observed at stress levels $\geq$ 26 MPa. For a given applied stress in the 13 to 30 MPa range, creep strain accumulation is highest in argon, followed by that in steam and in air. Creep strain rates range from 1.2 x 10^{-6} to 5.9 x 10^{-6} s^{-1} in air, from 7.4 x 10^{-6} to 8.4 x 10^{-3} s^{-1} in steam, and from 1.2 x 10^{-5} to 6.7 x 10^{-3} s^{-1} in argon. For creep stress levels $\leq$ 26 MPa creep rates are less than 2 x 10^{-5} s^{-1}. At 30 MPa, creep rates in steam and argon environments increase by nearly two orders of magnitude, while creep rates in air remain below 10^{-5} s^{-1}. At 32 MPa, creep rates produced in steam and in argon are about three orders of magnitude higher than the rates produced in air.

Creep run-out (set to 100 h) was achieved in all test environments for creep stress levels $\leq$ 20 MPa. At creep stresses > 20 MPa, creep performance was best in laboratory air and worst in steam. The run-out specimens exhibited an increase in stiffness and in strength. Prior creep significantly diminished composite's capability for inelastic straining. For applied stresses > 26

MPa, the presence of argon and especially the presence of steam drastically reduced creep lifetimes.

In all tests conducted in air, the failure occurs primarily through matrix damage and interplay delamination, with minimal fiber fracture. The same failure mechanisms are prevalent in tests of short duration (< 1 h) conducted in argon and in steam. Conversely, for test durations > 100 h in steam, the failure mechanism is dominated by fiber fracture. It is possible that the matrix undergoes additional sintering during the long-term tests. Additional sintering and consequently strengthening of the matrix may be behind the change in failure mechanisms.

REFERENCES

[1]F. W. Zok, "Developments in Oxide Fiber Composites," *J. Am. Ceram. Soc.*, **89**(11), 3309-3324 (2006).

[2]L. P. Zawada, J. Staehler, S. Steel, "Consequence of Intermittent Exposure to Moisture and Salt Fog on the High-Temperature Fatigue Durability of Several Ceramic-Matrix Composites," *J. Am. Ceram. Soc.*, **86**(8), 1282-1291 (2003).

[3]M. Parlier, M. H. Ritti, "State of the Art and Perspectives for Oxide/Oxide Composites," *Aerospace Sci. Technol.*, **7**, 211-221 (2003).

[4]M. A. Mattoni, J. Y. Yang, C. G. Levi, F. W. Zok, L. P. Zawada, "Effects of Combustor Rig Exposure on a Porous-Matrix Oxide Composite," *Int. J. Applied Cer Technol.*, **2**(2), 133-140 (2005).

[5]T. A. Parthasarathy, L. P. Zawada, R. John, M. K. Cinibulk, J. Zelina, "Evaluation of Oxide-Oxide Composites in a Novel Combustor Wall Application," *Int. J. Applied. Cer. Technol.*, **2**(2), 122-132 (2005).

[6]R. J. Kerans, R. S. Hay, N. J. Pagano, T. A. Parthasarathy, "The Role of the Fiber-Matrix Interface in Ceramic Composites," *Am. Ceram. Soc. Bull.*, **68**(2), 429-442 (1989).

[7]A. G. Evans, F. W. Zok, "Review: the Physics and Mechanics of Fiber-Reinforced Brittle Matrix Composites," *J. Mater. Sci.*, **29**:3857-3896 (1994).

[8]R. J. Kerans, T. A. Parthasarathy, "Crack Deflection in Ceramic Composites and Fiber Coating Design Criteria," *Composites: Part A*, **30**, 521-524 (1999).

[9]R. J. Kerans, R. S. Hay, T. A. Parthasarathy, M. K. Cinibulk, "Interface Design for Oxidation-Resistant Ceramic Composites," *J. Am. Ceram. Soc.*, **85**(11), 2599- 2632 (2002).

[10]K. M. Prewo, J. A. Batt. The Oxidative Stability of Carbon Fibre Reinforced Glass-Matrix Composites," *J. Mater. Sci.*, **23**, 523-527 (1988).

[11]T. Mah, N. L. Hecht, D. E. McCullum, J. R. Hoenigman, H. M. Kim, A. P. Katz, H. A. Lipsitt, "Thermal Stability of SiC Fibres (Nicalon)," *J. Mater. Sci.*, **19**, 1191-1201 (1984).

[12]J. J. Brennan. *Fiber Reinforced Ceramic Composites; Ch. 8.* Masdayazni KC ed. Noyes, New York, 1990.

[13]F. Heredia, J. McNulty, F. Zok, A. G. Evans, "Oxidation Embrittlement Probe for Ceramic Matrix Composites," *J. Am. Ceram. Soc.*, **78**(8), 2097-2100 (1995).

[14]R. S. Nutt, "Environmental Effects on High-Temperature Mechanical Behavior of Ceramic Matrix Composites," *High-Temperature Mechanical Behavior of Ceramic Composites*. S. V. Nair, and K. Jakus, editors. Butterworth-Heineman, Boston, MA, (1995).

[15]A. G. Evans, F. W. Zok, R. M. McMeeking, Z. Z. Du, "Models of High-Temperature Environmentally-Assisted Embrittlement in Ceramic Matrix Composites," *J. Am. Ceram. Soc.*, **79**, 2345-52 (1996).

[16]K. L. More, P. F. Tortorelli, M. K. Ferber, J. R. Keiser, "Observations of Accelerated Silicon Carbide Recession by Oxidation at High Water-Vapor Pressures," *J. Am. Ceram. Soc.*, **83**(1), 11-213 (2000).

[17]K. L. More, P. F. Tortorelli, M. K. Ferber, L. R. Walker, J. R. Keiser, W. D. Brentnall, N. Miralya, J. B. Price, "Exposure of Ceramic and Ceramic-Matrix Composites in Simulated and Actual Combustor Environments," *Proceedings of International Gas Turbine and Aerospace Congress*, Paper No. 99-GT-292 (1999).

[18]M. K. Ferber, H. T. Lin, J. R. Keiser, "Oxidation Behavior of Non-Oxide Ceramics in a High-Pressure, High-Temperature Steam Environment," *Mechanical, Thermal, and Environmental Testing and Performance of Ceramic Composites and Components*. M. G. Jenkins, E. Lara-Curzio, and S. T. Gonczy, editors. ASTM STP 1392, 210-215 (2000).

[19]J. A. Haynes, M. J. Lance, K. M. Cooley, M. K. Ferber, R. A. Lowden, D. P. Stinton, "CVD Mullite Coatings in High-Temperature, High-Pressure Air-H_2O," *J. Am. Ceram. Soc.*, **83**(3), 657-659 (2000).

[20]E. J. Opila, R. E. Hann Jr., "Paralinear Oxidation of SiC in Water Vapor," *J. Am. Ceram. Soc.*, **80**(1), 197-205 (1997).

[21]E. J. Opila, "Oxidation Kinetics of Chemically Vapor Deposited Silicon Carbide in Wet Oxygen," *J. Am. Ceram. Soc.*, **77**(3), 730-736 (1994).

[22]E. J. Opila, "Variation of the Oxidation Rate of Silicon Carbide with Water Vapor Pressure," *J. Am. Ceram. Soc.*, **82**(3), 625-636 (1999).

[23]E. E. Hermes, R. J. Kerans, "Degradation of Non-Oxide Reinforcement and Oxide Matrix Composites," *Mat. Res. Soc., Symposium Proceedings*, **125**, 73-78 (1988).

[24]A. Szweda, M. L. Millard, M. G. Harrison, *Fiber-Reinforced Ceramic-Matrix Composite Member and Method for Making*, U. S. Pat. No. 5 601 674, (1997).

[25]S. M. Sim, R. J. Kerans, "Slurry Infiltration and 3-D Woven Composites," *Ceram. Eng. Sci. Proc.*, **13**(9-10), 632-641 (1992).

[26]E. H. Moore, T. Mah, and K. A. Keller, "3D Composite Fabrication Through Matrix Slurry Pressure Infiltration," *Ceram. Eng. Sci. Proc.*, **15**(4), 113-120 (1994).

[27]M. H. Lewis, M. G. Cain, P. Doleman, A. G. Razzell, J. Gent, "Development of Interfaces in Oxide and Silicate Matrix Composites," *High-Temperature Ceramic–Matrix Composites II: Manufacturing and Materials Development*, A. G. Evans, and R. G. Naslain, editors, American Ceramic Society, 41–52 (1995).

[28]F. F. Lange, W. C. Tu, A. G. Evans, "Processing of Damage-Tolerant, Oxidation-Resistant Ceramic Matrix Composites by a Precursor Infiltration and Pyrolysis Method," *Mater. Sci. Eng. A*, **A195**, 145–150 (1995).

[29]R. Lunderberg, L. Eckerbom, "Design and Processing of All-Oxide Composites," *High-Temperature Ceramic–Matrix Composites II: Manufacturing and Materials Development*, A. G. Evans, and R. G. Naslain, editors, American Ceramic Society, 95–104 (1995).

[30]E. Mouchon, P. Colomban, "Oxide Ceramic Matrix/Oxide Fiber Woven Fabric Composites Exhibiting Dissipative Fracture Behavior," *Composites*, **26**, 175–182 (1995).

[31]P. E. D. Morgan and D. B. Marshall, "Ceramic Composites of Monazite and Alumina," *J. Am. Ceram. Soc.*, **78**(6), 1553–1563 (1995).

[32]W. C. Tu, F. F. Lange, A. G. Evans, "Concept for a Damage-Tolerant Ceramic Composite with Strong Interfaces," *J. Am. Ceram. Soc.*, **79**(2), 417–424 (1996).

[33]C. G. Levi, J. Y. Yang, B. J. Dalgleish, F. W. Zok, A. G. Evans, "Processing and Performance of an All-Oxide Ceramic Composite," *J. Am. Ceram. Soc.*, **81**, 2077-2086 (1998).

[34]J. B. Davis, J. P. A. Lofvander, A. G. Evans, "Fiber Coating Concepts for Brittle Matrix Composites," *J. Am. Ceram. Soc.*, **76**(5), 1249–57 (1993).

[35]T. J. Mackin, J. Y. Yang, C. G. Levi, A. G. Evans, "Environmentally Compatible Double Coating Concepts for Sapphire Fiber Reinforced γ-TiAl," *Mater. Sci. Eng.*, **A161**, 285–93 (1993).

[36]A. G. Hegedus, *Ceramic Bodies of Controlled Porosity and Process for Making Same*, U. S. Pat. No. 5 0177 522, May 21, (1991).

[37]T. J. Dunyak, D. R. Chang, M. L. Millard, "Thermal Aging Effects on Oxide/Oxide Ceramic-Matrix Composites," *Proceedings of 17th Conference on Metal Matrix, Carbon, and Ceramic Matrix Composites. NASA Conference Publication 3235, Part 2*, 675-90 (1993).

[38]L. P. Zawada, S. S. Lee, "Mechanical Behavior of CMCs for Flaps and Seals," *ARPA Ceramic Technology Insertion Program (DARPA), W. S. Coblenz WS, editor*. Annapolis MD, 267-322 (1994).

[39]L. P. Zawada, S. S. Lee, "Evaluation of the Fatigue Performance of Five CMCs for Aerospace Applications," *Proceedings of the Sixth International Fatigue Congress*, 1669-1674 (1996).

[40]T. J. Lu, "Crack Branching in All-Oxide Ceramic Composites," *J. Am. Ceram. Soc.*, **79**(1), 266-274 (1996).

[41]M. A. Mattoni, J. Y. Yang, C. G. Levi, F. W. Zok, "Effects of Matrix Porosity on the Mechanical Properties of a Porous Matrix, All-Oxide Ceramic Composite", *J. Am. Ceram. Soc.*, **84**(11), 2594-2602 (2003).

[42]F. W. Zok, C. G. Levi, "Mechanical Properties of Porous-Matrix Ceramic Composites," *Adv. Eng. Mater.*, **3**(1-2), 15-23 (2001).

[43]M. B. Ruggles-Wrenn, C. L. Genelin, "Creep of Nextel™720/Alumina-Mullite Ceramic Composite at 1200 °C in Air, Argon, and Steam," *Comp. Sci. Tech.*, **69**, 863-69 (2009).

[44]R. A. Jurf, S. C. Butner, "Advances in Oxide-Oxide CMC," *Eng Gas Turbines Power*, **122**(2), 202-5 (1999).

[45]L. P. Zawada, R. S. Hay, S. S. Lee, J. Staehler, "Characterization and High-Temperature Mechanical Behavior of an Oxide/Oxide Composite," *J. Am. Ceram. Soc.*, **86**(6), 981-90 (2003).

[46]J. A. Heathcote, X. Y. Gong, J. Y. Yang, U. Ramamurty, F. W. Zok, "In-Plane Mechanical Properties of an All-Oxide Ceramic Composite", *J. Am. Ceram. Soc.*, **82**(10), 2721-30 (1999).

[47]D. M. Wilson, L. R. Visser, "High Performance Oxide Fibers for Metal and Ceramic Composites," *Composites: Part A*, **32**, 1143-1153 (2001).

[48]S. Wannaparhun, S.Seal, "A Combined Spectroscopic and Thermodynamic Investigation of Nextel-720/Alumina Ceramic Matrix Composite in Air and Water Vapor at 1100°C," *J. Am. Ceram. Soc.*, **86**(9), 1628-30 (2003).

[49]C. X. Campbell, E. V. Carelli, K. L. More, P. Varghese, S. Seal, V. H. Desai, "Effect of High-Temperature Water Vapor Exposure on Nextel 720 in an Alumina-Matrix CMC," *Siemens Westinghouse Power Corporation Technical Document TP-02076*.

[50]R. J. Charles and W. B. Hillig WB, "The Kinetics of glass Failure by Stress Corrosion, *Symposium on Mechanical Strength of Glass and Ways of Improving It*. Florence, Italy. September 25-29 (1961). Union Scientifique Continentale du Verre, Charleroi, Belgium, 511-27 (1962).

[51]R. J. Charles and W. B. Hillig WB, "Surfaces, Stress-Dependent Surface Reactions, and Strength," *High-Strength Materials*. V. F. Zackey, editor. John Wiley & Sons, Inc., New York, 682-705 (1965).

[52]S. M. Wiederhorn, "Influence of Water Vapor on Crack Propagation in Soda-Lime Glass," *J. Am. Ceram. Soc.*, **50**(8), 407-14 (1967).

[53]S. M. Wiederhorn, L. H. Bolz, "Stress Corrosion and Static Fatigue of Glass," *J. Am. Ceram. Soc.*, **53**(10), 543-48 (1970).

[54]S. M. Wiederhorn, "A Chemical Interpretation of Static Fatigue," *J. Am. Ceram. Soc.*, **55**(2):81-85 (1972).

[55]S. M. Wiederhorn, S. W. Freiman, E. R. Fuller, C. J. Simmons, "Effects of Water and Other Dielectrics on Crack Growth," *J. Matl. Sci.*, **17**,3460-78 (1982).

[56]T. A. Michalske, S. W. Freiman, "A Molecular Mechanism for Stress Corrosion in Vitreous Silica," *J. Am. Ceram. Soc.*, **66**(4):284-288 (1983).

[57]T. A. Michalske, B. C. Bunker, "Slow Fracture Model Based on Strained Silicate Structures. J Appl. Phys., **56**(10), 2686-93 (1984).

[58]T. A. Michalske, B. C. Bunker, "A Chemical Kinetics Model for Glass Fracture," *J. Am. Ceram. Soc.*, **76**(10), 2613-18 (1993).

[59]E. A. V. Carelli, H. Fujita, J. Y. Yang, F. W. Zok, "Effects of Thermal Aging on the Mechanical Properties of a Porous-Matrix Ceramic Composite," *J. Am. Ceram. Soc.*, **85**(3), 595-602 (2002).

[60]Fujita H, Jefferson G, McMeeking RM, Zok FW. Mullite/Alumina Mixtures for Use as Porous Matrices in Oxide Fiber Composites," *J. Am. Ceram. Soc.*, **87**(2), 261-67 (2004).

[61]H. Fujita, C. G. Levi, F. W. Zok, G. Jefferson, "Controlling Mechanical Properties of Porous Mullite/Alumina Mixtures via Precursor-Derived Alumina," *J. Am. Ceram. Soc.*, **88**(2), 367-75 (2005).

FATIGUE BEHAVIOR OF AN OXIDE/OXIDE CMC UNDER COMBUSTION ENVIRONMENT

Shankar Mall[1] and Andrew R. Nye[1,2]

[1]Department of Aeronautics and Astronautics
Air Force Institute of Technology
[2]Materials and Manufacturing Directorate
Air Force Research Laboratory
Wright-Patterson AFB, OH 45433, USA

ABSTRACT

An oxide-oxide CMC with no interface, Nextel720[TM]/alumina, was investigated under tension-tension fatigue condition using the capabilities provided by a unique burner rig facility which simulated both the load and combustion conditions of hot-section components of gas turbine engines such as turbine blades and vanes. A set of fatigue tests performed using the burner rig (stress ratio, R = 0.05 and frequency = 1 Hz) provided S-N data and damage details, which were analyzed for the role and effects of oxidation on the failure and damage mechanisms. These test results were then compared with those obtained from fatigue tests (R = 0.05 and 1 Hz) in a standard furnace under laboratory air environement. The fatigue strength for 90,000 cycles, i.e. 25 hours was equal to 150 MPa, i.e. 90% of the ultimate tensile strength of the material under the combustion environment condition. Residual tensile strength of fatigue tested specimen at room temperature was almost equal to that of the virgin material which showed that there was no degradation in the test material from the combined fatigue and combustion exposure. Normalized fatigue strength, i.e. ratio of the fatigue strength and ultimate tensile strength at a certain number of cyles to failure, as well as damage mechanisms were almost similar in combustion and laboratory air environements. This suggested that the combustion environment condition was not damaging to the Nextel720[TM]/alumina.

INTRODUCTION

Several types of continuous fibers reinforced ceramic matrix composite (CMC) systems have been developed for their potential use in aircrafts/spacecrafts engines as well as for other high temperature applications over the last two decades or more. Oxidation is a major problem for the materials operating in several high temperature environments, such as found in gas turbine engines [1]. There are two main approaches for protection against oxidation in CMCs; the first one is to use inhibitors to slow the carbon-oxygen reaction rate, and the second one is to use barrier coatings to inhibit the ability of oxygen to reach the carbon [2]. An alternative is to have CMC with oxide fibers and oxide matrix with no engineered fiber/matrix interface due to their inherent resistance to oxidation. These damage-tolerant oxide/oxide CMCs have been characterized for applications in high temperature (>1100°C) environment [3-7]. Further oxide/oxide CMCs have inherent porosity in the matrix which provides the crack-deflecting behavior similar to that resulting from a weak fiber/matrix interface; this serves to enhance the damage tolerance [4, 7]. Several investigations have been conducted to characterize oxide/oxide CMCs at elevated temperatures (>1100°C) [8-10]. These studies have been performed under the ambient laboratory environment.

There have been limited numbers of studies to characterize the capabilities of oxide/oxide CMCs in the combustion environment. Parthasarathy et al. conducted modeling and burner rig testing of a combustor heat shield made from N720/A (Nextel[TM]720/alumina) and N720/AS (Nextel[TM]720/alumina silicate) oxide/oxide CMCs [11]. Thermo-mechanical analysis using a finite element method suggested that in-plane temperature gradients would have the greatest effect on the material stresses during service. Both the N720/A and N720/AS panels were tested in the burner rig combustor for ten hours. Temperatures at hot spots reached 1200 ~ 1250 °C during the burner rig test.

261

The N720/AS heat shield exhibited matrix cracking in a location along the ±45° direction while the N720/A heat shield did not have any crack [11]. Mattoni et al. conducted combustion rig tests of two N720/A heat shield [12]. The highest temperature seen by the heat shields were 1200 ~ 1225 °C, localized around the fuel nozzles. The first heat shield was exposed for 20 hours, and the second heat shield was exposed for 86 hours. The highest temperature seen by the heat shields were 1200 ~ 1225 °C, localized around the fuel nozzles. It was also noted that there was delaminations in both heat shields after burner rig exposure [12]. Tensile and Iosipescu specimens were machined from the heat shields in a 0°/90° orientation. Tensile testing revealed that the material exhibited a 10-20% decrease in tensile strength. However, the material exhibited 15% increase in shear strength. It was postulated that this was a result of matrix sintering, which increased the fiber-matrix bond strength [12].

Previous studies have thus shown that oxide/oxide CMCs have excellent fatigue properties at elevated temperature [9], and they appear to perform well under the burner rig condition [11, 12]. However, the experimental approaches taken in the previous studies did not involve gas turbine environment of a hot-section structural component, which experiences mechanical and thermal loadings simultaneously in the presence of oxidizing chemical species. To meet this need, a test apparatus, that provides a better link to realistic service environment, has been developed at the Air Force Institute of Technology (AFIT) in collaboration with the Materials & Manufacturing Directorate, Air Force Research Laboratory (AFRL). It combines mechanical loading capability with a real and controllable combustion environment to test the coupons under a variety of test scenarios involving realistic settings. Using this facility, an oxide/oxide CMC was tested under the tension-tension fatigue loading condition in a combustion environment using this AFIT/AFRL burner rig facility. Further, the results of the present study were compared to results of a previous study with the same CMC system under the same fatigue loading and temperature conditions but in a standard furnace involving laboratory environment [13]. This comparison highlighted the effects and role of oxidizing combustion environment on the fatigue response of the test material. These were the objectives of the present study.

EXPERIMENTS

AFIT/AFRL Burner Rig
    This burner rig facility has been named after the two organizations behind its development, namely AFIT and AFRL. Figure 1 shows this test facility that provides a unique capability of characterizating a coupon-size specimen under various simulated combustion and mechanical loading conditions of gas turbine engine components. It integrates a realist combustion environment to a mechanical test machine. An atmospheric pressure burner rig system mixes fuel and oxidizers to generate high temperature, high speed combustion flame, which is guided into the direction of a test specimen with the capability to vary the impingement angle in order to simulate an angled impingement for future studies. The specimen under the flame impingement simultaneously undergoes a controlled mechanical loading condition by an MTS servo-hydraulic material test system. The burner rig system is capable of producing hot combustion gas which travels into the downstream at a sub-Mach speed, providing extra usefulness to simulate even a thermal cycling condition of even the most advanced gas turbine engines. Thermal cycling is possible by a programmable mechanical actuation of the combustion rig in and out of the alignment with test specimen. The mechanical load up to 25 kN can be applied either in a sustained manner to study a creep scenario or a cyclic loading to investigate fatigue behavior. With its programmed thermal cycling as well as mechanical loading capabilities in both sustained and cyclic forms, virtually any combination of thermal cycling and mechanical loading (involving either creep, fatigue or both) scenarios experienced by a component in gas turbine engine can be accommodated with the AFIT/AFRL burner rig. The configuration of the AFIT/AFRL burner rig in the present study was configured such that a non-symmetric thermal field was rendered on the

test specimen by imposing the combustion condition on one side of the specimen only, while the other side was allowed to undergo natural convection with the ambient laboratory air. The thermal gradient induced by the combustion heating of such configuration accurately simulates conditions of air foils undergoing the impingement of hot gas stream on only one side during the operation of gas turbine engines. The flame impingement was perpendicular to specimen's width face in the present study. Further details of the burner rig are provided in [14].

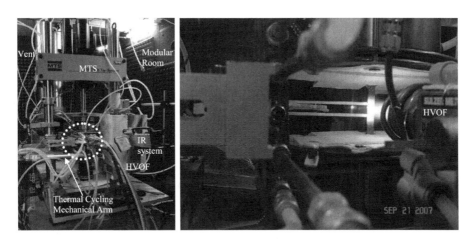

Figure 1: AFIT/AFRL burner rig; the right image represents the encircled area in the left

Material and Specimen

The test material was Nextel^{TM}720/alumina (N720/A) manufactured by COI Ceramics, San Diego, CA as 2.8 mm thick plate. Composite composed of uncoated Nextel^{TM}720 fibers (manufactured by 3M) in 8HSW, 0°/90° woven layers, with a density of about 2.78 g/cm^3 and fiber volume fraction of approximately 44%. Nextel^{TM}720 fiber is a meta-stable mullite having the chemical composition of Al$_2$O$_3$: 85% (by weight) and SiO$_2$ of 15% (by weight). This mullite fabric preform was infiltrated with the matrix precursor of alumina by a sol-gel process. Nextel^{TM}720/alumina CMC was made by a vacuum bag process under low pressure and low temperature followed by a pressureless sintering technique. There were numerous shrinkage matrix cracks, which were formed during processing due to mismatch of thermal expansion coefficient between matrix and fibers along with matrix porosity of about 24%. At room temperature, the ultimate tensile strength and elastic modulus of N720/A are reported to be equal to 170 ± 10% MPa and 60 ± 10% GPa [15].

Test specimens were machined from a single plate of N720/A having 12 plies of 8HSW fabric oriented at 0°/90°. Specimens were machined into a dog bone shape from the panel using a water jet. The specimens were cut to a length of 152 mm and a width of 12.70 mm. A dog bone section was cut with a gage length of 63 mm and a minimum width of 8.80 mm.

Burner Rig Testing

Prior to actual tests, the burner rig was calibrated for the test parameters which are provided in Table 1. This combustion condition was selected to simulate an application environment for hot section components such as turbine vanes and blades in modern gas turbine engines. Gas temperature was measured using R-type thermocouple. Due to the harsh combustion environment that carries aggressive chemical species and accompanies physical force by means of the flame thrust, it was not possible to devise a conventional contact method of measuring surface temperature. Instead the temperature measurement was made using a non-contact method involving the FLIR ThermaCAM P640 infrared (IR) thermal imaging system, which was calibrated for the emissivity by means of R-type thermocouple such that surface temperature can be monitored and recorded in real time during the test. The calibration was performed for the nominal surface temperature of 1235 ± 10 °C. The specimen was heated by the combustion flame on one side, while the other side was allowed to be heated through the natural convection with surrounding laboratory air. This mode of heating created a through-thickness thermal gradient, which involved a drop in surface temperature as much as 550 °C across the thickness from the front to back surface. Gas composition in the combustion was assumed as predicted by a computer code, i.e. Chemical Equilibrium with Applications (CEA) based on the calibration that showed a good match between the chemical contents predicted by the code and those experimentally collected using the TESTO 350 XL Gas Analyzer at lower temperature, i.e. less than 1000 °C [16]. The gas composition in the burner rig environment were estimated to be 26 vol% of CO_2, 14 vol% of O_2, 35 vol% of H_2O and approximately 25 vol% of other minor species including flame radicals such as OH [16].

Table 1: Test Parameters

Test Parameter	Condition	Calibration Tool(s)
Surface Temperature	~1235°C	Furnace, R-type TC & IR
Gas Temperature	< 1800°C	R-type TC
Gas Velocity	~ Mach 0.5	XS-4 High Speed Camera
Equivalence Ratio	~ 0.9	HVOF^{TM} Flow Controller
Gas Composition	H_2O, O_2, CO_2, CO, NOx	Testo XL 350 Gas Analyzer
Mechanical Loading	Fatigue (1 Hz & R = 0.05)	MTS
Test Duration	Up to 25 hours	N/A

Under the combustion environment, characterized by the parameters in Table 1, the test material was simultaneously fatigue loaded with a peak stress that was varied between each test, while the stress ratio (R) and cyclic frequency remained constant throughout the testing of all specimens at a stress ratio of R = 0.05 and frequency of 1 Hz. A run-out was set to the maximum number of cycles of 90,000 or 25 hours in the present study. If runout was reached, the test was terminated. Few specimens, which did not fail after the 90,000 cycles, were tested under monotonic tensile test condition to measure the residual strength at room temperature. The test data including applied peak stress, number of cycles at failure and the location of fracture were recorded. After the burner rig tests, microscopic analysis using Scanning Electron Microscopy (SEM) was carried out. Finally, to establish a baseline data, the monotonic tensile test of the test material was also conducted in the same combustion condition as used in the fatigue tests, i.e. combustion flame impinged the specimen on one side as in the fatigue test, and then it was tested under load control mode at a rate of 20 N/sec.

RESULTS AND DISCUSSION

A stress-strain relationship from a monotonic tensile test under the combustion test enviroment is shown in Figure 2. The ultimate tensile strength and failure strain were 167 MPa and 0.41%, respectively. The corresponding average values at 1200 °C under the ambient laboratory condition were 205 MPa and 0.39%, respectively . It appears that the tensile strength of the test material in a combustion environment decreased by about 20%. However, there is one important phenomenon which should be considered in this comparison, and it is the thermal stress from the one-sided heating of specimen. The non-uniform temperature distribution in the burner rig tested coupon resulted in a non-uniform stress distribution. Simple analysis suggests that the hot face, where the combustion gas directly impinged, will be in a compressive (or lesser tensile) state than the stress applied by the test machine. Estimation of this thermal stress is nontrivial due to both the three dimensional spatial temperature distribution as well as the bending deformation of the coupon, which will somewhat relieve the thermal stress. This would require a thermo-mechanical finite element analysis, which is outside the scope of the present effort.

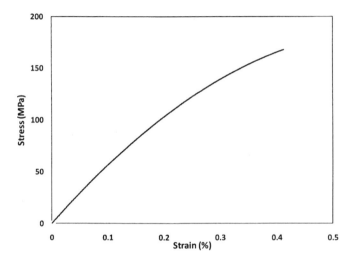

Figure 2: Monotonic tensile stress-strain curve under combustion condition

Fatigue life data in the combustion environment specimens are provided in Table I and also shown in Figure 3 where cycles to failure are plotted as a function of applied maximum cyclic stress. The applied maximum stress is normalized by the ultimate tensile strength in the combustion environment. For the comparison, fatigue life data under the standard furnace test, i.e. ambient laboratory condition at 1200 °C, are also plotted in this figure, and these were generated in a previous study [13]. The N720/A has excellent fatigue resistance under both test conditions when applied maximum stress was equal and less than 90% of the ultimate tensile strength. Further, this comparison shows that the fatigue strengths of the test material in the burner rig were equal to their counterparts in the ambient laboratory environment in terms of their respective ultimate tensile strength, since normalized fatigue strength indicates the percentage of knockdown in the ultimate tensile strength experienced by the test condition. As mentioned earlier, the ultimate tensile strength under the combustion environment was about 20% less than that in the ambient laboratory condition. Therefore, a similar reduction of about 20% in the fatigue strength is present up to 25 hours or 90,000 cycles. These comparisons clearly indicate that there was practically no or minimal damage on the N720/A from the harsh oxidizing condition of combustion. Therefore, it can be safely postulated that the reduction in the ultimate tensile strength and fatigue strength of N720/A under the harsh combustion environment was due to thermal stress from the one-sided heating of the specimen. This is also supported by variation of the secant modulus during cycling. Figure 4 shows the normalized secant modulus plotted against the logarithmic of fatigue cycles. The secant modulus was calculated from the stress-strain curves which were recorded during the fatigue tests. These curves were linear. The secant modulus was normalized by dividing it by its value during the first cycle. Figure 4 shows that the secant modulus changed very little over the course of the fatigue test, staying within ±10% of the modulus of the initial cycle. The minimal change in the modulus of the specimens further substantiates that there was little fatigue damage accumulation in the specimens tested in the combustion environment.

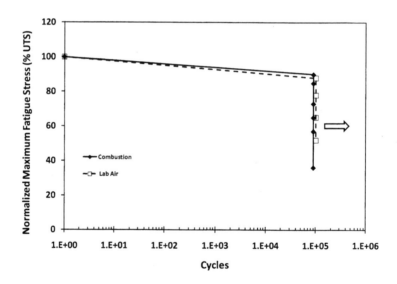

Figure 3: Fatigue diagram under combustion and laboratory environments

Table I: Test Results

Test Environment	Normalized Fatigue Stress (% UTS)	Cycles to Failure
Combustion	36	>90,000
Combustion	57	>90,000
Combustion	65	>90,000
Combustion	73	>90,000
Combustion	85	>90,000
Combustion	90	>90,000
Laboratory	88	>90,000
Laboratory	78	>90,000
Laboratory	65	>90,000
Laboratory	52	>90,000

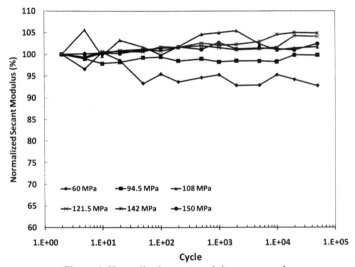

Figure 4: Normalized secant modulus versus cycles

To see what retained strength was present in the material after fatigue testing, the run-out specimens were monotonically tensile tested at room temperature. Figure 5 shows a typical stress-strain curve from a residual tensile strength test. The residual strength and Young's modulus after the run-out fatigue test in the combustion environment were within 10% of their counterparts from the virgin material which again substantiates that there was practically no damage accumulation from both fatigue and harsh combustion environment.

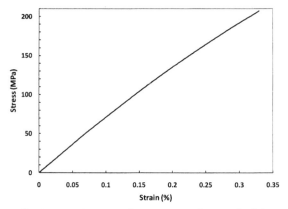

Figure 5: Monotonic tensile stress-strain curve of a run-out specimen under laboratory condition

Damage Mechanisms

Figure 6 shows surface of a test specimen before and after the fatigue test under combustion environment. This surface was exposed to the direct flame impingement. It is apparent from this comparison that there was no distinct feature which was generated from the harsh combustion environment. The pre-test and post-test photographs revealed that the matrix cracks originally present in the material did not change as the specimens were subjected to the fatigue and combustion conditions. To examine further, few specimens, tested under harsh combustion environment, were sectioned. A typical view from such sectioning is shown in Figure 7. The magnified views of internal surface from this figure are shown in Figure 8. Figure 8 (a) is from the region close to side where flame impingement occurred while Figure 8 (b) is from the back side of the specimen. From these higher magnification micrographs, it appears that the matrix condition is the same in the front (i.e. flame impingement side) as it is in the rear of the specimen. No densification of the matrix is apparent due to the combustion environment. Comparing these micrographs to the pristine specimen at the same magnification in Figure 9, the matrix condition and appearance look similar. These comparisons clearly show that combustion exposure had no noticeable effects on the oxide/oxide CMC, N720/A.

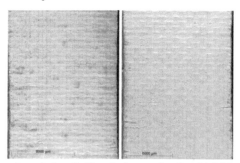

Figure 6. Specimen before (left) and after combustion (right) test (flame side)

Figure 7. Microstructural view of specimen after combustion test (flame side at left)

Figure 8. Magnified View of specimen shown in Figure 7; (a) is near flame impingement and (b) is on back side

Figure 9. Microstructural view of specimen before combustion test

CONCLUSIONS

Fatigue behavior of an oxide/oxide, CMC Nextel™ 720/Alumina (N720/A), was investigated in the combustion environment. The N720/A consisted of an 8-harness satin weave of Nextel™ aluminum oxide/silicon oxide fibers reinforced in alumina matrix. The combustion environment was generated using a High-Velocity Oxygen Fuel (HVOF) gun. The flame directly impinged the CMC specimen on one side as it was subjected to tension-tension fatigue at frequency of 1 Hz with stress ratio of 0.05, heating up the surface to 1235 ± 10 °C. All fatigue specimens survived 90,000 cycles, i.e. 25 hours when the applied maximum fatigue stress was equal or less than 150 MPa, i.e. 90% of the ultimate tensile strength of the material under the same combustion environment condition. Residual tensile strength of fatigue tested specimen at room temperature was almost equal to that of the virgin material which showed that there was no degradation in the test material from the combined fatigue and combustion exposure. Microstructural analysis also showed no effects of combustion environment on the test material. Comparison of fatigue behavior of the N720/A under the combustion and laboratory air environement showed that the normalized fatigue strength, i.e. ratio of the fatigue strength and ultimate tensile strength at a certain number of cyles, was almost equal in both these test conditions. This again suggests that the combustion environment condition was not damaging to the Nextel720™/alumina.

REFERENCES

1. Spriet P, Harbarou G. Applications of CMCs to turbojet engines: overview of the SEP experience. Key Engineering Materials, Switzerland: Trans Tech Publication, 1997; 1267-76.
2. Chawla KK. Ceramic Matrix Composites. New York: Chapman & Hall, 1993.
3. Lange FF, Tu WC, Evans AG. Processing of damage-tolerant, oxidation-resistant ceramic matrix composites by a precursor infiltration and pyrolysis method. Mater Sci Eng A 1995: A195:145–150.
4. Lunderberg R, Eckerbom L. Design and processing of all-oxide composites. In: Evans AG, and Naslain RG, editors. High-Temperature Ceramic–Matrix Composites II: Manufacturing and Materials Development. American Ceramic Society, 1995: 95–104.

5. Mouchon E, Colomban P. Oxide ceramic matrix/oxide fiber woven fabric composites exhibiting dissipative fracture behavior. Composites 1995; 26:175–182.
6. Morgan PED, Marshall DB. Ceramic composites of monazite and alumina. J Am Ceram Soc 1995, 78(6):1553–1563.
7. Tu WC, Lange FF, Evans AG. Concept for a damage-tolerant ceramic composite with strong interfaces. J Am Ceram Soc 1996; 79(2):417–424.
8. Staehler JM, Zawada LP, Performances of four ceramic-matrix composite divergent flap inserts following ground testing on an F110 turbofan engine. J. Am. Ceram. Soc., 2000; 83: 1727-38.
9. Steel S, Zawada LP, Mall S. Fatigue behavior of a Nextel 720/alumina (N720/A) composite at room and elevated temperature. Ceram. Eng. Sci. Proc., 2001; 22: 695-702.
10. Zawada LP, Hay RS, Lee SS, Staehler J. Characterization and high temperature mechanical behavior of an oxide/oxide composite. J. Am. Ceram. Soc. 2003; 86: 981-90.
11. Parthasarathy, TA, Zawada LP, John R, Cinibulk KM, Zelina J. Evaluation of oxide-oxide composites in a novel combustor wall application. Int. J. Appl. Cer. Techn. 2005; 2: 122-132.
12. Mattoni MA, Yang JY, Levi CG, Zok FW, Zawada LP. Effects of combustor rig exposure on a porous-matrix oxide composite. Int. J. Appl. Cer. Techn. 2005, 133-140.
13. Ruggles-Wrenn MB, Mall S, Eber CA, Harlan LB. Effects of steam environment on high-temperature mechanical behavior of Nextel[TM]720/Alumina (N720/A) continuous fiber ceramic composite. Composites Part A: Applied Science and Manufacturing, 2006; 37: 2029-2040.
14. Kim TT, Mall S, Zawada LP. Fatigue characterization of melt-infiltrated (MI) woven Hi-Nic-S/BN/SiC CMC using a unique combustion heating test facility. 33rd Int. Conf. Advanced Ceramics and Composites (ICACC), Manuscript ID 529430, 2009.
15. COI Ceramics, Unpublished Data.
16. McBride BJ, Gordon S. Computer program for calculation of complex chemical equilibrium compositions and application, NASA RP-1311, June 1996.

Erosion and Wear

PARTICLE EROSION WEAR BEHAVIOR OF NEW CONCEPTUAL SIC/SIC COMPOSITES

Author Name: Min-Soo Suh
Author Affiliation: Graduate School of Energy Science, Kyoto University
City, State, Country: Gokasho, Uji, Kyoto 611-0011, Japan

Author Name: Tatsuya Hinoki
Author Affiliation: Institute of Advanced Energy, Kyoto University
City, State, Country: Gokasho, Uji, Kyoto 611-0011, Japan

Author Name: Akira Kohyama
Author Affiliation: Institute of Advanced Energy, Kyoto University
City, State, Country: Gokasho, Uji, Kyoto 611-0011, Japan

ABSTRACT

The particle erosion of newly fabricated SiC/SiC composites by SiC particles was investigated at room temperature over a range of median particle sizes of 425–600 μm, velocities of 100 m/s and impact angles of 90° by weight loss measurements and scanning electron microscopy. A number of composite materials have been examined, commercial fiber-reinforced SiC/SiC by CVI and NITE route, and new conceptual NITE SiC/SiC composites. Microstructural observation was performed to examine the correlation between erosive wear behaviors and fabrication impurities. Erosive wear behavior was rather serious in prototypes of conceptual composite, which employed Pre SiC-fiber and phenolic resin. Result from the conspicuous defects in the prototype materials as the forms of porosity, fiber deformation, excess oxide, PyC deformation, interface cleavage, and, etc. Two dominant erosive wear mechanisms were observed, one is delamination of constituents mainly caused by erosive crack propagation. Another is detachment of constituents, which usually caused by an erosive impact. A unit size of delamination is the most decisive factor on wear volume. Bonding strength of each constituent is mostly affected by various forms of porosities. Therefore, fundamental cause and consequent results will be carefully elucidated. To reduce the dominant wear by improving the fabrication conditions, the microstructural defects will be also discussed. The final product of a conceptual composite show over double and half wear resistance in comparison to commercial CVI composite. Consequently, by controlling the fabrication impurities, the improvements have been successfully made for a new fabrication technique, as a result, the known defects are rarely observed in final product. Schematic wear model of SiC/SiC composites under particle erosion are proposed.

INTRODUCTION

Continuous SiC fiber-reinforced SiC matrix (SiC_f/SiC) composites are a promising candidate for a structural application, especially in severe conditions such as high temperature and wear resistant environments[1-6]. Wear cause durability and reliability issues, therefore, wear control i.e. design of operating conditions and selection of materials has become a strong need for the advanced and reliable technology of the future. It is well known that SiC has one of the highest hardness of all single-phase ceramics, moreover, possesses fascinating performance in various conditions[2-4]. Brittle behavior is the dominant failure factor for most ceramics, but SiC_f/SiC composites have solved most of the inherent brittle issues. It is also known as one of the most advanced composite material systems due to the commercial availability as well as promising mechanical properties in elevated temperature. The use of SiC_f/SiC composites such as combustor liners and turbine vanes provides the potential of improving next-generation turbine engine performance, through lower emissions and higher cycle efficiency, relative to today's use of super alloy in high temperature components. The demand of the new advanced materials is increasing, especially in the recent high-efficiency advanced energy system and/or severe environments, at the same time the understanding of wear mechanisms can lead to improve material performance and also essential to design advanced tribosystems.

SiC$_f$/SiC composite is composed mainly matrix, fiber, and matrix-fiber interface as a constituent part. In case of the highly qualified grade SiC$_f$/SiC composite developed in Kyoto University, the Tyranno SA^{TM} grade fiber and the pyrolytic carbon (PyC) interface by CVD route are employed as fiber-reinforcement and fiber-interface, respectively. The mechanical properties are promising due to near-exact microstructures e.g. near-full dense, on-demanded accurate thickness of the interface, and no residual carbon or silicon area (see figure 1). State of the art in SiC/SiC composites has solved most of the potential issues of near-net shaping and machining, joining technology, and evaluation technology. There are no doubts that SiC/SiC composites are at the forefront of advanced materials technology and prime candidates for structural material in severe environments. However, it still has issues of brittle reliability and extremely high cost due to the complication of many complex processes for production.

Figure 1. High densification and crystallization of high-qualified grade SiC$_f$/SiC composite.

A novel cost conscious fabrication on SiC/SiC composite has been developed for high erosion resistance. To accomplish both cost issues and foreign object damage (FOD) resistance, Pre-SiC fiber was employed as fiber-reinforcement with adopting a new phenolic resin as fiber-interface, in particular to substitute CVD coated PyC interface. A number of conceptual SiC/SiC composites were fabricated in order to investigate the dominant wear correlative to crucial parameters of material fabrication. This study intended to improve not only erosion performance but material properties by following methods. Optimizing the fabrication conditions through the controlling of microstructural defects observed in prototype SiC/SiC composites, and examining the wear behaviors for matrix, fiber and interface as a constituent of composites.

EXPERIMENTAL PROCEDURES

Material preparation

Nano-powder Infiltration and Transient Eutectoid (NITE) process[6] has been selected to fabricate SiC/SiC composites with various fabrication conditions. It is a highly optimized liquid phase sintering (LPS) process, which available in commercial. Figure 2 shows the brief concept of NITE process. Fabrics are prepared as a form of prepreg sheet and pre-coated by methods explained in figure 3 to employ the PyC interface around the filament. The prepreg is impregnated with the matrix precursor by passing through the slurry. A powder mixture of over 90 wt% SiC and under 10 wt% of sintering additives was prepared by milling process for raw material of slurry. Finally, hot-pressed into a dimension of 40 mm x 20 mm sized plate under Ar atmosphere at 1850 °C, applying 20 MPa of load by a graphite mold. Figure 3 illustrates schematic differences in fiber coating method.

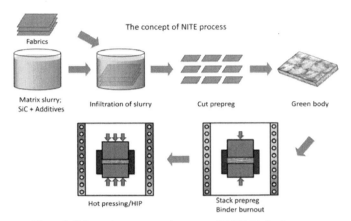

Figure 2. Schematic concept and process of NITE fabrication route.

Figure 3. Schematic comparison of fabric preparation.

Conceptual composites were fabricated in different conditions by modifying the amount of employed PyC for interface-control and fiber heat treatment for bonding-control among the fibers. Sintering conditions (e.g. temperature, applied pressure, and holding time) and fabrication compositions (e.g. amount of SiC powder and sintering additives) are remained as constant. Fabrication characteristics of the various composite materials used in this test are listed in table 1.

Table I. Fabrication characteristics of various SiC/SiC composites

SiC$_f$/SiC composites	Fabrication process	Employed coating	Fiber coating method	Fiber type	Amount of employed PyC	Fiber heat treatment
Commercial product #1	NITE	PyC	CVD	Cef-NITE™	500 nm	N/A
Commercial product #2	CVI	PyC	CVD	Tyranno SA™	80 nm	N/A
Prototype (LOT#11)	NITE	PyC	Pre-pyrolysis	Pre-SiC fiber	Large	500 °C
Prototype (LOT#12)	NITE	PyC	Pre-pyrolysis	Pre-SiC fiber	Small	500 °C
Final product (LOT#17)	NITE	PyC	Pre-pyrolysis	Pre-SiC fiber	Small	Delicately Controlled

Erosion wear test
        Erosive wear test was carried out by impinging SiC powder (green carborundum #36) on the surface of a test specimen. Using jets to utilize repeated impact erosion with a small nozzle delivering a compressed gas containing those abrasive particles, which coincident with ASTM G76 as shown in figure 4. The particle impact velocity of 100 m/s was used to evaluate the composites, weight of impacted particles is not in discuss since the purpose of this study is to understand the wear behavior

versus fabrication parameters of new SiC/SiC composites. The related data, which can be a reference to compare with the enormous data in previously published works is explained in the author's other works. Table 2 shows the experimental conditions for the test. It is well known that the effect of impingement angle on wear rates is the greatest close to perpendicular for brittle materials.

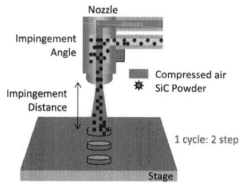

Figure 4. Schematic illustration of particle erosion environment.

Table II Experimental conditions of particle erosion wear test

Test parameters	Erosion method	Erodent	Step of erosion	Impingement conditions		
				Angle	Distance	Jet pressure
Value	Solid particle compressed air	SiC powder [a]GC #36	3 steps	90 °	40 mm	5 atm

[a]Green carborundum (GC) is a bulk grinding compound generally for polishing purpose
Distribution of particle size for #36 is 600 ~ 425 μm

Protection seal was adopted to protect the partial surface of a test specimen from particle impingement while the erosion tests. Detachment of the protection seal and removal of debris by air-gun proceeds, and then surface topography was measured by a surface profilometer after each step of erosion. The average worn depth was calculated by the difference in topography between the baseline of uneroded surface and eroded surface. Erosive wear rate was calculated by the average volume of worn surface divided by the total area of eroded surface.

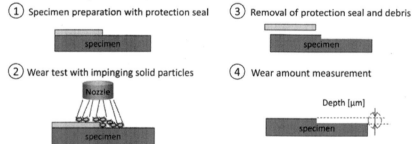

Figure 5. Procedure of wear amount measurement after each erosion step

RESULT AND DISCUSSION

Erosion performance and material analysis

Erosion wear surface of various SiC/SiC composite specimens were observed by direct-eye as shown in figure 6. The result shows that newly fabricated conceptual SiC/SiC composite has reasonable wear resistance comparing with two commercial composites (see figure 7). Erosion performance of conceptual NITE composites, especially for the final product (LOT#17) was superior almost twice more resistance against commercial product made by CVI (chemical vapor infiltration) route. Cef-NITE^{TM} fiber-reinforced NITE composite shows the best erosion resistance while final product (LOT#17) shows only 20 % inferior to Cera-NITE^{TM} performance.

| Commercial NITE product #1 | Commercial CVI product #2 | Prototype LOT#11 | Prototype LOT#12 | Final product LOT#17 |

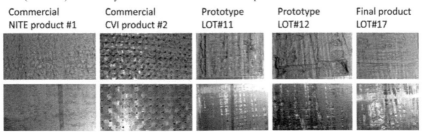

Figure 6. Bare eye observation on the worn surfaces of composites before and after the erosion test.

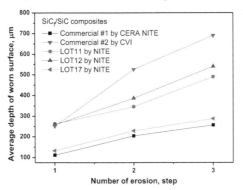

Figure 7. Wear rate under solid particle erosion test for various SiC/SiC composites.

Prototype material LOT#11 has employed volume fraction of PyC about treble more than final product LOT#17. Figure 8 shows the amount of employed PyC among the fibers. SE micrograph in back scattered electron imaging (BEI) mode indicates fibers and PyC in color of grey and black, respectively. Severe fiber deformation and micro-pore formation are observed in prototype LOT#11. Increase of PyC amount results in severe fiber deformation and/or pore generation, which plays an important role for the erosive wear performance. Figure 9 shows conspicuous defects in prototype material; a) irregular fiber deformations, b) insufficient densification of the matrix, c) distribution of residual oxide, d) micro-pores, e) deformation and crack formation on PyC interface, f) weak bonding strength between fiber and interface, and g) melted-flowed oxide phase around the fiber.

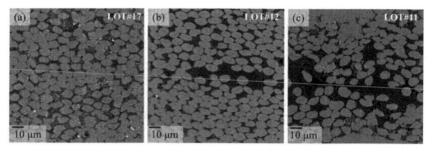

Figure 8. Employed amount of PyC, in (a) Final product, and in prototype materials (b), and (c).

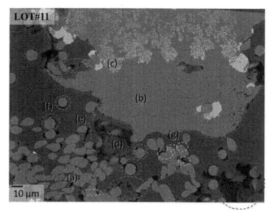

Figure 9. Conspicuous defects observed in typical region of prototype specimen LOT#11.

Wear Behavior

Two dominant erosive wear behaviors were observed in this study, one is delamination of constituents. Another is detachment of constituents, the former usually caused by erosive crack propagation and later caused by an erosive impact.

A unit size of delamination is determined by crack propagation length, mostly along the interface of constituents. It is the most decisive factor on amount of wear. When a crack reaches to the interface a reflection along the interface simultaneously occurs. In case of the strong bonded interface, the crack propagation neither reach to next crack nor results in any delamination. However, in the opposite case, crack propagated to the end of confluent crack, and consequently, causing huge delamination. It is observed that most of the erosive crack propagates through the weakest area. Evidential proof was confirmed in two prototype materials that the detachment of matrix and fiber were conspicuously observed in plenty of areas, which results from weak bonding strength. The final product has relatively strong bonding strength so that those matrix and fiber detachments were hardly observed. Comparatively, detachment of matrix and fiber was rarely occurred by an erosive impact.

Strong bonding between fiber and interface absorbs the impact energy but erosive impact gradually worn the constituents.

Wear on matrix

Eroded surface of conceptual composites was examined by SEM. Different erosive wear behavior was observed as shown in figure 10. It shows the great difference between the coarse matrix and dense matrix. Due to inadequately applied pressure, phenomena of insufficient densification of matrix and scarce grain growth were observed in prototype LOT#11. Grain size of final product was relatively larger according to sufficient grain growth. Both grain pull-out and intergranular fractures by erosive impact were observed in eroded surface of final product (see figure 10-d).

Figure 10. SE micrographs of eroded surface on (a) coarse matrix of prototype LOT#11, and a typical eroded region of (b) prototype LOT#11, (c) LOT#12, and (d) final product LOT#17.

Microscopic observation on the cross section of prototype materials also confirmed the porosities in the matrix. As shown in figure 11, (b) shows pores between remnant agglomerates compared with (a) pores in the dense matrix. These inadequately pressurized areas and insufficient grain growth are one of the reasons of severe matrix detachment.

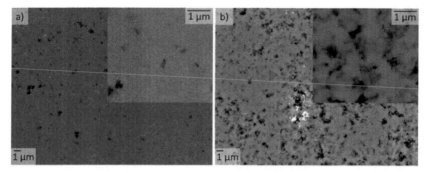

Figure 11. SE micrographs (BEI mode) of typical porosities in prototype LOT#11
(a) pores in dense matrix and (b) coarse matrix.

Wear on fiber and interface
Detachment of fiber was conspicuously observed in prototype LOT#11. As a form of aperture between fiber and PyC, porosity caused easy fiber detachment by a single impact. It was mainly generated by volume contraction of Pre-SiC fiber itself during the crystallization process. On PyC interface, for the evidence, exact shaped and sized traces of fiber detachment were observed (see figure 12-a). Interface cleavage was also observed, which was generated while the hot-pressing process.
Bonding strength between fiber and interface was relatively strong that most of the wear was caused by gradual eroding (see figure 12-b). In addition, behaviors such as sudden detachment of constituents were rarely observed in final product LOT#17.

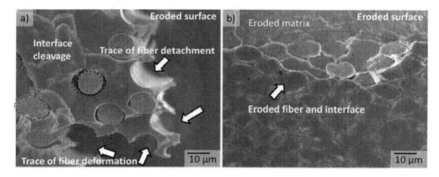

Figure 12. SE micrographs of fiber detachment and deformation observed in
(a) prototype LOT#11, and (b) final product LOT#17

These porosities influenced on a bonding strength of each constituent, which results in severe detachment of fiber and matrix by an erosive impact (also shown in figure 13). Plenty of fiber detachments were observed with eroding of all constituents in prototype composite.

Figure 13. Erosive wear on each constituent observed in prototype LOT#11.

Erosion wear models

Unit length of erosive crack propagation and confluence of cracks dominate the unit size of constituent delamination. Erosive crack can propagate further and joining to other cracks also often occurred in the coarse matrix. Minor erosion occurs when the erosive crack joined to next confluent crack. Huge erosion occurs when propagation of erosive crack reaches to the interlayer of constituents where the bonding interface is fragile, and after that the crack branches along the matrix-interface.

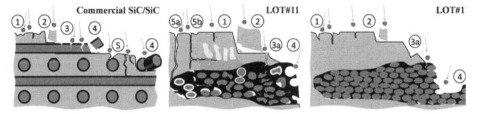

Figure 14 Cross sectional view of the erosive wear on SiC/SiC composites

Figure 14 shows the dominant erosive wear mechanisms on various SiC/SiC composites. Each number represents following erosive behaviors: 1. Matrix eroding and crack initiation, 2. Large and small unit of matrix detachment, 3. Interface and/or fiber eroding and cracking, 3a. Interface cleavage by eroding, 4. Fiber detachment, 5. Crack propagation along the interface and/or through the matrix, 5a. Through a weak interface between matrix and PyC layer, and 5b. Through a weak coarse matrix.

CONCLUSION

Erosive wear test was carried out on a number of SiC/SiC composites. Schematic wear models are proposed with following details. Two dominant erosive wear mechanisms were observed for new conceptual composites, one is delamination of constituents, which mainly occurred by crack propagation through the weakest area. Another is detachment of constituents by an erosion impact. A unit size of delamination dominated by the erosive crack propagation length, which mostly determine the size of wear volume.

The fabrication improvements have been successfully made for a new conceptual composite, as the result, the known defects are rarely observed in final product. Crucial fabrication factors for erosion wear resistance were; 1. Weak bonding strength of constituent parts, 2. Poor densification of matrix and scarce grain growth, 3. Interface cleavage with micro-pores, and 4. Aperture along the fiber-interface

ACKNOWLEDGEMENT

The author would like to thank the Ministry of Education, Culture, Sports, Science and Technology (MEXT) of Japan for scholarship and the 21 Global COE program of Kyoto University for partial-financial support.

REFERENCE
[1]K. Komeya, M. Matsui, in Materials Science and Technology Vol. 11, ed. M.V. Swain (VCH, Weinheim, 1994), pp. 517–565.
[2]M.-S. Suh, et al., "Friction and Wear Behavior of Structural Ceramics Sliding against Zirconia", Wear **264** (9-10), 800-806 (2008).
[3]Suh, M.-S., Kohyama, A.: Special Issues on "in situ" Crystallized SiC/SiC Composites. Proceeding of ISAE2009, 439-442 (2009).
[4]Suh, M.-S., Kohyama, A.: Effect of Porosity on Particle Erosion Wear Behavior of Lab. Scale SiC$_f$/SiC Composites. Int. J. of Modern Phys B, under press (2009)
[5]J.A. DiCarlo, et al., in Handbook of Ceramic Composites, ed. P.B. Narottam P. Bansal (Kluwer Academic Publishers, Boston, 2005), p. 33.
[6]A. Kohyama, et al.: Development of SiC/SiC composites by nano-infiltration and transient eutectoid (nite) process. Ceramic Eng. and Sci. Proc., **23**, 311-318 (2002).

THRESHOLD OF RING CRACK INITIATION ON CVD-SiC UNDER PARTICLE IMPACT

Min-Soo Suh
Graduate School of Energy Science, Kyoto University
Gokasho, Uji, Kyoto 611-0011, Japan

Sang-Yeob Oh
School of Mechanical & Automotive Engineering, Kyungpook National University
386 Gajang-dong, Sangju, Gyung-buk 742-711, South Korea

Tatsuya Hinoki
Institute of Advanced Energy, Kyoto University
Gokasho, Uji, Kyoto 611-0011, Japan

Chang-Min Suh
School of Mechanical Engineering, Kyungpook National University
1370, Sankyuk-dong, Puk-ku, Daegu 702-701, South Korea

Akira Kohyama
Institute of Advanced Energy, Kyoto University
Gokasho, Uji, Kyoto 611-0011, Japan

ABSTRACT
This study examines the threshold of ring crack initiation to provide the fundamental and morphological cracking behavior of impact damage on pure β-SiC structure in comparison of other brittle material such as soda-lime glass. Particle impact test was carried out by using 2 mm spherical steel ball as a projectile at room temperature. For the polished CVD-SiC, the diameter of outermost ring crack increase linearly with the increase of impact velocity. Cracking patterns were examined and schematically established for the test materials versus impact energy. Threshold of ring crack initiation on polished CVD-SiC was about 0.7E-03 J. Radial cracks were observed over 2.15E-02 J at the first time in both rough and smooth surface of CVD-SiC. The result of outermost ring crack in two surface finish were almost similar to a certain impact velocity but dramatic behavior was also observed. The threshold of ring crack initiation is important significance in order to prevent the fracture of components.

INTRODUCTION

Silicon carbide (SiC) has many of fascinating properties required for severe environments such as high strength at elevated temperatures, low density, low coefficient of thermal expansion, and high wear and oxidation resistance for high-temperature components[1-9]. A major problem in using brittle materials for structural applications is the localized damage/cracking when subjected to impact by foreign objects[5-18]. Ceramics, because of their brittle nature, are susceptible to fracture toughness, which is still a hot issue despite the superior characteristics it has. Particle impact can be a major source of cracks in ceramics, which can cause substantial degradation of strength[5-13]. To ensure the fundamental reliability and durability issues of materials foreign object damage is essential. Therefore, foreign object damage (FOD) associated with particle impact have to be considered when ceramic materials are designed for structural applications[5-18]. Recently, due to the tremendous efforts given to the development and improvement of fabrication technology on continuous SiC fiber-reinforced SiC matrix (SiC/SiC) composites[2-6], simultaneous FOD researches are ongoing on a new advanced SiC/SiC composites for structural material usage[5-9]. In particular, SiC/SiC composites are considered as a promising material for next generation aerospace, space propulsion systems, and advanced energy systems, because of their high temperature strength, high wear and creep resistance, high thermal conductivity, and graceful failure under loading[1-6]. These composites are fabricated by various processing approaches: chemical vapor infiltration (CVI), liquid phase sintering (LPS), polymer infiltration and pyrolysis (PIP), and melt infiltration (MI). The Nano-powder Infiltration and Transient Eutectoid (NITE) process is one of the most successful LPS processes which have been developed in our research group[2-6]. It was aimed for reduced porosity, advanced matrix quality, and strong fiber-matrix interface. Therefore the superior characters are near-zero porosity, no residual Si and C, and excellent mechanical and thermal properties comparing with other process.

There are quite a number of parameter which affects to the FOD performance in the view point of fracture mechanism, such as grain size, inherent micro-cracks, porosity, and sintering/bonding additive. Sintering additives is necessary although the excess amount will act as a defect, however without additives it is difficult to densify because of covalent nature of Si-C bonding and low self diffusion coefficient. To understand the complex nature of SiC/SiC composite under impact damage, three constituent parts have to be evaluated separately. Beta structured SiC crystalline, matrix-fiber PyC interface, and fabrication impurities such as sintering oxides, and residual Si or C. For those reasons, CVD-SiC is prepared for solid particle impact test due to its promising high chemical purity and stoichiometric homogeneity, and absence of micro-cracks/pores. Soda-lime glasses were also prepared as a reference material to understand the impact damage behaviors of brittle material. Numerous efforts has been made to understand the impact behaviors of SiC ceramic materials under particle impact[5-18], however the result of literature survey shows that impact behavior of CVD-SiC in the focus of crack initiation and cracking behaviors is novel. This study intended to provide morphological impact behavior on pure β-SiC matrix focused on the crack initiation. And to examine the impact damage resistance on standard SiC structure to solve the complex nature of SiC/SiC composites.

EXPERIMENTAL PROCEDURES

Material preparation

Commercial CVD-SiC (Chemical vapor deposition silicon carbide) and soda-lime glass were prepared for particle impact test. CVD-SiC has purity greater than 99.9995%, and stoichiometric homogeneity with negligible porosity and inherent microcracks. Typical properties are listed in table I. In addition, pure β-SiC (3C-SiC) structure promises the best stiffness to weight ratios, and wear and corrosion resistance. Dimension of 20x20x3 mm plate, polished to two different average roughness of 0.02 and 0.2 μm, are subjected to particle impact damage (see figure 1). Commercial soda-lime glass was selected as a reference material due to the excellent transparency. Mechanical properties of subjected materials are shown in table II.

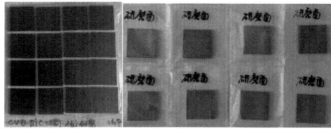

Figure 1 CVD-SiC with 0.02 μm as mirror surface finish (left) and 0.2 μm as rough surface (right)

Table I Typical Properties of **CVD Silicon carbide**[TM] ceramic material

Properties	Typical values[1]
Crystal structure (face-centered cubic β-phase)	FCC, polycrystalline
Sublimation Temperature (C)	~2700
Grain Size (μm)	5
Density (g cm[-3])	3.21
Hardness (kg mm[-2]) Knoop (500 g load) Vickers (500 g load)	2540 2500
Chemical Purity[2]	≥99.9995 % SiC
Flexural Strength, 4-point[3] @ RT (MPa/Ksi) @ 1400°C (MPa/Ksi)	415/60 575/84

(1) Average values at room temperature.
(2) Total metallic impurities
(3) Flexure beams had a 0.5μm RMS surface finish

Table II Mechanical properties of specimen and projectile

	Young's modulus (GPa)	Hardness (Hv)	Relative density	K_{IC} (MPa√m)	Poisson's ratio (ν)
Soda-lime glass	74	560	2.53	0.75	0.2 ~ 0.27
CVD-SiC	466	2500	3.21	3.3	0.21
Steel ball	208	880	7.83	-	0.3

Particle impact test

The experimental apparatus is schematically shown in figure 2. A partially stabilized 2 mm dia. of steel ball was fixed on the end of a ball carrier, and settled as a sabot in a gun barrel. Air compressor was introduced to propel gas pressure to the specified level in the reservoir depending on prescribed impact velocity. The solenoid valve instantaneously opened, the gas released toward to the pistol, accelerating a steel-ball projectile through the gun barrel to impact the target specimen. One impact was made under each set of conditions at room temperature. Impact velocity of each projectile was determined using oscilloscope based on the time-of-flight principle. The steel ball projectiles were purchased from AKS, Amatsuji Steel Ball mfg., Japan.

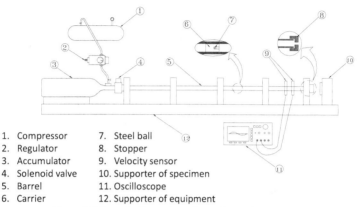

1. Compressor
2. Regulator
3. Accumulator
4. Solenoid valve
5. Barrel
6. Carrier

7. Steel ball
8. Stopper
9. Velocity sensor
10. Supporter of specimen
11. Oscilloscope
12. Supporter of equipment

Figure 2 Schematic illustration of experimental equipment

After the impact test, the trace of impact damage was tracked down by a laser surface profilometer. Figure 3 shows an example of data acquisition by non-contact profilometer, which has vertical resolution of nanometer level. Although the rough surface of CVD-SiC sample as R_a=0.2 μm, the trace of ring crack was clearly observed. The scan speeds of the non-contact profilometer are dictated by the light reflected from the surface and the speed of the acquisition electronics. It also can reduce possible surface damage by wear or any other chances of contamination. Optical microscope (VHX-200, Keyence, Japan) was used to characterize the trace of impact damage as shown in figure 4. After measurement, AFM (atomic force microscope VN-8010, Keyence, Japan) was employed to analyze the surface topography, impact damage behavior, and also for data verification.

RESULT AND DISCUSSION

Schematic cracking patterns on soda-lime glass versus impact velocity of 2 mm steel ball projectile are shown in figure 5. It was observed that near 10 m/s (KE=1.36E-04 J) is the threshold of ring crack initiation for commercial soda-lime glass. As the increase of impact velocity, formation of cone crack was sequentially observed. Second ring and multi-rings were also observed in certain level of impact energy. Initiation of lateral crack was subsequently held on, radial cracks and lateral cracks were observed in series. High impact velocity results a crushed impact site with a chipped region and a complex damage of nature in soda-lime glass.

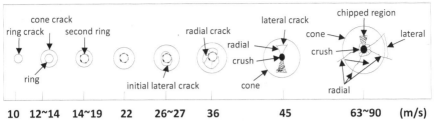

10   12~14   14~19   22   26~27   36   45   63~90   (m/s)

Figure 3 Schematics of sequential crack patterns versus impact velocity on soda-lime glass

An example of surface measurement by laser surface profilometer is shown in figure 4, it confirms the formation of ring crack under the 63 m/s of low impact velocity on CVD-SiC with average roughness 0.2 μm. Considering the roughness of area inside the ring crack diameter, it seems to imply that the asperities of specimen surface was pushed down by a impact of projectile. And around the impact site the generation of crater is observed. It seems that the crater is generated by any impact of particle[10] without distinction of inherent impurities and defects the test specimen has.

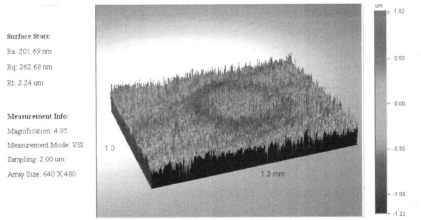

Figure 4 An example of surface analysis result by laser surface profilometer on CVD-SiC

Optical microscope (VHX-200, Keyence, Japan) was used to characterize the trace of impact damage as shown in figure 5. After the measurement, AFM (atomic force microscope VN-8010, Keyence, Japan) was employed to analyze the surface topography, impact damage behavior, and also for data verification.

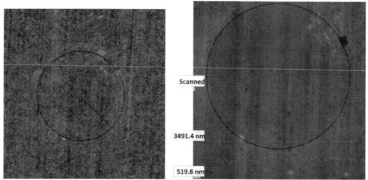

Figure 5 Ring crack on rough CVD-SiC after impact damage was evaluated by video microscope image (left), and atomic force microscope (right)

Test result of particle impact on CVD-SiC polished to 0.02 μm are shown in figure 6. In order to examine the quantitative effect of impact damage on a surface of specimen, the outermost diameters of ring cracks were measured and plotted. This figure shows the relation between the outermost diameter of ring crack and impact velocity. It shows that the outermost diameter of ring crack linearly increases with proportional increase of impact velocity. Three different regimes are marked in the diagram. Regime A is presumed as the threshold of ring crack initiation because below this level there was no sign of ring crack. Only the ring cracks including second ring and multi-rings are observed on the surface in regime B. Regime C shows transition from ring crack to other crack such as radial crack, at least two radial cracks were observed in this regime.

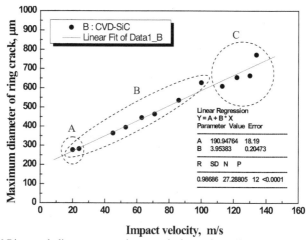

Figure 6 Ring crack diameter versus impact velocity on CVD-SiC with mirror surface

Optical micrographs of impact damages on CVD-SiC polished to 0.02 μm are shown in figure 8. Threshold of ring crack initiation is about 20 m/s (KE=0.70E-03 J), which shows a shape of spiral. It shows that at the end of outer crack it connects to a relatively thin crack trying to close the path as a ring but it seems that the energy was not enough (see figure 7, 0.7E-03J 100 mag.). Shortly, second ring was formed in the range of 24 ~ 53 m/s (KE=1.00E-03 ~ 6.70E-03 J) before multi-rings appears. Radial cracks were observed over 110 m/s (KE=2.15E-02 J) of impact but the maximum diameter of ring crack in over this point was relative larger than rough CVD-SiC. Diameter of innermost ring increases up to 53 m/s but almost constant over the tested impact velocity, since then multi-rings were observed. As it is shown, outer ring cracks were produced successively from inner to outer during the sequential impact process with certain dynamic behavior, which is that the outermost diameter increased with increasing impact velocity. It is assumed that the impact energy and the inertia increase with the increase of impact velocity. In addition, the diameters of produced ring cracks are varied by the material properties of projectile. It is also supposed that radial cracks will affect the material strength of after the impact occurred. The number of radial crack cumulate more as the increase of impact velocities but in this study it was below the number of three.

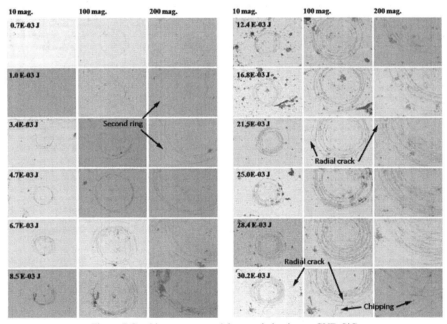

Figure 7 Cracking patterns and fracture behavior on CVD-SiC

Generally, commercial soda-lime glass has surface roughness of 0.00 μm like mirror surface, but other materials have mostly rougher surface. Particle impact test result on CVD-SiC with rough surface Ra=0.2 μm was also examined to know how surface roughness affects on impact damage (see figure 8). Outermost diameter of ring crack is formed similar to the polished CVD-SiC but the values are slightly larger than smooth surface, however marking 122 m/s as a turning point, the diameter of ring crack dramatically decrease as the increase of impact velocity. Particle impact test was retaken for the smooth surfaced CVD-SiC to verify the previous results. Radial cracks were observed in marked area for both two surface conditions. It seems that the formation of radial cracks results the non-uniform energy dissipation in pure initiation of ring crack, consequently the outermost ring crack have irregular behaviors in these level of impact energy.

Threshold of ring crack initiation was observed at 63 m/s (KE=6.70E-03 J), it was hard to seek for the minute impact damage below this level due to the surface roughness. It is assumed that there was energy dissipation when the projectile impacts on the rough surface. The deformation of surface asperities absorbs impact energy comparatively larger then mirror finished surface.

SE micrographs of impact damage on CVD-SiC polished to 0.2 μm are shown in table 3. Mostly, impact site shows multi-rings, and the radial crack appeared over 100 m/s of impact for the first time as similar as polished CVD-SiC. Three radial cracks are formed in case of 118 and 144 m/s velocity of impact.

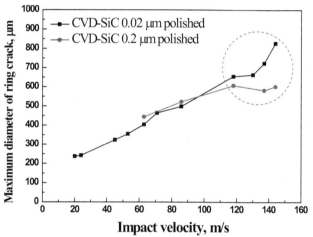

Figure 8 Ring crack versus impact velocity on CVD-SiC with two different surface finish

Table 3 Impact velocity versus magnification of SE micrographs on 0.2 μm polished CVD-SiC

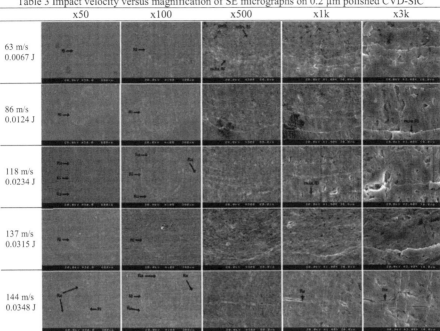

	x50	x100	x500	x1k	x3k
63 m/s 0.0067 J					
86 m/s 0.0124 J					
118 m/s 0.0234 J					
137 m/s 0.0315 J					
144 m/s 0.0348 J					

It is obvious that ring crack is formed even a low impact velocity, and almost no degradation in mechanical properties but damage on the surface. By building the database of threshold of ring crack initiation, the database can be one of the useful parameter to decide fracture damage under FOD.

Threshold is important that at the view point of initial damage, below this level can be will be safe from any other damages. In order to prevent the fracture of components the study on the initial ring crack damage subjected by particle impact has an important significance.

CONCLUSION

Threshold of ring crack initiation was examined for two different roughness of CVD-SiC. Outermost diameter of ring crack linearly increase as the increase of impact velocity in both surface conditions. Surface roughness does not seems to affect on formation of ring crack below 118 m/s of impact velocity, hence the results were similar despite the surface finish. However, the ring crack in rough surface dramatically decrease above this level, where the radial cracks were observed. The dominant reasons is possible chance of energy dissipation due to the deformation of asperities and irregular energy dissipation due to the formation of the radial cracks. Morphological cracking pattern was examined and established for CVD-SiC.

ACKNOWLEDGEMENT
The author would like to thank the Ministry of Education, Culture, Sports, Science and Technology (MEXT) of Japan for scholarship, and the 21 Global COE program of Kyoto University for partial-financial support.

REFERENCE
[1]M.-S. Suh, et al.: Friction and Wear Behavior of Structural Ceramics Sliding against Zirconia, *Wear* **264** (9-10), 800-6 (2008).
[2]M.-S. Suh, et al.: Special Issues on "in situ" Crystallized SiC/SiC Composites. *Proc. of ISAE2009*, 439-42 (2009).
[3]A. Kohyama, et al.: Development of SiC/SiC composites by nano-infiltration and transient eutectoid (nite) process. *Ceramic Eng. and Sci. Proc.*, **23**, 311-8 (2002).
[4]Y. Katoh, et al.: SiC/SiC composites through transient eutectic-phase route for fusion applications. *J. of Nuclear Mat.*, **329-333** (1-3 PART A), 587–91 (2004).
[5]M.-S. Suh, and Kohyama, A.: Effect of Porosity on Particle Erosion Wear Behavior of Lab. Scale SiC_f/SiC Composites. *Int. J. of Modern Phys B*, under press (2009)
[6]M.-S. Suh, et al.: Particle Erosion Wear Mechanism of New Conceptive SiC/SiC Composites, *Tribology Letters*, under press (2010)
[7]Sung R. Choi: Foreign object damage phenomenon by steel ball projectiles in a sic/sic ceramic matrix composite at ambient and elevated temperatures. *J. Am. Ceram. Soc.*, **91** (9), 2963-8 (2008).
[8]R.T. Bhatt, et al.: Impact resistance of uncoated SiC/SiC composites. *Mater. Sci. and Eng. A*, **476** (1-2), 20-8 (2008).
[9]S. R. Choi, R. T. Bhatt, J. M. Perrira, and J. P. Gyekenyesi, Foreign Object Damage Behavior of a SiC/SiC Composite at Ambient and Elevated Temperatures, ASME Paper No. GT2004-53910 (2004).
[10]I. Maekawa, et al.: Damage Induced in SiC by a Particle Impact. *Eng. Fracture Mech.* **40** (4-5), 879-86, (1991)
[11]A. G. Evans, Strength degradation by projectile impacts, *J. Am. Ceram. Soc.* 56 405-9 (1973).
[12]D. A. Shockey, D. C. Erlich and K. C. Dao, Particle impact damage in silicon nitride at 1400°C, *J. Mater. Sci.* 16, 477-82 (1981).
[13]Y. Akhnune, Y. Kanato and K. Matoda, Spherical-impact damage and strength degradation in silicon nitrides for automobile turbocharger rotors, *J. Am. Ceram. Soc.*, **72** (8), 1422-8 (1989).
[14]D. A. Shockey, D. J. Rowcliffe, K. C. Dao and L. Seaman, Particle impact damage in Silicon Nitride, *J. Am. Ceram. Soc.*, **73** (6), 1613-9 (1990).
[15]A. M. Rajendran and J. L. Kroupa, Impact Design Model for Ceramic Materials, *J. Appl. Phys.*, **66** (8) 3560–5 (1989).
[16]A. G. Evans and T. R. Wilshaw, Dynamic Solid Particle Damage in Brittle Materials: An Appraisal, *J. Mater. Sci.*, 12, 97–116 (1977).
[17]B. M. Liaw, A. S. Kobayashi, and A. G. Emery, Theoretical Model of Impact Damage in Structural Ceramics, *J. Am. Ceram. Soc.*, 67, 544–8 (1984).
[18]M. van Roode, et al., "Ceramic Gas Turbine Materials Impact Evaluation," *ASME Paper* No. GT2002-30505 (2002).

ADVANCED CERAMIC-STEEL PAIRINGS UNDER PERMANENT SLIP FOR DRY RUNNING CLUTCH SYSTEMS

Prof. Albert Albers
IPEK – Institute of Product Development, KIT – Karlsruhe Institute of Technology
Karlsruhe, Germany

Dipl.-Ing. Michael Meid
IPEK – Institute of Product Development, KIT – Karlsruhe Institute of Technology
Karlsruhe, Germany

ABSTRACT

Running clutches under permanent slip offers multiple applications regarding vibration damping or torque distribution in drivetrains, for instance. Regarding state-of-the-art applications, only lubricated clutches are used. Advanced engineering ceramics show specific benefits in wear behavior and thermal resistance and are therefore representing an interesting chance for running also unlubricated clutches under permanent slip conditions. The emphasis of this analysis is the basuc characterization of the tribological system behavior of the non-oxide ceramic/steel friction pairing SSiC/C45E regarding friction coefficient and wear. As influencing factors the sliding speed and contact pressure between the friction surfaces, as well as the rotational speed level, leading to different convectional heat transfer rates, are varied and analyzed. Additionally, the resultant temperatures are used for characterization of their influence. The organic state-of-the-art facing Valeo 820DS is used for as a reference. The analysis results of running advanced ceramics under permanent slip are very promising concerning friction coefficient level and stability as well as wear behavior.

INTRODUCTION

More than ever, the run on innovative and efficiency enhancing technologies in automotive vehicles is accelerating, forced by regularization of penalty taxes for high carbon dioxide emission, rising fuel costs and last but not least the enforced interest to reduce the effect on the climate change. Furthermore, increased vehicle requirements with regards to comfort and dynamics have to be fulfilled. A common approach is downsizing of the engine and using turbochargers. This results in lower weight and better efficiency among others because of an improved torque characteristic of the engine, for instance. However, higher vibrations of the engine torque are generated affecting the comfort of the vehicle and being considered as a negative effect. Therefore vibration reducing drivetrain components have to be integrated and optimized in order to damp these vibrations to a tolerable level.

A common system is the dual mass flywheel (DMF) and in a high end version its combination with a centrifugal force pendulum. Regardless of the advantages, a DMF has also disadvantages. Additional costs, increased weight and needed design space could be mentioned. Another approach could be running the clutch under controlled permanent slip in order to reduce drive train vibrations. Additionally, implementation of a clutch running under permanent slip may lead to a further increase of comfort in combination with other vibration damping approaches.

Lubricated multiple disk clutches are already used under permanent slip conditions in several applications. Possible automotive applications are the dynamic vibration damping between engine and drivetrain, torque distribution between front and rear axles in all-wheel drivetrains or torque distribution between the wheels of an axle. The reason for using lubricated clutches instead of unlubricated clutches is the heat input caused by energy dissipation of a clutch running under permanent slip.

In comparison to lubricated clutches, lower energy losses and lower system costs are citable for dry running clutches. Until now, common organic facings are not resistant enough for running longer time with high energy input without being destroyed immediately by getting deflagrated or without a huge drop in lifetime. The compensation of the increased wear could be problematic with a state-of-the-art SAC (self adjusting clutch). Advanced ceramic-steel friction pairings with monolithic ceramics could allow running a clutch under permanent slip due to a better wear behavior and the higher thermal resistance, whose characterization is the emphasis of this research.

FRICTION MATERIALS AND EXPERIMENTAL METHODS
In order to characterize the tribological system behavior, experimental analyses have been performed at testing level IV on the dry friction test bench of the IPEK[1]. Within testing level IV, the component clutch disk is tested gathering information about its tribological behavior including influences of cushion spring characteristics as well as integration concepts for ceramics concerning fixation and bearing of pellets. The experiments have been performed using an advanced engineering ceramic/steel friction pairing and also a conventional state-of-the-art organic facing running against cast iron as a reference.

Clutch disk with advanced engineering ceramics
As ceramic/steel friction pairing the non-oxide ceramic SSiC and the steel C45E in normalized condition is used. This steel is selected in collaboration with the Institute of Material Science II of the KIT, considering tribological characteristics. In previous research, the friction pairing SSiC/C45E has shown a good tribological behavior concerning wear characteristics, coefficient of friction and gradient of coefficient of friction. Beyond that a good hybrid compound of SSiC and metal is realizable[3].

The hardness of the steel C45E is about 200 HV05 whereas the hardness of the engineering ceramic SSiC is much higher (2540 HV05). The ceramic pellets are integrated in "as-fired" state with a roughness $R_z \cdot$ 2.6 µm and the steel disks have a roughness $R_z \cdot$ 6.8 µm.

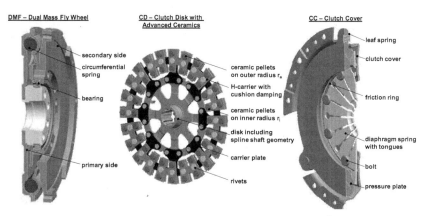

Figure 1. Clutch system with dual mass flywheel and clutch disk prototype[2] with the advanced ceramic/steel friction pairing EkasicF® (SSiC)/C45E

The used prototype has been optimized with regards to increased transmissible torque, improved wear resistance and weight (Figure 1). The clutch disk is composed of a weight optimized disk including spline shaft geometry for the transmission input shaft of the experimental vehicle of the IPEK, a carrier ring for mounting the H-shaped cushion spring components with the advanced engineering ceramics/steel compound. During clutch modulation, the cushion springs generate a designated contact pressure adaptation up to the operation point.

The cushion spring design improves the wear balance between the different ceramic pellets, so that the pellets can be used "as-fired" leading to lower production costs. Additionally, the cushion spring interacts with the deterioration improved pellet design. The pellets have a spherical friction surface to avoid wear of edges leading to an enormous increased abrasive wear characteristic of metallic friction partners[4]. The radius of the spherical friction surface of the pellets is chosen dependently to the angle of the single cushion springs.

Conventional clutch disk with organic facing

As a reference, the series clutch disk with the organic facing Valeo 820DS is used running against pressure plates of compacted graphite iron (CGI) GJV-300. The organic facings are cushioned with flat springs acting distributed on the whole surface in contrast to the previously described prototype with advanced ceramics. The organic facings are fastened by rivets on the cushion springs. The friction surface is incorporated with grooves in order to transport loose particles out of the friction contact (Figure 2).

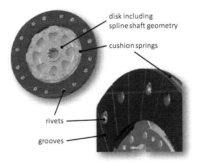

Figure 2. State of the art clutch disk with an organic facing

TEST BENCH SETUP

The dry friction test bench consists of two high dynamic rotary current servo motors, which are connected by the clutch disk (Figure 3). The clamping force is provided with a step motor acting on a spindle integrated in the sledge.

The clutch disk is mounted on a spline shaft, equivalent to the transmission input shaft, which offers a degree of freedom in axial direction. The pressure plate on the side of the prime mover (pressure plate 1) is fixed in axial direction and the second pressure plate (pressure plate 2) is fixed in axial direction on the sledge, which can be moved by the step motor. The attachment of the pressure plates is designed removable to allow a replacement of them. The transmitted torque of the second pressure plate is transmitted over three stiff linear bearings to the first pressure plate, also synchronizing them[3].

MEASUREMENT INSTRUMENTATION

As measured values the rotating speed, the transmitted torque, the axial clamping force and the temperature of the first pressure plate are considered as relevant for basic characterization of the tribological system behavior.

As measurement equipment a rotary torque transducer with an effective range up to 1000 Nm and an axial force transducer with an effective range up to 10.000 N, a type K thermocouple and rotary encoders with a high resolution at each e-machine are applied (Figure 4). The torque and axial force transducers as well as the thermocouple communicate contactless.
The roughness of parts is measured separately with a tactile surface measurement instrument.

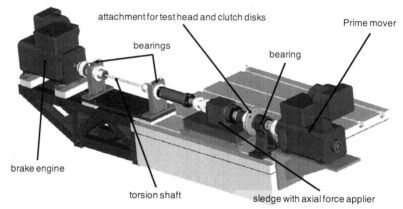

attachment for test head and clutch disks

Prime mover

bearings

bearing

brake engine

torsion shaft

sledge with axial force applier

Figure 3. Dry Running Test Bed of IPEK – Set-up for the clutch disk prototype 3

ENERGY INPUT
The dissipated energy of slipping clutches lead to a heat generation in the friction contact and a heat transfer into the pressure plates and the ceramic pellets. The amount of dissipated energy and the corresponding power is dependent on the transmitted torque and the rotational speed difference between the pressure plates and the clutch disk. Comparing the energy input of the different operating points, a wide range from circa 0.1 kW up to 1 kW is covered during the analysis.

TEST PLAN
In this analysis the operating parameters transmitted torque T_c, rotational speed n_{cd} and rotational speed difference between the pressure plates and the clutch disk n_d are varied. Hence the analysis references to the characteristic tribological parameters sliding speed, contact pressure and temperature, which are determined or calculated out of the measured variables.
The variation of the operating parameters is done in this analysis with the Design of Experiments (DoE) method "one factor at a time" (OFAT), which allows visualizing the influences of the different operating parameters under comparable conditions. The combination of the operating parameters n_{cd} = 800 rpm, n_d = 35rpm and T_c = 100Nm is used as reference point for the OFAT analysis. These operating parameters lie between the lower and upper end of the ranges, so the reference point builds the intersection point of the several OFAT variations. The reference point will also be used to analyze the running-in-characteristic of the friction pairing.
In order to achieve statistical data, several test cycle runs are made in series at each operating point. Beyond that, a temperature check is done after each test cycle to receive comparable data. If the temperature exceeds the chosen value of 50°C after a test cycle, the next test cycle will not be started until the measured temperature drops under this chosen temperature limit. The reason for running the

clutch at such a low temperature is that the test bench is an open system and it isn't possible to reach higher temperatures at all operating points.

Figure 4. Dry Running Test Bed – Setup with measurement instrumentation

TEST CYCLE

Several test cycles have been performed in order to analyze the tribological behavior under permanent slip conditions (Figure 5). The reason for using several short test cycles is the possibility to interrupt the tests due to unexpected problems or temperature limits.

Each test cycle starts at standstill of the test bench and with a clearance between the pressure plates and the clutch disk. At the beginning of a test cycle (Point 1, Figure 5) the pressure plates and the clutch disk are accelerated synchronously from standstill to the clutch disk rotational speed n_{cd} (Point 2, Figure 5).

Then the engine speed is increased to a higher speed level n_e in order to generate a speed difference n_d between the friction surfaces (Point 3, Figure 5). After reaching the desired rotational speed difference n_d the step motors actuated sledge begins to close the pressure plates (Point 4, Figure 5) until the desired transmitted torque T_c is reached due to the applied clamping force F_{cl}. After that, the test bench runs with a constant torque under permanent slip for the time t_{PS}, which has been chosen to 120s. Therefore the step motor of the sledge is actuated dynamically by the torque control in order to adjust the clamping force F_{cl} to the necessary value.

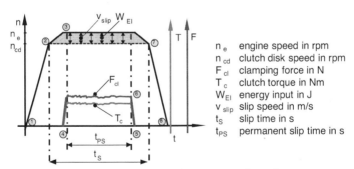

n_e engine speed in rpm
n_{cd} clutch disk speed in rpm
F_{cl} clamping force in N
T_c clutch torque in Nm
W_{EI} energy input in J
v_{slip} slip speed in m/s
t_S slip time in s
t_{PS} permanent slip time in s

Figure 5. Test procedure for permanent slip cycles

After the permanent slip time t_{PS} is over (Point 6, Figure 5) the sledge opens the pressure plates and the transmitted torque T_c and the clamping force F_{cl} decreases to zero (Point 5, Figure 5). At the end of a test cycle the rotational speed difference n_d is reduced to zero (Point 7, Figure 5) and then both engines are stopped synchronously (Point 8, Figure 5). Finally the temperature check is activated to delay the start of the next cycle in case of an exceeded temperature limit.

MEASUREMENT ANALYSIS PROCEDURE
The characteristic tribological parameters sliding speed v_s, contact pressure p_{Cl} and the coefficient of friction μ have to be calculated with the measured values axial force F_{Cl}, transmitted torque T_c and the static component variable average friction radius r_m.
For calculating the contact pressure, a closer look on the friction system has to be taken. It isn't allowed to use the theoretical friction area of the pellets or the organic facing, called nominal friction area, for analyses, because this will result in deviations. The detection of the effective friction area is only necessary for the calculation of the contact pressure and not for other characteristic values.
Concerning the organic facing, the not wearing groove and rivet areas have to be subtracted from the nominal friction area in order to receive the effective friction area. In contrast, the detection of the prototypes effective friction area is more complicated. The ceramic pellets don't wear over the whole surface (Figure 14), because all 96 friction contacts underlie a tolerance chain of ceramics and of single cushion springs. Some pellets show good wear pattern from the beginning, but other pellets only wear slightly. This effect gets reduced during a longer running-in and the friction coefficient gets stabilized due to better wear distribution.
The effective wear surface is detected by summation of all single friction areas. It is to mention, that this detected area isn't constant over lifetime due to wear of the pressure plates and wear of the ceramic pellets. Prior analysis showed that the effective friction area can be assumed with a good approach as 25% of the nominal friction area. Hence, the contact pressure can be calculated independent of the regarded clutch disk using the corresponding effective friction area.

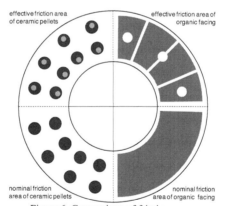

Figure 6. Comparison of friction areas

For characterization of wear behavior two different methods accomplished:
Linear wear intensity W_I in $\mu m/km$:
This value relates the wear height h_w to the sliding distance. It is based on the assumption of an average wear over propagation and describes therefore a linear correlation between wear and

sliding way. In order to calculate this value, the distance under slip conditions has to be summated and the height of the wear has to be measured.

$$W_l = \frac{h_w}{2 \cdot \pi \cdot r_m \cdot \sum_{i=1}^{z} \frac{n_d}{60} \cdot t_{ps,i}}$$

Function wear W_F in mm³/MJ:
This value relates the volumetric value of wear to the energy input due to friction. To calculate this value, the energy input has to be summated and the wear volume V_w has to be detected:

$$W_F = \frac{V_w}{\sum_{i=1}^{z} W_{EI,i}}$$

Concerning the organic clutch disk the wear volume V_w can be calculated out of the wear height h_w and the effective friction surface A_{eff} of the organic facing.

Concerning the clutch disk prototype, the wear volume V_w has to be detected out of the wear marks on the pressure plates (Figure 14), because of the significant higher wear of steel instead of ceramics. The wear volume V_w is calculated out of several measurements with a surface measurement instrument. Therefore the surface profiles of the wear marks are compared with a linear interpolated line forming original surface profiles in order to receive the wear profile (Figure 7). For calculating the wear volume V_w the product of one incremental wear height h_w and its corresponding circumference is integrated over the relevant range of measurement points.

$$V_w = \int 2 \cdot \pi \cdot r \cdot h_w$$

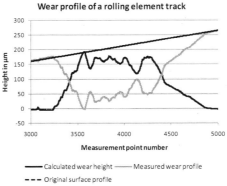

Figure 7. SSiC/C45E wear profile

RESULTS AND DISCUSSION
The goal of this permanent slip analysis is firstly the basic characterization of the tribological behavior of the SSiC/C45E friction pairing and secondly the gathering of data of the system behavior. Concerning the tribological behavior, different influencing factors on the friction coefficient are

analyzed and the wear behavior is analyzed as a whole. Concerning the system, the thermal behavior and heat transfer is interesting regarding longer permanent slip phases, which are planned in the future. The tests with the organic facing Valeo 820DS are only performed for comparison purposes.

The analysis is performed by running test cycles in sequence with a test bench standby for cooling down between the test cycles, if the temperature is too high. For the analysis the sequence of test cycles is split and each test cycle is evaluated separately.

The test cycle, being exemplary discussed now, is measured at the reference point n_{cd} = 800 rpm, n_d = 35 rpm and T_c = 100 Nm (Figure 8). The visualized measurement data has been post-processed to eliminate high frequency vibrations in the signal. You can also see oscillations with a frequency about 0.53 Hz in the post-processed signal data most likely caused by the torque control of the test bench.

The clamping force shows a decreasing behavior and simultaneously the transmitted torque remains constant while running under permanent slip. Thus implicates, that the friction coefficient has to increase at the same time (Figure 8).

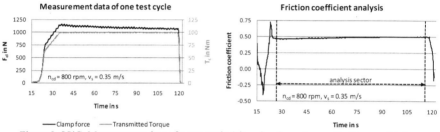

Figure 8. SSiC: Measurement data of a test cycle (clamping force, transmitted torque, friction coefficient)

Until beginning and after ending of the permanent slip time the measured test data isn't exploitable for further analysis. An example is the physically impossible negative value of the friction coefficient. At this point, there isn't any contact between the friction surfaces and the measured values are therefore nearby zero. Thus could lead to meaningless values during calculation of the friction coefficient. Hence, for characterization of the tribological behavior only the illustrated analysis sector with permanent slip conditions is used (Figure 8).

The measurement data of the following discussion is quoted in two kinds:
If nothing is explicitly mentioned, the quoted values are based on mean values of an analysis sector, when comparing several test cycle runs at one operating point. Respectively, when comparing different operating points, the quoted values are mean values of all analysis sectors of an operating point.
If it is separately quoted in a diagram, single measurement points of different test cycles are used in order to reduce the influence of the temperature.

Running-in characteristic

It is very important to pay attention to the running-in characteristic when discussing influences of variables on the friction coefficient. Between the measured values at the beginning and at the end of the test a considerable difference is remarkable (Figure 9). Both measurements have been performed at the reference point with the friction pairing SSiC/C45E. The difference of the total energy input between the two runs is circa 3 MJ up to 4 MJ. For the analysis of the influencing factors, the run with the closer basis to the other operating points is used.

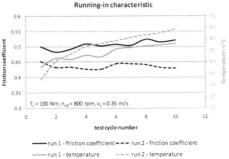

Figure 9. SSiC: Running-in characteristic

Influence of the contact pressure

The measurements show, that the friction coefficient is stable up to the measured contact pressure of 1.65 Mpa (Figure 10). At the operation point with a contact pressure of 2.12 Mpa a highly increased friction coefficient has been measured. A hypothesis for that could be an also highly increased abrasive wear behavior. For determination of the wear phenomenon separate tests have to be performed, because within this analysis, only the total wear is analyzed. In contrast to the organic reference facing, the friction coefficient of SSiC is significant more stable.

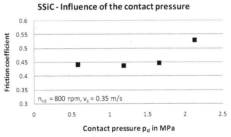

Figure 10. SSiC: Contact pressure

Influence of the sliding speed

The coefficient of friction seems to be almost constant over the sliding speed, if a constant temperature is considered (Figure 11). This constancy is only valid for small sliding speeds, for higher sliding speeds the friction coefficient has a negative gradient[6]. Only the beginning of the measurements at this operating point had a higher temperature level, which could be compared to other operating points. The reason for that is the small energy input, compared to the heat losses at this rotational speed level. Hence, the mean value of the friction coefficient is much lower due to the influence of the temperature.

The characteristic of advanced ceramics and the organic reference facing is almost equivalent, but there's a significant difference between the levels of the friction coefficient (Figure 11).

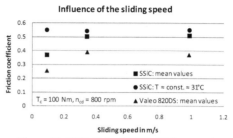

Figure 11. SSiC, Valeo 820DS: Sliding speed

Influence of the temperature

The influence of the temperature can be detected by analyzing the behavior of the friction coefficient over an entire test cycle sequence at one operating point (Figure 12). The result is that the friction coefficient is rising with an increasing temperature within the examined temperature range.

It is to mention, that the measured temperature doesn't match the friction contact temperature. The reason therefore is that there is a small distance between the friction contact and the position of the thermocouple in the pressure plate. Additionally, the measured temperature of the pressure plate has a time delay due to the heat capacity of the pressure plate.

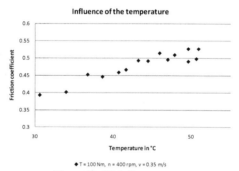

Figure 12. SSiC: Temperature

Influence of the rotational speed

Higher rotational speeds are increasing the convectional heat transfer from the friction surfaces, the clutch disk and the pressure plates to the surrounded air. This leads to lower temperatures in the friction contact area and to lower pressure plate temperatures at steady state conditions. It's an influence of the rotational speed on the friction coefficient remarkable (Figure 13). Actual, the rotational speed level should not have any direct influences on the tribological behavior since no relevant tribological parameters are changed. The most likely reason for that is a highly influenced friction contact temperature. The advanced ceramic SSiC has a higher conduction of heat in contrast to the pressure plates of steel. It is supposable, that the heat losses of the pellets are highly increasing with higher rotational speeds due to the cliffy design of the clutch disk leading to high turbulences of the air.

Influence of the rotational speed

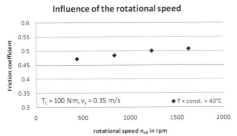

Figure 13. SSiC: Rotational speed

Wear characteristic

The characteristic wear values for the SSiC/C45E friction pairing are calculated with the theoretical energy input and sliding way extracted out of the test plan.

- Linear wear intensity: W_I = 1.01 μm/km
- Function wear: W_F = 31.99 mm³/MJ

The linear wear intensity marks a very good value, which is definitely necessary for running a clutch under permanent slip. The function wear characteristic lies in contrast a little bit over measured values of synchronization tests due to the long sliding distance.

During friction, the pellets are getting polished, because of the roughness change during testing. The roughness changed from a median $R_z = 2.19$ μm before testing to $R_z = 0.51$ μm of worn areas after testing.

For detection of the wear characteristics of the organic facing further tests have to be performed. At the moment the organic facing does not yet wear over the whole friction surface.

Figure 14. Ceramic pressure plates after testing and wear behavior of the clutch disk

Abrasive particles

During the permanent slip tests with the advanced ceramics SSiC, brown particles emerged within the friction contact, whose components are unknown yet. Until now, no comparable particles have been generated in synchronization tests with the same friction pairing so far. You can see the brown discoloration of the wear marks and the particles in Figure 14. One possible reason could be low energy input and low sliding speeds, similar to frictional corrosion. The analysis of the particle composition and the reason for the particle generation will continue.

CONCLUSION

The first experimental results of running a clutch disk with the advanced engineering ceramic SSiC under permanent slip showed very promising results. The friction coefficient is on a high level and in reference to the organic facing more stable with a considerable difference in the level of the friction coefficient. In contrast to the organic facing, the convectional heat transfer is much higher with the prototype providing longer running times or higher energy input under permanent slip conditions. The linear wear intensity is on a very good level, which is also necessary for running a clutch under permanent slip conditions. In contrast to synchronization tests the value represents only the sixth part. The function wear is about 10% above synchronization tests but still at a very good level.

OUTLOOK

Based on these promising results, further tests for running advanced ceramics under permanent slip are planned. Concerning the discussed friction pairing, analysis with longer running times are considered to increase the rate of wearing pellets and to verify the transferability of the achieved results.

Additionally, material analyses on abrasive particles, the pressure plates and some ceramic pellets have to be carried out to detect the appeared wear mechanisms. Therefore images with microscope and SEM (scanning electron microscope) will be taken and additionally the consistence of the particles will be determined with an SEM or an EDX (energy dispersive X-ray).

It is also planned to perform permanent slip analysis with another advanced ceramic/steel pairing. Therefore another prototype with aluminum oxide has been build up. Within this prototype the hybrid compound of metal and advanced ceramic is soldered in contrast to the tested prototype with SSiC.

In order to review the transferability of the results between different testing levels, experiments with a modular testing head will be performed. It offers the possibility to run single pellets with a controlled pressure distribution and at different friction diameters. The method of testing friction pairings at a high degree of abstraction leads to a cost reduction, if it is possible to transfer the results to higher testing levels.

Another outlook is to rise the temperatures during the analysis by installation of a cover around the test head or an adapted test plan.

ACKNOWLEDGEMENTS

The authors would like to thank the Deutsche Forschungsgemeinschaft (DFG) for financial support within the frame of the Collaborative Research Centre CRC 483 "High performance sliding and friction systems based on advanced ceramics"
http://www.sfb483.uni-karlsruhe.de

NOMENCLATURE

n_e	[rpm]	engine speed		μ	[-]	coefficient of friction
n_{cd}	[rpm]	clutch disk speed		t_s	[s]	sliding time
n_d	[rpm]	speed difference between engine and clutch disk		t_{ps}	[s]	permanent slip time
T_c	[Nm]	transmitted torque		W_{EI}	[kJ]	energy input
F_{cl}	[N]	clamping force		A_w	[m²]	wear area

v_s	[m/s]	sliding velocity	h_w	[μm]	wear height
A_{nom}	[m²]	nominal friction area	V_w	[mm³]	wear volume
A_{eff}	[m²]	effective friction area	W_I	[μm/km]	wear intensity
p_{cl}	[N/mm²]	contact pressure	W_F	[mm³/MJ]	function wear
r_m	[m]	median friction radius			

REFERENCES

[1] Czichos, H., Habig, K.-H.: Tribologie-Handbuch: Reibung und Verschleiß, 2. Auflage, Vieweg, Wiesbaden 2003, ISBN 3-528-16354-2.

[2] Albers, A.; Mitariu, M; Ott, S.: Effiziente Leistungssteigerung im Antriebsstrang durch innovative Reibkupplungen auf Basis ingenieurkeramischer Friktionswerkstoffe. GfT - Gesellschaft für Tribologie, Tagungsband 49. Tribologie-Fachtagung, Göttingen, 22. - 24. September 2008, 35/1ff

[3] Albers, A., Ott, S., Mitariu, M.: Tribologische Systemuntersuchungen an Fahrzeugkupplungen mit integrierten ingenieurkeramischen Komponenten. VDI-Berichte Nr. 1943, Getriebe in Fahrzeugen, Friedrichshafen, 27. Juni 2006, S.445 – 468.

[4] Albers, A., Ott, S., Mitariu, M.: Innovative trockenlaufende Reibkupplungen auf Basis ingenieurkeramischer Friktionswerkstoffe - Tribologische Charakterisierung auf Systemebene. GfT - Gesellschaft für Tribologie, Tagungsband Tribologie-Fachtagung 2007 vom 24. bis 26. September 2007 in Göttingen, 52/1ff.

[5] Albers, A., Mitariu, M., Ott, S.: System- und Friktionsverhalten eines prototypenhaft umgesetzten Kupplungssystems mit ingenieurkeramischen Friktionswerkstoffen. GfT - Gesellschaft für Tribologie, Tagungsband 50. Tribologie-Fachtagung, Göttingen, 21.-23. September 2009, 75/1ff.

DEVELOPMENT AND VALIDATION OF LUBRICATED MULTI-DISK CLUTCH SYSTEMS WITH ADVANCED CERAMICS

Prof. Albert Albers
IPEK – Institute of Product Engineering, KIT – Karlsruhe Institute of Technology
Karlsruhe, Germany

Dipl.-Ing. Johannes Bernhardt
IPEK – Institute of Product Engineering, KIT – Karlsruhe Institute of Technology
Karlsruhe, Germany

Dipl.-Ing. Sascha Ott
IPEK – Institute of Product Engineering, KIT – Karlsruhe Institute of Technology
Karlsruhe, Germany

ABSTRACT
Lubricated frictional systems and their frictional materials have to fulfil high standards concerning mechanical and thermal load capacity, wear and coefficient of friction. Engineering ceramics show high potential to increase the power density of these systems.
The main goal is to develop methods and tools to design highly loaded lubricated frictional systems. Representative for highly loaded frictional systems with low pressure and high sliding speed, a lubricated multi-disk clutch has been chosen as a system for demonstrating purpose. This system is being used to deepen the knowledge about the tribological systems behaviour.

INTRODUCTION
To fulfill increasing demands concerning efficency and environmental impact new powertrain systems especially in motor vehicle applications are required. In most cases the powertrain of vehicles uses clutch systems to enable gearshift and start-up. The clutch system has an high impact on the dynamic systems behaviour, power density and efficency of the powertrain. Depending on the powertrain concept there are different requirements on clutch systems. New powertrain concepts are maybe using more than one clutch to connect and disconntect different engines and ancillary units depending on the operating condition. Multi-disk clutch systems which are often used in the powertrain of vehicles as well as in industrial plants, have an high impact on power density and dynamic behaviour of the whole system. A safe operation of the whole system has to be guaranteed under strongly variing conditions combined with the need of increasing power density.

SYSTEMS ANALYSIS OF THE LUBRICATED MULTI-DISK CLUTCH SYSTEM

CLUTCH SYSTEM
Figure 1 shows a lubricated multi-disk clutch system used in passenger vehicles. The system enables the start-up procedure and in this function it has to be considered as a torque converter replacement. The powertrain shows higher efficiency compared to a system using a torque converter.

Fig 1: Lubricated multi-disk clutch

The abstract description of a multi-disk clutch system according to the C&CM (Contact and Channel Model [1]) (Figure 2) clarifies the relevant working surface pairs and Channel and Support Structures. By the use of this type of description it is possible to analyze the function of a multi-disk clutch system. During operation the clutch system is being actuated by applying a clamping force into axial direction. This results in a contact pressure acting in the frictional contact (WSP_{fric}) of the clutch system allowing to transmit a frictional torque. From the frictional contact the torque is transmitted via the inner and outer carrier to the in- and output shafts. During sliding operation heat is generated. To avoid thermal overstraining the clutch system is convection cooled. Usually oil is used as cooling medium.

Fig 2: C&CM-analysis of multi-disk clutch

The detailed analysis of the frictional contact (WSP$_{fric}$ (Figure 3)) shows the different types of friction. Depending on the load, the mikrotexture of the frictional surface, the material of the friction pairing and the lubricant there are areas with hydrodynamic, or boundary friction.

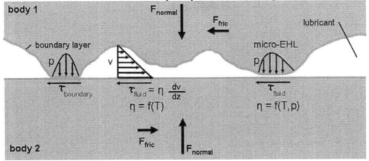

Figure 3: Mixed friction

Depending on the operating conditions the system is being damaged continuously. Based on the systems analysis there are the following causes of damage:
- Interaction between friction plate, counter plate and lubricant
- Interaction with the systems environment

Based on experimental investigations using state of the art lubricated multi-disk clutch systems Hauser [2] identifies the thermal load of the oil as the main influence concerning damaging of the frictional contact of a lubricated clutch system.

ENERGY EQUATION OF MULTI-DISK CLUTCH SYSTEM

Figure 4 shows a black-box description of a multi-disk clutch system.

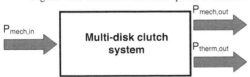

Figure 4: Energy equation of multi-disk clutch system

During operation there is a mechanical and thermal power transmitted through the systems boundary. Depending on the operating conditions the proportion between mechanical and thermal power output is varying.

$$P_{mech,in} - P_{mech,out} - P_{therm,out} = \frac{dQ_{clutch}}{dt}$$

$$P_{mech,in} - P_{mech,out} = \frac{dQ_{clutch}}{dt} + P_{therm,out}$$

Equation 1: Energy equation multi-disk clutch

The difference between mechanical power in- and output is being transformed into heat which heats up the clutch system respectively is transferred to the cooling medium. The heating of the oil results in accelerated oil aging. That is the reason why the transfer of frictional heat to the cooling medium has to be seen as an indicator concerning power density of a lubricated multi-disk clutch system.

LUBRICANT
The Arrhenius equation describes the connection between temperature and speed of a chemical reaction.

$$v = v_0 e^{-\frac{E_A}{kT}}$$

$v = speed \quad of \quad reaction$

$v_0 = constant \quad of \quad proportionality$

$E_A = activation \quad energy$

$k = Boltzmann \quad constant$

$T = temperature$

Equation 2: Arrhenius equation

The following equation describes the proportion between identical chemical reactions at different temperature levels. The proportion of reaction speeds is called acceleration coefficient:

$$\frac{v_2}{v_1} = e^{-\frac{E_A}{k}\left(\frac{1}{T_2}-\frac{1}{T_1}\right)} \qquad \frac{v_2}{v_1} = A \qquad A = acceleration \quad coefficient$$

Equation 3: Acceleration coefficient

The acceleration coefficient allows to compare two identical volumes (dV_1, dV_2), which are only different in temperature, concerning the proportion of reaction speeds.

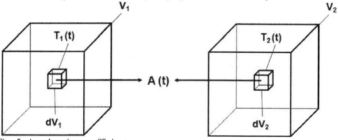

Fig: 5: Acceleration coefficient

The temperature distribution in a volume is depending on time and on position. To compare the chemical reaction between the two volumes V_1 and V_2 the acceleration coefficient has to be integrated with local temperature and volume. The resulting factor is called extended acceleration coefficient A^*.

$$A^* = \int_V \int_t A \, dt \, dV$$

$V = volume$

$t = time$

Equation 4: Extended acceleration coefficient

MULTI-DISK CLUTCH SYSTEM

In the framework of the research of CRC 483 a prototype of a lubricated multi-disk clutch has been developed and built-up (Figure 6) [3]. This prototype system uses advanced ceramics to realize a ceramic steel material pairing in frictional contact. The advanced ceramic material shows much higher mechanical and thermal strength compared to usually used organic facings. Higher strength of the friction pairing allows a reduced contact area to improve heat transfer for reduced thermal load of the frictional system.

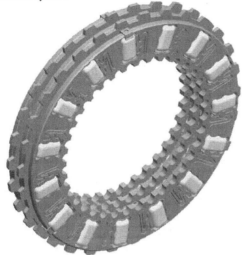

Figure 6: Multi-disk clutch system using advanced ceramics

Using the high strength of the ceramic material combined with a suitable design of the clutch system allows reducing the thermal load of the clutch system and therefore improving the power density.

SIMULATION

Compared to state of the art systems the prototype (Figure 6) uses a smaller friction area. Reducing the friction area generates higher local load. The calculation of oil aging (Equation 3) shows that increased temperature generates much higher oil aging. A result of reducing friction area could be an improved heat transfer with lower mass temperature but very high local temperature which accelerates oil aging.

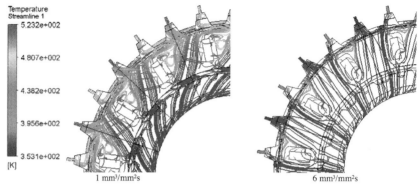

$n_{input} = 1090$ 1/min, $n_{output} = 140$ 1/min, $q_{fric} = 0{,}19$ W/mm^2

Figure 7: Oil flow in multi-disk clutch system

Figure 7 shows the result of a simulation based analysis of the prototype clutch system using CFD. The clutch system is in permanent sliding operation with a specific friction power of 0,19 W/mm^2 with a different specific oil flow (1 or 6 mm^3/mm^{2}s)

	Multi-disk clutch prototype		State of the art
Specific friction power in W/mm^2	0,19	0,19	0,093
Specific oil flow in mm^3/mm^{2}s	1	6	1
Medium counter plate temperature in °C	187	98	280
Maximum counter plate temperature in °C	235	229	-
Medium oil temperature outlet in °C	177	96	129
A* (1 mm^3/mm^{2}s / 6 mm^3/mm^{2}s)	12,28		-

The comparison shows that by increasing oil flow the medium counter plate temperature could be reduced by 89 K and the maximum counter plate temperature by 6 K. Based on this analysis a calculation of the extended acceleration coefficient has been carried out to determine the oil aging as a result of the thermal load during operation. The calculation shows that increasing the oil flow has a high impact on oil aging.

Furthermore the simulation results have been compared to experimental results using a state of the art clutch system of the same dimensions. Although this clutch system has been loaded by 50 percent compared to the prototype system it shows much higher (93 K) medium counter plate temperatures. Even the maximum counter plate temperature of the prototype system is lower (45 K) than the medium temperature of the state of the art system. This will result in a much higher oil aging speed of the state of the art clutch system. The improvement in power density of the prototype system shows the potential of using advanced ceramics as friction materials and the need of suitable methods as well.

CONCLUSION AND OUTLOOK

Within the presented research topic a systems analysis has been carried out to deepen the knowledge about the relevant interaction between load and the damaging mechanisms. Based on this analysis some possibilities to improve power density of the system have been identified. The integration of advanced ceramics as a friction material combined with a suitable systems design is a very powerful possibility. To realize a prototype different simulation methods, especially concerning fluid flow and heat transfer, have been used and integrated into development process.

The next step is to analyse and optimize the fluid flow through the frictional contact to improve the tribological behaviour depending on the specific operating conditions. Furthermore the simulation models have to be validated by experimental investigations.

ACKNOWLEDGEMENTS

The authors would like to thank the Deutsche Forschungsgemeinschaft (DFG) for financial support within the frame of the Collaborative Research Centre CRC 483 "High performance sliding and friction systems based on advanced ceramics"

REFERENCES

1   Matthiesen, S.: Ein Beitrag zur Basisdefinition des Elementmodells „Wirkflächenpaare und Leitstützstrukturen" zum Zusammenhang von Funktion und Gestalt technischer Systeme, Forschungsberichte des Instituts für Produktentwicklung, Band 6, Karlsruhe 2002.

2   Hauser, C.: Einfluss der Ölalterung auf Reibcharakteristik und Reibschwingverhalten von Lamellenkupplungen, Dissertation FZG TU München 2007.

3   Albers, A., Ott, S., Bernhardt, J.: Systems development of lubricated multi-disc clutch systems with advanced ceramics, Nordtrib 2008, 13th Nordic Symposium on Tribology, Tampere, Finland, June, 10.-13., 2008.

Modeling

VIRTUAL TESTING AND SIMULATION OF MULTIPLE CRACKING IN TRANSVERSE TOWS OF WOVEN CMCs

Pierre Pineau, Guillaume Couégnat, Jacques Lamon
CNRS/University of Bordeaux, Laboratory for ThermoStructural Composites
Pessac - France
[pineau, lamon,couegnat]@lcts.u-bordeaux1.fr

ABSTRACT

The present paper develops a method of virtual testing with a view to investigating the local response of tows within textile Ceramic Matrix Composite (CMC) under various loading conditions. The transverse tows contain matrix, voids and fibres which act as stress concentrators. Mesh for Finite Element analysis is constructed from micrographs of composite cross sections. Cracks were introduced into the mesh. Multiple cracking was simulated. Transverse tow tensile behavior and flaw density functions were derived from finite element computations of stress-state.

INTRODUCTION

Stress-induced damage plays a significant role in the potential use of Ceramic Matrix Composites (CMCs) as structural components. It involves matrix cracking and it depends on several factors including loading conditions, mechanical properties of constituents and composite structure. From this point of view, the 2D woven CMCs like the SiC-matrix ones display several interesting features, such as a multiple length scale structure (from millimetre to micrometer) and ceramic constituents sensitive to size effects and stochastic fracture in terms of initiation and location of cracks. Microstructure/properties relations are therefore of primary importance with a view to designing high-performance materials as well as components.

The present paper proposes a virtual testing procedure to simulate multiple cracking in the weft tows of 2D woven CMCs. Tow microstructure is reproduced and cracks are introduced in the Finite Elements mesh.

The approach has been developed on 2D woven SiC matrix composites, made via CVI (chemical vapour infiltration) and reinforced by carbon or SiC fibers. But it can be applied to various multidirectional CMCs. In addition to differences in constituent properties, composite structure can be different in terms of amount of fibres in tows or defects distribution. But, at microscopic and mesoscopic scales these composites exhibit a similar pattern: fibres, intra-tow matrix and voids in the tows, and then, tows and inter-tow matrix with voids at upper scale (figures 1 and 2).

Damage affects composite stiffness in the loading direction. Three main damage modes which appear in succession as the load increases have been identified [1]:

- First, cracking in the inter-tow matrix

- Second, cracking in the transverse tows (tows perpendicular to the loading direction: figure 1): matrix cracks are parallel to the tow direction. They are initiated by fibres, voids or interfaces (figure 2). They propagate through the intra-tow matrix. Deflection occurs at the interface with longitudinal tows.

- Third, cracking in the longitudinal tows (tows parallel to the loading direction): after saturation of previous damage, the longitudinal tows carry the loads. Intra-tow matrix cracks bridged by fibres develop. Fibres breaks then lead to tow failure.

The distinction made between longitudinal and transverse tows is relative to on-axis loading direction: for a damaged piece of material that has been 90° rotated with respect to the loading direction and damaged, the former transverse tows are now the longitudinal ones and all the tows contain transverse and longitudinal cracks.

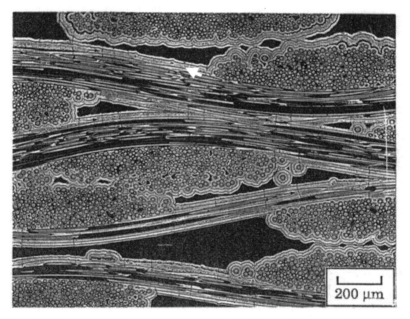

Figure 1: Micrograph of 2D woven SiC/SiC showing the microscopic structure of tows, and the mesoscopic structure (tows and inter-tow matrix) of composite.

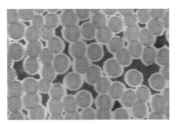

Figure 2: Detail of tow microstructure: fibers, interphases and voids.

Matrix multiple cracking and fragmentation have been studied characterized and modelled for unidirectional composites [2], when the matrix cracks are perpendicular to the loading direction and to fibres axis. Multiple cracking under increasing load requires two conditions: statistical distribution in matrix failure data and reloading of uncracked fragments. In unidirectional composites the random distribution of matrix flaws induces statistical features and the fibres permit reloading of the undamaged parts of matrix. Concerning the transverse tows in 2D woven composites, the statistical features result from the presence of fibres and voids in the intra-tow matrix which cause stress concentrations and from variability in fiber/matrix bonding and matrix strength, whereas the longitudinal neighbouring tows are the reloading elements. These features have to be introduced explicitly in the meshes of the virtual testing process.

VIRTUAL TESTING: HETEROGENEOUS OBJECT ORIENTED APPROACH

Finite Elements Analysis (Fea): cells

The FEA mesh is constructed from micrographs. Images of transverse tows are extracted from composite cross sections (figure 3) and fibres and voids are identified. The meshing procedure starts by the numerical analysis of the image of a tow (figure 3). First of all, contrast smoothing is done to obtain the same colour levels per constituent. The fibres are then modelled as ellipses, the centre position, minor and major axes and orientation angles of which are identified. The frontiers of voids are described using a list of points. These data are transformed into a mesh. The accuracy of the mesh around the pores was adapted to avoid irregularities in the meshed interface that would generate stress concentrations or overestimations.

In order to take into account the presence of mesopores between tows in 2D woven CMC (figure 1), the frontier of the transverse tow under consideration was identified as the envelop curve of the peripheral fibres dilated to the appropriate width (figure 4).

The tow was then placed between two reinforcing elements which represent the longitudinal tows (figure 4). They consist of an elastically equivalent material whose properties are obtained using classical mixtures law.

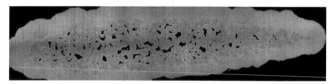

Figure 3: Image of a transverse tow extracted from a micrograph of SiC/SiC specimen.

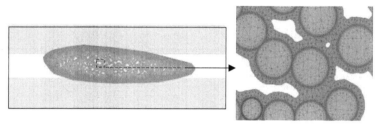

Figure 4: Simulation cell (left) and mesh details (right).

Crack initiation: failure criterion

Fracture of ceramic constituents in CMCs is brittle except in some cases of environmental activated delayed failure of fibres. The microstructure of the transverse tows revealed by cross sections (figures 2-3) displays heaps of fiber sections and voids. Both are regarded as inclusions with Young's modulus respectively null and E_f. They act as stress concentrators. In that way they are considered as defects which may induce cracks.

A stress criterion of crack initiation was selected. Both fibres and voids populations are in competition. For each defect a post-processing operation is used to find the maximum stress along defect/matrix interface. This stress defines the defect critical stress σ_{Cf} for a fibre and σ_{Cp} for a void.

$$\sigma_C(i) = \max_{interface(i)} (\sigma_I) \quad (1)$$

where σ_I is the principal stress in the interface.

The failure criterion is simply expressed as:

$$\sigma_R \le \sigma_C \quad (2)$$

where σ_R is the failure stress.

The failure stress σ_R was taken first to be unique, and then to have a statistical distribution. In the first step, multiple cracking results from scatter in critical stress values σ_C.

Crack propagation: meshing process

Matrix cracks in the transverse tows are parallel to fiber direction and orthogonal to longitudinal tow direction (fig.1-3). Because of matrix brittleness, crack propagation is instantaneous through the tow. Crack deflection occurs between inter-tows matrix and longitudinal tows. In our case the deflection of cracks along longitudinal tows could not be correctly controlled as the meshes were not adapted to this issue. However these debond cracks can be arbitrarily integrated into the meshes.

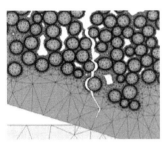

Figure 5: Example of transverse crack in a mesh of transverse tow. Note that the crack was arrested at longitudinal tow frontier.

The cracks are introduced into the meshes by duplicating nodes once the fracture inducing defect and the location of the largest critical stress have been identified. Crack orientation is chosen orthogonal to the loading direction. Nodes along the crack are selected up to the longitudinal tow frontier using an iterative procedure. The crack can follow fibre/matrix interfaces and also pass through voids. These nodes are then duplicated and reassigned in the elements located on the other side of the crack (figure 5).

SIMULATION OF MULTIPLE CRACKING

Several virtual tests of multiple cracking can be carried out by adjusting the loading conditions or scatter in failure sress, which leads to various cracking patterns.

Loading conditions and stress state

Uniaxial tension of longitudinal tows was obtained using loading conditions based on displacements which were applied to the opposite right and left sides of the cell (fig. 4). The elastic constants required for computations are listed in table I for SiC/SiC tows [3]. Longitudinal tow properties were estimated using mixtures law.

Table I: Elastic constants of constituents of SiC/SiC transverse tows.

	E	v
SiC$_{Fibres}$	200 GPa	0.12
SiC$_{Matrix}$	350 GPa	0.2
Longitudinal tow	270 GPa	0.15

Stress analysis shows that fibres and voids are stress concentrators (fig. 6). Further examination of stress state confirms that critical stresses σ_C are located at defect poles and that they coincide with σ_{11} (stress component parallel to loading direction). The two defects populations are characterized by critical stress values σ_{C_f} and σ_{Cp} (figure 7). Critical stress σ_{C_f} can be identified at two locations: either at the interface or in the matrix close to the interface. The value of stress in the matrix was $1.177^{+0.043}_{-0.007}$ times as high as that at the interface. It is worth pointing out that accuracy for the interface value was limited by discontinuity of certain stress values on both sides.

Plots of cumulative densities show that the voids are more severe than the fibres. SiC matrix (SiC$_m$) strength is estimated to be around 275 ± 25 MPa. The SiC$_f$/SiC$_m$ fiber/matrix interface is generally strong. The opening stress has been estimated to be larger than 420MPa [4]. Thus, it can be concluded from figure 7 that the void population is probably that one which generates the cracks on those composites with strong interfaces. But the situation will be different with weak interfaces. The frontier between both trends needs to be identified.

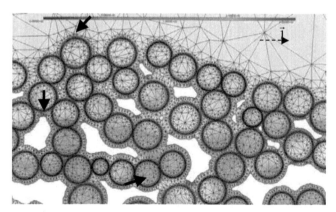

Figure 6: σ_{11} stress field: stress concentrations (red colour and arrows).

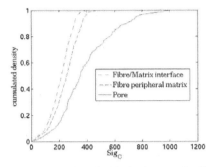

Figure 7: Cumulative density of critical values σ_c for fibres and voids

Multiple cracking process

A single value of SiC_m strength (275MPa) was used first. The first crack was induced by the most severe void under a 0.026% strain (figure 8). A crack generates a circular domain of stress relaxation (figure 8), where stresses are very low so that defects in this area can be regarded as innocuous. It is worth mentioning that a similar circular domain is obtained in homogeneous media. Outside this circular domain with diameter determined by tow width, stress concentrations will generate additional cracks (black arrows figure 8). The crack also causes a significant stress concentration in longitudinal tows as they carry the entire load (grey arrows figure 8).

An iterative procedure identifies the voids that generate the successive cracks. Damage proceeds up to 0.1% strain (figures 9 and 10). The resulting stress/strain behaviour and the evolution of defect population during multiple cracking are shown by figure 10. These results agree with available experimental results obtained on 2D CVI SiC/SiC composite [1].

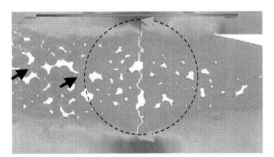

Figure 8: Crack unloading zone and stress concentrations.

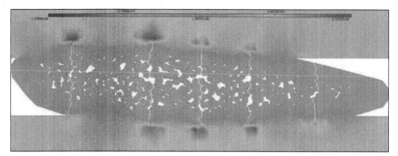

Figure 9: σ_{11} stress field in the damage tow with 5 cracks at 0.1% strain.

In a second step, scatter in σ_R was introduced through a normal distribution function ($\sigma(\sigma_R) = 275, \mu(\sigma_R) = 10$). 99% of σ_R values are higher than 250MPa and lower than 300MPa. When σ_R is different from 275MPa strains and stresses at onset of cracking are different from above (table II) but the crack inducing defects and, consequently, cracks location stay the same. The strain level for the first crack is in good agreement with the elastic limit (0.035) which is measured on current 2D SiC/SiC composite.

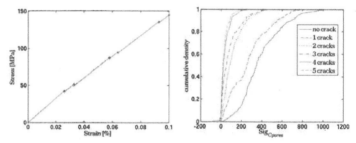

Figure 10: Stress/strain behaviour (left) and successive defects populations (right).

Table II: Strain dependence on failure stress dispersion

Number of crack	1	2	3	4	5
$\sigma_R = 275$ MPa	0.0261	0.0329	0.0583	0.0640	0.0930
$\sigma_R = norm(275,10)$ MPa #1	0.0263	0.0340	0.0570	0.0638	0.0942
$\sigma_R = norm(275,10)$ MPa #2	0.0264	0.0338	0.0570	0.0584	0.0923

Reproducibility in the cracking scheme can be explained by the relatively low density of crack inducing big voids. Scatter in σ_R would be more efficient in the case of C/SiC with higher contrast in elastic properties of constituents and weaker fibre/matrix interfaces. Fibres would initiate cracks and their population would be denser than the voids one.

Number of cracks

Several saturation criteria can be applied to determine the maximum number of cracks. The key point is that debond cracks at longitudinal tow frontier were not taken into account although they have a predominant action on load transfers in the structure. Stress relaxation zones around cracks can be considered to be the exclusion zones for further crack initiation. So, the exclusion zones are circles with a diameter equal to crack length. In that way, the saturation crack number will be overestimated. It is given by the number of overlapping circles within the tow. At this stage, the minimum crack spacing distance equals circle radius and half tow height. Thus, an upper bound for the number of cracks at saturation is determined by tow height h versus longitudinal/transverse load transfer length L:

$$N_{max} = E(\frac{2L}{h} + 1) \qquad E(.) \text{ stands for integer part} \qquad (3)$$

$N_{max} = 6$ or 5 cracks was obtained using equation (3). This number is close to the results of simulations, and it is slightly larger than the experimental ones currently observed on SiC/SiC composite (3 or 4). This overestimation of experimental data is reasonable.

CONCLUSION

A virtual testing procedure was proposed to investigate the behavior of heterogeneous parts selected within woven composites and was applied to SiC/SiC composites. It uses finite element meshes constructed from micrographs. The meshes mimic the microstructure. The fibers and the voids are reproduced. Simulation of cracking is based on the stress-state. Cracks are introduced at the location of maximum stresses, with respect to matrix and interface strengths, which can be either constant or scattered.

The tensile behavior as well as multiple cracking was simulated for a transverse tow in SiC/SiC composite. The crack pattern was found to be in agreement with current observations on SiC/SiC composites. The stress-strain curves for transverse tows as well as flaw density were derived from simulations

The maximum number of cracks was computed when the transverse tow is strongly bonded to the longitudinal ones. Exclusion zones consist of circular domains of stress relaxation around cracks. It was shown that an upper bound for the number of cracks at saturation can be derived in simple form in terms of tow dimensions. This expression shows the influence of tow geometry on multiple cracking. Further refinements will introduce the influence of tow bonding.

ACKNOWLEDGEMENTS

The authors would like to thank French Ministry of Research and Education, CNRS, C.E.A. and SNECMA Propulsion Solide.

REFERENCES

[1]J. Lamon, "Handbook of Ceramic Composites – Chap.3: Chemical Vapor Infiltrated SiC/SiC Composites (CVI SiC/SiC)", 55-76. Edited by N.P.Bansal, Klewer Academic Publishers (2005).

[2]N. Lissart, J. Lamon, "Damage and failure in ceramic matrix minicomposites: experimental study and model", *Acta mater*. **45**, 1025-1044 (1997).

[3]Bobet J.L., Lamon J., "Thermal residual stresses in ceramic matrix composites - I: Axisymetrical model and finite element analysis" *Acta. Metall. Mater*. **43**, 2241-2253 (1995).

[4]S. Pompidou, "Déviation des fissures par une interface ou une interphase dans les composites et les multicouches" (Crack deflection by interfaces and interphases in composites and multilayers), PhD Thesis, University of Bordeaux 1, N°2694, 2003.

ISOTHERMAL CHEMICAL VAPOR INFILTRATION MODELING BY RANDOM WALKS IN CMT 3D IMAGES AT TWO SCALES

G. L. Vignoles[1], I. Szelengowicz[1], W. Ros[1,2], C. Mulat[1,2], C. Germain[2]

1. University Bordeaux 1
LCTS – Lab. for ThermoStructural Composites
3, Allée La Boëtie
F33600 PESSAC

2. University Bordeaux -ENITAB
IMS – Lab. for Integration from Materials to Structures
351 Cours de la Libération
F 33405 TALENCE Cedex

ABSTRACT: The production cycle of high-quality Ceramic-Matrix Composites (CMCs) often involves interphase or matrix deposition by Chemical Vapor Infiltration (CVI). This costly step has motivated many modeling approaches, in order to provide guidelines for process control and optimization.

In this context, numerical tools for direct modeling of isothermal, isobaric CVI in complex 3D images of the composite architecture, acquired e.g. by X-ray Computerized Microtomography (CMT) have been developed. To address inter- and intra-bundle length scales inherent to a composite with a woven textile reinforcement, a numerical strategy has been set up, based on two numerical tools. They solve diffusion-reaction equations and handle simultaneously the progressive evolution of the porous structure. They involve distinct random walk methods and image handling routines.

The small-scale program uses Pearson random walks simulating rarefied gas transport; the fluid/solid interface is explicitly represented as a set of triangles through a Simplified Marching Cube approach. Direct simulation of CVI in intra-bundle pores is possible with such a tool. Effective laws for the evolution of porosity, surface and transport properties as infiltration proceeds are inferred from these simulations by averaging and are considered as inputs for the next modelling step.

The large-scale solver uses Brownian motion simulation; the porous medium is considered as a continuum with locally heterogeneous and anisotropic diffusivities, and the deposition reaction is handled through a survival probability computation. Simulation of the infiltration of a whole composite material part is possible with this program.

Validation of these tools on test cases, as well as some examples on actual materials, are shown and discussed.

INTRODUCTION

Thermostructural materials, like Ceramic Matrix Composites (CMCs)[1] and Carbon-Fiber Reinforced Carbons (CFRCs) [2,3], already in use in very demanding technologies like aerospace, are promising candidates for new applications in the field of aeronautic propulsion[4] and nuclear power engineering[5], among others. They consist of a fiber arrangement, woven or not, filled with a matrix. The particular nature of the matrix makes it necessary to use specific insertion techniques. One of the most efficient ones is Chemical Vapor Infiltration (CVI), by which a gas precursor is allowed to diffuse through the fiber arrangement (preform) at moderately high temperatures; they are cracked and eventually lead to a solid deposit. The fibers are thus progressively coated by the matrix; some residual porosity is present. This process gives the best matrix quality, but is expensive and difficult to control in terms of deposit homogeneity throughout the preform thickness. The quality of materials prepared by CVI relies on processing conditions (such as vapor precursor concentration, temperature and pressure), as well as on intrinsic properties of the preform. Experimental determination of the conditions which lead to an optimal infiltration is time-consuming and expensive. That is the reason why a global modeling of CVI is of great interest to optimize the final density and homogeneity of the composites[6-11]. A long-term goal is the development of a virtual material toolbox, which allows going in a computer from the design of a composite architecture to its full property assessment, without actually constructing the material.

The keys of the model are: (i) the determination of geometrical characteristics and transport properties of the preform at various stages of infiltration[12], namely: the effective gas diffusivity, either in continuum or in rarefied regime, the gas permeability to viscous flow in the case of pressure-driven CVI[13], and the heat conductivity, in the case of thermal-gradient modifications of CVI[14], and (ii) the simulation of the infiltration itself.

The aim of this paper is to present a 3D image-based modeling approach for isothermal, isobaric CVI (aka I-CVI). The architecture of the fibrous arrangement, in which fibers are frequently grouped into bundles or tows, requires that the study pay attention to at least two length scales: inter-bundle and intra-bundle. Numerous past works have treated I-CVI modeling at both scales[15]; the specificity of the work presented here is the ability to account for precise geometric details of the preform structure, as obtained from high-resolution X-ray tomographic imaging or equivalent methods, and to deal with two different scales inside the material with two distinct random-walk techniques.

The first part will briefly present the image acquisition procedure in the case of C/C composite. In the next part, the fibre-scale simulation, which has been developed since several years, will be recalled; then, the newer large-scale simulation will be presented. Application to actual 3D media will be shown and discussed in the last part of the document.

IMAGE ACQUISITION

In order to have a good representation of the C/C composite architecture, CMT scans should be performed at various scales: indeed, the diameter of a single fiber is roughly 8 μm, while the space period of the textile arrangement may span several millimeters. Fortunately, all these scales are accessible to X-ray CMT, using classical X-ray sources for the largest ones and Synchrotron Radiation X-ray CMT (XRCT) for the smallest. However, in this last case, one has not a direct access to the density distribution in the material, since the tomographs are acquired in phase contrast or "edge-detection mode". They require special image processing algorithms to successfully separate the solid from the void phase[16] and the fibers from the matrix[17,18].

Details on the experimental procedure have been given in previous papers[19,20]. The samples were raw and partly infiltrated C fiber preforms made of stacked satin weaves held together by stitching; they have been scanned with a 0.7 μm voxel edge size resolution, using the setup of the ESRF ID 19 beamline. Lower resolution scans (6.7 μm/voxel) were also made on the same samples, in order to connect with a maximal confidence the fiber-scale and larger scales, like the Representative Elementary Volume (REV) scale. Figure 1a) displays a lower resolution image of the whole sample, the upper part of which has been scanned with higher resolution.

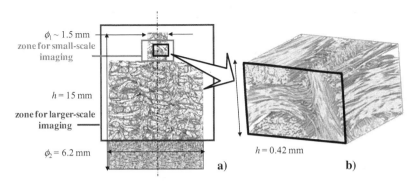

Figure 1. a)Low-resolution scan of a C/C sample for CT acquisition at 2 distinct resolutions. Sample outer diameter is 6 mm. b) 3D rendering of partly infiltrated C fibers in a preform, after segmentation from an edge-detection high–resolution tomographic scan.

FIBRE-SCALE SIMULATION

The fibre-scale infiltration simulation is based on a simplified representation of gas kinetics in a fluid/solid binarized medium. The discretization of the fluid/solid interface follows a Simplified Marching Cube scheme[21,22], which is a compromise between the accuracy of the Marching Cube and computational resource savings.

The random walk scheme is inspired from Burganos & Sotirchos[23], allowing for rarefied regime, transition regime and ordinary diffusion. The walkers are handled according to the algorithm sketched at fig. 2. The orientation distribution of the steps is isotropic in the bulk fluid, and follows Knudsen's cosine law on the walls. The free path (distance between two molecule-molecule collisions) is sampled from a decreasing exponential distribution. This method has been validated for the computation of effective diffusion coefficients [24,25].

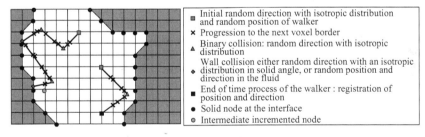

Figure 2: Scheme of the gas-kinetic random walker displacements in a 2D example using SMC surface discretization, with deposition modeling.

Chemical reaction with the walls is represented by the sampling of sticking events; the probability is fixed and is a function of the molecular velocity and of the linear heterogeneous reaction rate constant[26]. The handling of walkers after a collision may be performed in several ways. First, if the walker is randomly in the fluid with constant probability, one represents a situation in which the active species responsible for solid growth appears by the slow decomposition of a source gas, as occurs in the case of pyrocarbon CVD from methane[27]. This kind of simulation has been validated against an analytical solution in a straight slit pore[26].

Situations where the active species comes from outside the considered material sample are managed by connecting the 3D image to reservoirs with constant walker concentrations (*i.e.* it is a grand-canonical Monte-Carlo simulation). Concentration differences between the reservoirs simulate an overall concentration gradient throughout the material sample. Again, in this situation, it is possible to test the method against an analytical result in the case of a straight slit pore. From Figure 3 it is seen that the agreement is excellent for the concentration profile. Figure 4 shows that the effective reaction constant of the porous structure may be recovered as a function of the local Thiele modulus,

$$\Phi = L\sqrt{\frac{kS_v}{D}} \tag{1}$$

where k is the local heterogeneous reaction rate constant, d_p is the pore diameter, and D is the gas diffusion coefficient; the results are in excellent agreement with the analytical solution.

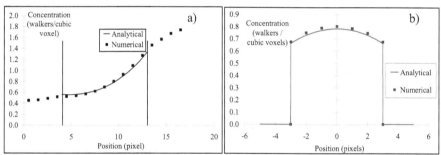

Figure 3. Validation of the code with respect to the case of a straight slit pore connected to two reservoirs with fixed concentrations. a) Longitudinal concentration profile, b) Transverse concentration profile.

The handling of the surface alteration is carried out by a discrete VOF technique, in which the amount of deposited solid is stored in the fluid node closest to each sticking event position; when this amount exceeds a threshold, the node is converted to solid and the local interface is modified on-the-fly. The surface evolution scheme has been validated against analytical solutions in the case of composite materials ablation[28].

The results of the program are:
(i)   a computation of the effective diffusivity and averaged reaction rate.
(ii)   A prediction of the morphological evolution of the medium, through the production of image sequences, from which any morphological quantity may be recomputed, like e.g. the internal surface area, pore diameter statistics, etc …

The evolution laws for the effective diffusion and reaction coefficients as infiltration proceeds are taken as inputs for the large-scale model, which is described next.

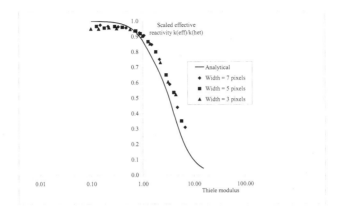

Figure 4 : Effective deposition rate constant in the case of a flat pore in a concentration gradient – validation of numerical simulations with respect to the analytical solution

LARGE-SCALE SIMULATION

The large-scale computer code is based on a distinct random-walk method. Here, sharp fluid/solid interfaces are not usual; instead, every voxel contains a certain fiber density which can be associated to a given diffusivity tensor. A preliminary processing step of the program is to determine the direction of the fibers by image processing, namely, finding the eigenvectors and eigenvalues of the structure tensor, *i.e.* the Hessian of the image[29]. The applied filter was based on a gradient mask optimized with respect to cylinder detection[30]. Then, every voxel is associated to a full local diffusivity tensor, the amplitude of which is given by the grayscale level and the principal directions of which are given by the detected fiber orientation.

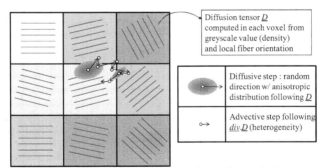

Figure 5: 2D Scheme of the Ito-Taylor numerical scheme for random walks in large-scale images (here, 3x3 voxels), in which the fiber direction has been assigned.

For the resolution of the Fokker-Planck transport problem, it has been chosen to implement a particle solver, through an Itō-Taylor numerical scheme. The solver is inspired from advection-dispersion solvers[31]; it allows accounting for the local anisotropy of the medium in each voxel by biasing the orientation distribution according to the local fiber orientation. The code also accounts for local heterogeneities of the diffusion tensor by adding to the diffusive step an advective step in which the drift is computed from the image gradient.

The program has been validated against various simple cases in which analytical solutions are available: homogeneous diffusion tensor aligned with the image axes or off-axis, linear gradient of the diffusivity; image with a discontinuity of the diffusion coefficient. All results were excellent with a moderate number of walkers (approximately 800).

In addition to transport, the program has to simulate deposition inside the fiber bundles. This is achieved by the computation of a survival probability at each time step with size δt according to:

$$P_s = \exp(-k_{\text{eff}} . \delta t) \qquad (2)$$

where k_{eff} is the effective (homogenized) reaction rate constant; then, a random number is sampled from a uniform distribution in $[0;1]$. If it exceeds the survival probability, then the walker is swallowed, and the local infiltration progress variable is incremented.

The program has been validated against a 1D diffusion-reaction case for which the traditional hyperbolic cosine is the solution of the concentration field. Figure 5 shows the excellent agreement between the analytical and numerical solution.

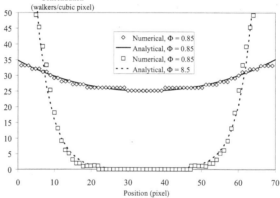

Figure 6: Local concentration profiles in a 1D diffusion-volume reaction problem – validation of the numerical simulation with respect to the analytical solution. (Φ = Thiele modulus)

APPLICATION TO ACTUAL 3D IMAGES OF COMPOSITES: RESULTS AND DISCUSSION

An example simulation is shown in Figure 6. A 400x400x350 voxels image (0.28x0.28x0.245 mm^3) has been infiltrated in continuum regime ($Kn = 0.035$) and a sticking probability equal to 0.16. From the rendering of the deposited matrix, it is clearly seen that the thickness is not constant: indeed, fibers lying close to a bundle periphery receive a much larger deposit than in the bundle center.

The evolution of the composite may be monitored through the evolution of the scaled Thiele modulus:

$$\Phi / \Phi_0 = \sqrt{\frac{\sigma_v \; D_0}{D \; \sigma_{v0}}}$$

(3)

as the pore volume decreases. The Thiele modulus is a measure of the reaction/diffusion ratio; high values are synonymous of a difficult infiltration. This quantity is roughly proportional to the inverse square root of the average pore diameter: accordingly, it is prone to increase when infiltration proceeds[32]. Figure 7a) displays such an evolution for a 100x100x100 block. The increase is firstly linear; then, it accelerates when getting closer to the percolation threshold, which in this case is extremely low.

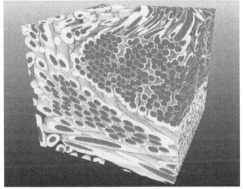

Figure 6: Rendering of a simulated matrix with parameters $Kn = 0.035$ and $P_c = 0.16$.
Volume fractions : Fibers : 28%, matrix : 36%

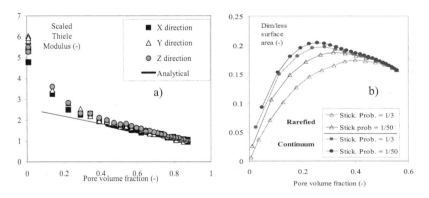

Figure 7: Evolution of some quantities during infiltration. a) Scaled Thiele modulus; b) Scaled surface area for various values of Kn and P_c.

Such an evolution is shown to depend not only on the structure of the porous medium, but also on the initial values of the reaction/diffusion ratio and on the Knudsen number. Differences between a purely geometrical dilation – which corresponds to the kinetic limit – and a diffusion-reaction competition case have been illustrated in Figure 7b). There are distinct evolutions of the same initial image under deposition with various conditions. The case of low sticking probability and continuum diffusion regime leads to the highest values of the surface area, corresponding to an optimal infiltration; when the Knudsen number or the sticking probability increase, the infiltration is less efficient.

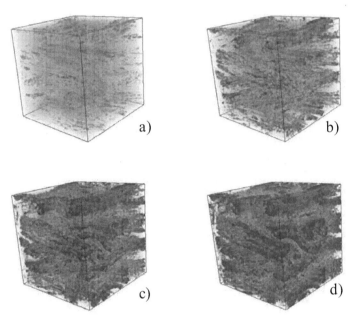

Figure 8: Simulated infiltration of a C/C composite sample at macro-scale. The image is 70x70x70 pixels in size.

The results of the small-scale infiltrations have been cast into evolution laws $\underline{D}_{eff} = f(\varepsilon,\varepsilon_0)$ and $k_{eff} = f(\varepsilon,\varepsilon_0)$. For sake of simplicity, and retaining enough realism, we use the laws given by Rikvold & Stell[33] for unidirectional beds of partially penetrable random cylinders, for which an excellent agreement has been found for C/C composites[20]. The initial radius of the fibers is denoted r_0 and is constant for all fibers; on the other hand, the initial porosity ε_0 may vary from place to place. A growth parameter is defined as:

$$\delta = \frac{r}{r_0} \tag{4}$$

At each capture event, the infiltration parameter d is increased in the following way :

$$\Delta\delta = \frac{C_{gas}M_s}{(N_w/V_{cell})M_{gas}} \cdot \frac{1}{S_v r_0}$$

(5)

where (N_w/V_{cell}) is the « numerical concentration of walkers », so that the first ratio in the right-hand side is the amount of solid volume brought by a single walker.

Then, porosity, internal surface area and pore diameter may be computed as :

$$\varepsilon = \varepsilon_0 \exp\left[\left(\delta^2 - 1\right)\left(1 - \frac{1}{\varepsilon_0}\right) - (\delta - 1)^2\left(1 - \frac{1}{\varepsilon_0}\right)^2\right]$$

(6)

$$S_v = \frac{2\varepsilon}{r_0}\left(1 - \frac{1}{\varepsilon_0}\right)\left(\frac{1-\delta}{\varepsilon_0} - 1\right)$$

(7)

$$d_p = \frac{4\varepsilon}{S_v}$$

(8)

As shown in eq. (1), the Thiele modulus is also a function of the structural parameters. The local diffusion tensor components (parallel and perpendicular) are given by:

$$1/D_i^{eff} = \left(1/D_i^b\left(\varepsilon,\varepsilon_0,\Phi\right) + 1/D_i^K\left(\varepsilon,\varepsilon_0,d_p,\Phi\right)\right)$$

(9)

The dependence of the Knudsen and binary coefficients to the porosity has been obtained by diffusion simulations[25]. All these laws have been plugged in the large-scale simulator.

Figure 8 displays an example infiltration with a low initial value of the global Thiele modulus ($\Phi_0 = 0.05$). The fiber bundles are becoming progressively denser while the inter-bundle space (containing some isolated fibers) evolves in a much slower fashion. These first qualitative results match well the conclusions of the morphological analyses[20] which have been carried out on the same samples at various stages of infiltration.

CONCLUSION AND OUTLOOK

This contribution has presented two numerical tools, based on distinct random walk methods, adapted to simulating diffusion/reaction problems involved in chemical vapor infiltration. The tools have been validated, and application to actual 3D images produced by X-ray CMT is demonstrated. More detailed post-processing on the simulation output, and application to various materials are the next objectives of the working program. The codes constitute a toolbox for multi-scale modeling adapted to real composite material architectures; moreover, they can also be used on ideal images and help the materials engineer to design fibrous arrangements with better infiltration capabilities.

REFERENCES
[1] W. Krenkel, *Ceramic Matrix Composites*, Wiley-VCH, Weinheim, 2008.
[2] G. Savage, *Carbon/Carbon composites*, Chapman & Hall, London, 1993.
[3] E Fitzer and L. Manocha, *Carbon reinforcements and carbon/carbon composites*, Springer, Berlin, 1998.
[4] E. Bouillon, P. Spriet, G. Habarou, C. Louchet, T. Arnold, G. Ojard, D. Feindel, C. Logan, K. Rogers and D. Stetson, Engine test and post engine test characterization of self sealing ceramic

matrix composites for nozzle applications in gas turbine engines, in *Proceedings of ASME Turbo Expo*, June 14-17, Vienna, GT2004-53976, 2004.

[5] H.C. Mantz, D.A. Bowers, F.R. Williams, M.A. Witten, A carbon-carbon panel design concept for the inboard limiter of the Compact Ignition Tokamak (CIT), in *Proceedings of the IEEE 13th Symposium on Fusion Engineering* vol. 2, IEEE-89CH2820-9, 1990, 947.

[6] T. L. Starr, A. W. Smith, Advances in modeling of the forced chemical vapor infiltration process, *Mat. Res. Soc. Symp. Proc.* **250**, 207-14 (1992).

[7] P. McAllister, E. E. Wolf, Simulation of a multiple substrate reactor for chemical vapor infiltration of pyrolytic carbon within carbon-carbon composites, *AIChE J.*, **39**, 1196-209 (1993).

[8] G. L. Vignoles, C. Descamps, N. Reuge, Interaction between a reactive preform and the surrounding gas-phase during CVI, *J. Phys. IV France* **10**, Pr2-9-Pr2-17 (2000).

[9] N. Reuge, G. L. Vignoles, Modeling of isobaric–isothermal chemical vapor infiltration: effects of reactor control parameters on a densification, *J. Mater. Proc. Technol.* **166**, 15-29 (2005).

[10] D. Leutard, G. L. Vignoles, F. Lamouroux, B. Bernard, Monitoring density and temperature in C/C composites elaborated by CVI with radio-frequency heating, *J. Mater. Synth. and Process.* **9**, 259-73 (2002).

[11] I. Golecki, Rapid vapor-phase densification of refractory composites, *Mater. Sci. Eng.* **R20**, 37-124 (1997).

[12] J. Y. Ofori, S. V. Sotirchos, Structural model effects on the predictions of CVI models, *J. Electrochem. Soc.* **143**, 1962-73 (1996).

[13] T. L. Starr, Gas transport model for CVI, *J. Mater. Res.* **10**, 2360-6 (1995)

[14] G.L. Vignoles, J.-M. Goyhénèche, P. Sébastian, J.-R. Puiggali, J.-F. Lines, J. Lachaud, P. Delhaès, M. Trinquecoste, The film-boiling densification process for C\ C composite fabrication: From local scale to overall optimization, *Chem. Eng. Sci.* **61**, 5336-53 (2006).

[15] G. L. Vignoles, Modelling of CVI processes, *Adv. Sci. Technol.* **50**, 97-106 (2006).

[16] G. L. Vignoles, Image segmentation for phase-contrast hard X-ray CMT of C/C composites, *Carbon* **39**, 167-73 (2001).

[17] J. Martín-Herrero, C. Germain, Microstructure reconstruction of fibrous C/C composites from X-ray microtomography, *Carbon* **45**, 1242-53 (2007).

[18] C. Mulat, M. Donias, P. Baylou, G. L. Vignoles, C. Germain, Axis detection of cylindrical objects in 3-D images, *J. Electronic Imaging* **17**, 0311081-89 (2008).

[19] O. Coindreau, P. Cloetens, G. L. Vignoles, Direct 3D microscale imaging of carbon-carbon composites with computed holotomography, *Nucl. Instr. and Meth. in Phys. Res. B* **200**, 295-302 (2003).

[20] O. Coindreau, G. L. Vignoles, Assessment of structural and transport properties in fibrous C/C composite preforms as digitized by X-ray CMT. Part I : Image acquisition and geometrical properties, *J. Mater. Res.* **20**, 2328-39 (2005).

[21] G. L. Vignoles, Modelling binary, Knudsen, and transition regime diffusion inside complex porous media, *J. Phys. IV France* **C5**, 159-66 (1995).

[22] M. Donias, G. L. Vignoles, C. Mulat, C. Germain, Simplified Marching Cubes, submitted to *IEEE Trans. on Visualization & Computer Graphics* (2009).

[23] V. N. Burganos , S. V. Sotirchos, Knudsen diffusion in parallel, multidimensional, or randomly oriented pore structures, *Chem. Eng. Sci.* **44**, 2451-62 (1989).

[24] O. Coindreau, G. L. Vignoles, and J.-M. Goyhénèche, *Ceram. Trans.* **175**, 71-84 (2005).

[25] G. L. Vignoles, O. Coindreau, A. Ahmadi, D. Bernard, Assessment of geometrical and transport properties of a fibrous C/C composite preform as digitized by X-ray computed micro-tomography. Part II: Heat and gas transport, *J. Mater. Res.* **22**, 1537-50 (2007).

[26] G. L. Vignoles, C. Germain, O. Coindreau, C. Mulat, and W.Ros, *ECS Trans.* **23**, 449-454 (2009).

[27] S. Middleman, The interaction of chemical kinetics and diffusion in the dynamics of chemical vapor infiltration, *J. Mater. Res.* **4**, 1515-24 (1989).

[28] J. Lachaud, G. L. Vignoles, A Brownian motion technique to simulate gasification and its application to C/C composite ablation, *Comput. Mater. Sci.* **44**, 1034-1041 (2008).

[29] F. Michelet, J.-P. Da Costa, O. Lavialle, Y. Berthoumieu, P. Baylou, C. Germain, Estimating Local Multiple Orientations, *Signal Processing* **87,** 1655-69 (2007).

[30] C. Mulat, M. Donias, P. Baylou, C. Germain, G. L. Vignoles, Optimal orientation estimators for detection of cylindrical objects, *Signal, Image and Video Processing* **2**, 51-58 (2008).

[31] E. M. LaBolle, G. E. Fogg, A. F. B. Thompson, Random-walk simulation of transport in heterogeneous porous media: Local mass-conservation problem and implementation methods, *Water Resources Res.* **32**, 583-593 (1996).

[32] G. L. Vignoles, O. Coindreau, C. Mulat, C. Germain, J. Lachaud, Benefits of X-ray CMT for the modelling of C/C composites, *Advanced Engineering Materials*, to appear (2010).

[33] P. A. Rikvold, G. Stell, « d-dimensional interpenetrable-sphere models of random two-phase media : microstructure and an application to chromatogaphy », *J. Coll. Int. Sci.* **108**, 158-162 (1985).

ADSORPTION AND SURFACE DIFFUSION OF OXYGEN ON 3C AND 2H-SiC POLYTYPES: NUMERICAL STUDIES

Junjie Wang[1], Litong Zhang[1], Qinfeng Zeng[1], G. L. Vignoles[2]

1. Northwestern Polytechnical University
National Key Laboratory for ThermoStructural Composites
Xi'An
China

2. University Bordeaux 1
LCTS – Lab. for ThermoStructural Composites
3, Allee La Boetie
F33600 PESSAC
France

ABSTRACT - Silicon carbide oxidation is an important technological issue either in semiconductor or in thermostructural composite applications. Since this phenomenon starts with oxygen adsorption and surface diffusion, these first steps have been investigated by theoretical chemistry methods, in order to provide some insights. Density Functional Theory (DFT) calculations have been performed for the determination of binding energies of single oxygen atoms on different surface sites (Si-terminated and C-terminated) of cubic (3C) and hexagonal (2H) polytypes of SiC. Atomic movements of adsorbed oxygen have been tracked; diffusion paths and energy barriers have been found. The influence of the oxygen coverage ratio has also been investigated, showing a synergetic effect between oxygen adatoms. Finally, the differences between 3C and the more reactive 2H polytypes are also discussed.

INTRODUCTION

Silicon carbide (SiC), which occurs in a large variety of polytypes, is an interesting candidate material for use in high-temperature and high-power applications, such as hot structural components of gas turbines, heat exchanger tubes for industrial furnaces[1] and high temperature, high power, and high frequency electronic devices and sensors[2].

Future nuclear power technological concepts rely heavily on SiC as a high-temperature, high neutron fluence resistant material[3]. The hexagonal 2H, 3C, 4H and 6H-SiC polytypes are the most commonly grown and studied. The polytypes differ by the stacking sequence periodicity of the Si and C atomic planes, which leads to different electronic properties such as a band gap ranging from 2.39 eV (3C-SiC) to 3.33 eV (2H-SiC)[4].

When silicon carbide is used as thermal structure material, it can be oxidized in the working environment. On the contrary, silicon carbide designed for electronic applications is intentionally oxidized during processing to form a thin layer of SiO_2. The oxidation of SiC is therefore an important issue in practically all its applications. So, the oxidation of silicon carbide surfaces has attracted much attention as one of the most important processes in current and future SiC technology.

A precise control of the surface oxidation and an improvement of oxidation resistance require a detailed scientific knowledge of the initial reactions of oxygen with SiC surfaces on atomic scale. Many groups have extensively studied chemical reactions at SiC surfaces and subsequent oxidation processes by using various experimental techniques[5-18] and first-principles calculations[19-21]. However, atomic structures of oxide complexes and initial oxidation processes even at submonolayer coverage remain not fully understood yet, especially for 3C-SiC(111) and 2H-SiC(0001) surfaces that are technologically relevant.

In this paper, a detailed theoretical study of the initial oxidation process for 3C-SiC(111) and 2H-SiC(0001) ideal surfaces is presented. The adsorptions of O atoms on these surfaces have been firstly studied through ab initio DFT calculations. By means of the Nudged Elastic Band (NEB)[22] method, the diffusion pathways between stable adsorption sites and their energy barriers have been achieved. Approximate potential energy surfaces can be built from these results. Finally, in order to investigate the non-linearity arising from high oxygen coverage ratios, the adsorption of O atom pairs is also studied. The emphasis of the discussions is set on the comparison between the 3C and 2H polytypes of SiC.

NUMERICAL METHODS

The SiC surface models (illustrated in Fig. 1) in the present calculations are built with a periodic supercell containing a vacuum width of 20 Å and a slab consisting of twelve layers of Si(C) atoms with a 3×3 lateral unit cell (9 primitive surface cells per supercell). In addition, each broken sp_3 bond at the bottom layer atoms in each supercell is saturated with one hydrogen atom. The lowest three layers of atoms were kept fixed in order to hold the characteristics of a more realistic surface, while the rest of the unit cell was allowed to relax during the geometry optimizations, with or without oxygen.

Side view

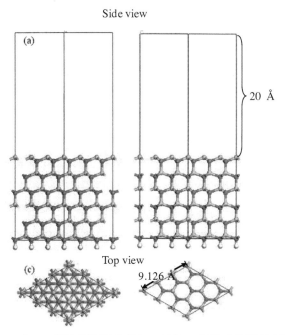

Top view

Figure 1. Side and top views of 3C-SiC (111) ((a) and (c)) and 2H-SiC (001) ((b) and (d)) Si-terminated surfaces. In this and all following figures related to the surface structure, gray, yellow and white spheres respectively indicate carbon, silicon and hydrogen atoms.

The oxygen atom(s) adsorption energies are calculated using DMOL^3 from Accelrys[23,24]. This code is especially useful for calculations involving large, periodic surfaces of a material, as in the slab-supercell modeling approaches. It is a density functional theory (DFT) code, with implementations of the local density approximation (LDA) and generalized gradient approximation (GGA) frameworks. Here, the DND basis set, which is comparable to the Gaussian 6-31G* basis sets, and the GGA-PW91[25] are used. The real space cutoff radius is of 4.0 Å for Si and C. All-electron basis sets are used for all the elements. A Fermi smearing of 0.01 hartree is employed to improve computational performance. The convergence criteria for energy, gradient, and displacement are respectively 2×10^{-5} hartree, 4×10^{-3} hartree/Å, and 5×10^{-3} Å. Accurate Brillouin zone sampling is ensured by summing over a finite set of k-points according to the Monkhorst-Pack scheme with a grid spacing of 0.05 Å^{-1} (i.e. a 3×3×1 k-point setting).

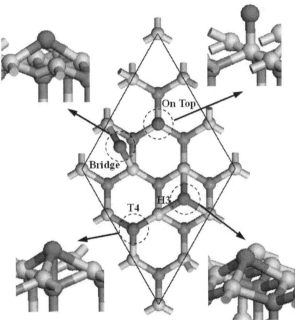

Figure 2. The four adsorption sites for oxygen adsorption on Si-terminated 2H-SiC. The same sites are considered for the 3C polytype and for C-terminated faces. Red spheres denote oxygen atoms.

Four adsorption sites are considered in this work. Figure 3 shows different O adsorption configurations, which are: (i) a single-coordinated "On Top" site (OT in the following), (ii) a twofold-coordinated "Bridge" site (BR in the following), (iii) a threefold-coordinated "Hollow" site (H3 in the following) and (iv) a fourfold-coordinated site (T4 in the following).

The adsorption energies can be calculated using the following equation:

$$\Delta E_{ads} = \left(\frac{N}{2} E(O_2) + E(slab) - E(total) \right) / N \qquad (1)$$

where N is the number of oxygen atoms adsorbed on the SiC surface, $E(slab)$ and $E(total)$ are the calculated total energies of the surface with and without oxygen, respectively. $E(O_2)$ is the total energy of an isolated dioxygen molecule.

To determine the activation barrier for the O atom diffusion among adsorption sites, the complete linear synchronous transit (LST) and quadratic synchronous transition (QST)[26] search methods are employed, followed by transition state confirmation through the nudged elastic band (NEB) method[22].

ADSORPTION ENERGIES

Table 1 summarizes the energies of adsorption for oxygen atoms on Si and C- terminated 2H(0001) and 3C(111) SiC surfaces. On the Si-terminated side, the largest adsorption energy corresponds to the BR situation, in which the oxygen atom is in a siloxane conformation and satisfies the octet rule. This situation is preferred to the OT situation -- because Si is not likely to form a double bond with O -- and to the H3 and T4 situations, because the excess of valence electrons confers them a marked antibonding character.

Table 1. Oxygen adsorption energies on Si- and C- terminated 2H(001) and 3C(111) SiC surfaces. (*) This site has only marginal stability.

Configuration	E_{ads} (eV)		
	2H(0001)	3C(111)	Difference
Si-terminated			
BR	4.17	3.98	0.19
H3	4.09	3.78	0.31
OT	3.72	3.82	-0.09
T4	3.05	(*)2.72	0.34
C-terminated			
OT	1.70	1.72	-0.02

On the C-terminated side, the only stable site is the OT site, in which the oxygen atom is in a carbonyl conformation. The adsorption energy is neatly lower, because the very large valence electron density reduces strongly the bond order.

The 2H surface displays a neatly larger reactivity with respect to oxygen. This could be expected from its slightly lower formation energy. Indeed, there is a larger amount of ionic character in the chemical bonding of 2H-SiC, which is the cause of the wider band gap. This makes adsorption of the very electronegative O atoms more favorable.

The OT site is the most insensitive one to polytype nature, because it corresponds to the actual crystallographic site of the Si or C atoms of the next layer; moreover, it is the site in which the O atom is most distant from its second neighbors.

Finally, the T4 site has a markedly distinct energy and stability when comparing the two surfaces. Indeed, in 3C, the T4 site has only saddle-point stability, i.e. it is stable with respect to vertical displacements of the O atom, but not with respect to lateral movements. In 2H, the stability is full, and the adsorption energy is much larger. This is probably due to the presence of staggered 3rd-neighbor carbon atoms in 3C, which are more repulsive than in eclipsed configuration.

SURFACE DIFFUSION OF ADSORBED OXYGEN

The LST/QST methods followed by NEB confirmation have been applied to O adatoms located on all kinds of sites and moving to all possible neighboring sites. A list of transition states was produced for all possible path segments. The results have been collected under the form of approximate potential energy surfaces (PES), as illustrated in figure 3. The BR site energy is taken as reference. The stars indicate the position of the localized activation states. It becomes possible to define from these PESs a minimal energy path for diffusion over the whole surface, which starts obviously from any BR site and reaches another equivalent BR site. It appears that the minimal path differs between 3C and 2H. The path over 2H is BR-> H3->BR->T4->BR, while over 3C it is: BR->H3->BR-> BR. Indeed, in the latter case, the T4 site has no lateral stability. So, the diffusion path avoids it, as can also be seen on Figure 4, which is a plot of the system energy as a function of position along the minimal energy path for both surfaces, taking again the BR site energy as a reference. The energy barriers are markedly lower for 3C-SiC; this comes from the lesser adsorption energy in BR conformation.

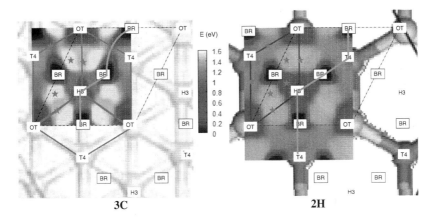

Figure 3. Approximate 2D PES of oxygen atom diffusion on 3C-SiC(111) surface (left) and 2H-SiC (001) surface (right), interpolated from the energies of the sites and transition states. The stars indicate the locations of identified transition states. The green lines with arrows represent the lowest-energy continuous diffusion paths on both surfaces.

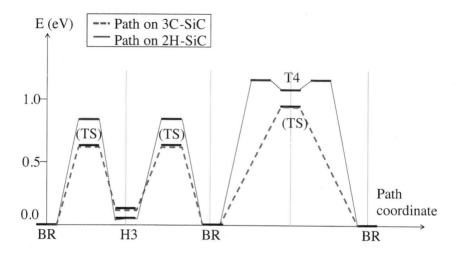

Figure 4. Energy profile of diffusion paths on 3C and 2H Si-terminated SiC surfaces. The BR site is taken as energy reference.

INFLUENCE OF THE COVERAGE RATIO : ADATOM PAIR INTERACTION

To investigate the influence of the coverage ratio, adsorption of O atom pairs has been simulated. Starting configurations with various site pair possibilities have been chosen and the systems were allowed to minimize their energy. The configurations which have been found stable are listed in Table 2, with their optimized O-O distances and the pairing energy, i.e.:

$$E_{pairing} = ½ (\Delta E_{ads}(\text{atom pair}) - \Delta E_{ads} (\text{atom } 1) - \Delta E_{ads} (\text{atom } 2)) \qquad (2)$$

A positive pairing energy indicates an extra stabilization of the adatoms by their interaction.

Table 2. List of site pairs found stable from the study, with O-O distances and pairing energy (eq. (2)).

Conformation		3C		2H	
		Distance (Å)	Pair Energy (eV)	Distance (Å)	Pair Energy (eV)
BR-BR	I	2.768	-0.122		
	II	3.051	0.022		
	III	3.686	0.072	3.51	0.17
	IV	4.069	0.079	4.02	0.28
	V			4.38	-0.26
	VI	5.342	-0.067	5.18	-0.10
BR-H3	I			3.571	0.221
BR-OT	I	3.055	-0.167	3.124	-0.018
	II	3.108	-0.078	3.171	-0.177
	III			3.826	0.041
	IV	4.438	-0.037	4.417	0.012
	V	4.649	-0.08	4.608	0.015
H3-H3	I			5.256	-0.557
H3-OT	I			3.29	-0.064
	II	3.612	0.019		
	III	4.731	-0.063	4.677	0.12
OT-OT	I	3.082	-0.197		
	II			3.698	-0.125
	III	5.298	-0.041	5.185	0.045

Table 2 shows that there is an optimal disposition in which the shortest O-O distance is about 4 Å for the 2BR configurations. For the 2OT configurations, the adsorption energy increases with the shortest O-O distance: the largest O-O distances are favored.

Figure 5 is a plot of the pairing energy against the O-O distance. It appears that the general tendency is a repulsion at the shortest distance, an attraction at medium distances (3.5 − 4.5 A), and again a dominant repulsion at larger distances. The absolute value of the pairing energies is not very large, though non negligible (up to 10% of the adsorption energy). The 2H surface displays larger interaction energies than the 3C surface. In particular, the BR-BR III, BR-BR IV and BR-H3 configurations display the largest positive pairing energy values.

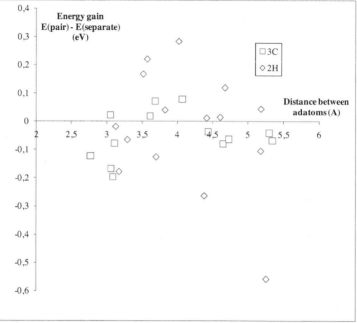

Figure 5. Relative energy of adsorbate pairs as a function of their separation on Si-terminated 3C and 2H SiC surfaces.

From these results, we can conclude that the adsorption of oxygen on 3C-SiC (111) surface favors low coverage situations. However, in some cases, the adsorption of oxygen favors a moderate O-O distance of about 3 to 4 Å: these are the 2BR-II, III and IV and the OT-H3-I configurations, whose adsorption energies are larger than those reference energies. This reveals that there are some electronic effects between the O adatoms, and the electronic effects are related with the O-O distance (between 3 and 4 Å), like favorable orbital overlaps. Moreover, in situations of attraction, the oxygen adatom charge $q(O)$ has been found to be lower than $0.95e$, as opposed to the other cases.

For 2H-SiC (0001) surface, the adsorptions of 2BR, BR-H3 and BR-OT configurations favor a moderate O-O distance of 3.5 to 4 Å. However, the adsorptions on 2H3, H3-OT and 2OT sites favor an O-O distance larger than 4.5 Å.

CONCLUSION AND OUTLOOK

This work focused on some issues for accurate numerical simulations of O atom adsorption on 2H-SiC (0001) and 3C-SiC (111) surfaces.

Among the possible sites on 2H-SiC (0001) surface, BR and H3 are the most stable on the Si side, while OT is preferred on the carbon side. The first ones are like siloxane bridges, while the last ones are like carbonyl groups. However, the most stable adsorption configurations on the Si-terminated 3C-SiC (111) surfaces are BR and OT. This difference could be explained by the presence of staggered 3rd-neighbor carbon atoms in 3C, which are more repulsive than in eclipsed configuration. Similarly, the O atom prefers to adsorb at OT site on C-side of 3C-SiC (111) surface.

Atomic movements of adsorbed oxygen have been tracked. By means of the NEB method, the energy barriers of oxygen atom diffusion pathway among different stable adsorption sites on 2H (0001) and 3C (111) SiC surfaces are achieved. The path over 2H is BR-> H3->BR->T4->BR, while over 3C it is: BR->H3->BR-> BR.

The influence of the oxygen coverage ratio has also been investigated, showing a synergetic effect between oxygen adatoms. This kind of electronic effect between the O adatoms is related with the O-O distance (between 3 and 4 Å) and oxygen adatom charge.

Several related investigations about SiC are under development, with the main goal to get the dynamic knowledge of SiC oxidation process.

In summary, further investigations include:

- The relaxation and oxygen adsorption behavior on Si- and C-terminated cubic SiC (111) and wurtzite SiC (0001) surfaces.
- The first-principles molecular dynamic simulation of oxygen molecule adsorption on SiC surface.
- The stability of $Si_{1-x}C_xO_2$ at SiC/SiO_2 interface and the diffusion activation energies of O_2 and CO in SiO_2 and $Si_{1-x}C_xO_2$.

REFERENCES

[1] N. S. Jacobson, J. Am. Ceram. Soc. **76**, 3 (1993).

[2] M. A. Capano, R. J. Trew, Mater. Res. Soc. Bull. **22**, 19 (1997).

[3] A. Kohyama, Y. Katoh, S.M. Dong, T. Hino and Y. Hirohata, FT/P1-02, 19th IAEA Fusion Energy Conference Lyon, France (2002)

[4] N. T. Son, O. Kordina, A. O. Konstantinov, et al., Appl. Phys. Lett., 65, 3209 (1994).

[5] V. M. Bermudez, J. Appl. Phys. **66**, 6084 (1989).

[6] W. Voegeli, K. Akimoto, T. Urata, S. Nakatani, K. Sumitani, T. Takahashi, Y. Hisada, Y. Mitsuoka, S. Mukainakano, H. Sugiyama, X. Zhang, H. Kawata, Surf. Sci. **601**, 1048 (2007).

[7] X. Xie, K. P. Loh, N. Yakolev, S. W. Yang and P. Wu, J. Chem. Phys. **119**, 4905 (2003).

[8] J. A. Schaefer and W. Göpel, Surf. Sci. **155**, 535 (1985).

[9] H. Ikeda, K. Hotta, T. Yamada, S. Zaima, and Y. Yasuda, Jpn. J. Appl. Phys. Part 1 **34**, 2191 (1995).

[10] C. Virojanadara and L. I. Johansson, Surf. Sci. **505**, 358 (2002).

[11] C. Virojanadara and L. I. Johansson, Phys. Rev. B **71**, 195335 (2005).

[12] D. G. Cahill and Ph. Avouris, Appl. Phys. Lett. **60**, 326 (1992).

[13] H. Ikegami, K. Ohmori, H. Ikeda, H. Iwano, Sh. Zaima, and Y. Yasuda, Jpn. J. Appl. Phys., Part 1 **35**, 1593 (1996).

[14] F. Amy, P. Soukiassian, Y. K. Hwu and C. Brylinski, Phys. Rev. B **65**, 165323 (2002).

[15] P. Soukiassian and F. Amy, J. Electron Spectrosc. **144-147**, 783 (2005).

[16] Y. Hoshino, R. Fukuyama and Y. Kido, Phys. Rev. B **70**, 165303 (2004).

[17] C. Radtke, I. J. R. Baumvol, B.C. Ferrera and F. C. Stedile, Appl. Phys. Lett. **85**, 3402 (2004).

[18] D. Schmeißer, D. R. Batchelor, R. P. Mikalo, P. Hoffmann and A. Lloyd-Spetz, Appl. Surf. Sci. **184**, 340 (2001).

[19] E. Wachowicz, R. Rurali, P. Ordejón and P. Hyldgaard, Comp. Mater. Sci. **33**, 13 (2005).

[20] A. G. P. Déak, J. Knaup, Z. Hajnal, Th. Frauenheim, P. Ordejón, J.W. Choyke, Physica B **340-342**, 1069 (2003).

[21] M. Di Ventra and S. T. Pantelides, Phys. Rev. Lett. **83**, 1624 (1999).

[22] G. Henkelman and H. Jonsson, J. Chem. Phys. **113**, 9901 (2000).

[23] B. Delley, J. Chem. Phys. **92**, 508 (1990).

[24] B. Delley, J. Chem. Phys. 113, 7756 (2000).

[25] J. P. Perdew, J. A. Chevary, S. H. Vosko, K. A. Jackson, M. R. Pederson, D. J. Singh, C. Fiolhais, Phys. Rev. B. **46**, 6671 (1992).

[26] N. Govind, M. Petersen, G. Fitzgerald, D. King-Smith and J. Andzelm, Comp. Mater. Sci. **28** 250 (2003).

EFFECTIVE SURFACE RECESSION LAWS FOR THE PHYSICO-CHEMICAL ABLATION OF C/C COMPOSITE MATERIALS

G. L. Vignoles[1], J. Lachaud[1,2], Y. Aspa[1,3], M. Quintard[3,4]

1. University Bordeaux 1
LCTS – Lab. for ThermoStructural Composites
3, Allee La Boetie
F33600 PESSAC

2. NASA Ames,
MS 230-3,
Moffett Field, CA 94035

3. University Toulouse
IMFT – Institute of Fluid Mechanics in Toulouse
1, Allee Prof. Camille Soula
F31000 TOULOUSE

4. CNRS
IMFT – Institute of Fluid Mechanics in Toulouse
1, Allee Prof. Camille Soula
F31000 TOULOUSE

ABSTRACT :
The thermostructural parts which have to suffer the most severe temperatures in a rocket nozzle or in a Thermal Protection System for atmospheric re-entry are frequently Carbon/Carbon composites. They principally undergo physico-chemical ablation by oxidation and sublimation. Fibers, matrices and interphases have marked differences in their ablation resistance, which usually leads to typical surface roughness features.

A comprehensive study of the relations between material parameters, physico-chemical conditions, and the surface recession leads to the possibility to reproduce numerically any surface roughness pattern. In addition, the effective recession rates are provided by the approach, in transient and steady states.

In this work, we discuss the results of this modeling approach with an emphasis on the « composite laws » that can be inferred from them, as a practical design tool for the engineer. For instance, a « weakest-link » law has been put forward: it is shown in what physico-chemical regime such an assumption is true. Extensions to other regimes are also provided.

INTRODUCTION
Materials for atmospheric re-entry body protection are mostly ablative Thermal Protection Systems (TPS); among them, carbon/carbon (C/C) and carbon/phenolic resin (C/R) composites are of common use[1,2], because of their excellent compromise between thermal, thermochemical and mechanical properties[3]. The principle of thermal protection is that an appreciable amount of the received heat flux is converted into an outwards mass flux through endothermic processes, like sublimation and chemical etching: this induces surface recession[4]. Surface roughening then appears: this unavoidable phenomenon has several consequences of importance in the case of atmospheric re-entry. First, it increases the chemically active surface of the wall; and second, it contributes to the laminar-to-turbulent transition in the surrounding flow[5,6]. Both of these modifications to the physico-chemistry lead to an increase in heat

transfer, resulting in an acceleration of the surface recession[7]. The TPS thickness design has to account for this rather strong effect.

Another space technology application for the same class of materials is the fabrication of rocket nozzle throats and inner parts. Here again, the acquisition of surface roughness during rocket launch is a critical issue, not because of the laminar-to-turbulent transition, but principally because of its impact on surface recession velocity, and on the possibility of triggering mechanical erosion[8].

For both applications, if general phenomenological tendencies are predictable, the understanding of the interaction between the flow and the material has to be improved. One of the key points of ablative material design is the understanding of the relation between the global effective recession rate of the composite and the individual recession rates of its constituents.

In the past years, a large effort has been made at modeling ablation from a material point of view[9-11]. The observation of the material morphology at all scales has led to the conclusion that morphological features of the surfaces were linked to the contrast between constituents, as well as to the reaction/diffusion (interfacial/bulk transfer) ratio. A modeling study has been designed and performed[12]; its results have largely confirmed the first hypotheses. Analytical models have also been exploited for a better understanding of the role of the parameters[13].

In addition to the successful results on morphology reproduction, there is also another interesting output of the study, which is the prevision of effective ablation resistances (or reactivities) from the constituent individual properties. The aim of this paper is to summarize and discuss these results, which can be presented as "composite laws".

The document is organized as follows. A first part will recall the essentials of the approach, and the main results on morphology. Second, the effective laws will be presented for initial (flat) surfaces, then steady-state surfaces, and, finally, transient surfaces. A final discussion will conclude the document.

MODELLING METHODS

From the morphological study, it appears that the modeling of the roughness onset on ablative composites should feature the following elements[11]:

(i)- Surface recession, under the action of oxidation or of sublimation. The surface recession velocity at any point depends on surface orientation, and on the rate of mass transfer. This latter quantity is a function of surface temperature and of local concentration of reactant gas or the local partial pressure of sublimed species, compared to the equilibrium value, e.g. through a Knudsen-Langmuir relationship.

(ii)- The local gas concentration is attained by solving a mass balance equation featuring consumption or production by the surface, and transport in the bulk of the gas phase.

(iii)- Similarly, the local surface temperature is evaluated by a heat balance equation featuring transport in the solid gas phases, as well as interfacial heat consumption.

(iv)- When necessary, the chemical reactivity will be a function of space, in order to translate the possible material heterogeneity.

Point (i) is modeled by a Hamilton-Jacobi equation which describes the propagation of the surface under the action of a Hamiltonian which depends *a priori* on surface concentration and temperature. Points (ii) and (iii) may be modeled by transport equations for gas species and temperature; gas-phase transport should feature diffusion (possibly multi-component), and convection. This last element requires knowledge of the gas phase velocity field, in possibly turbulent flow conditions, which may be extremely difficult to obtain in a realistic fashion.

A first model has been built using a set of rather restrictive assumptions:

- Isothermal conditions. This is suited to two kinds of situations: (1) isothermal oxidation tests; (2) micro-scale simulations where it appears that temperature differences across a characteristic roughness feature length are small enough to be neglected.

- Gas transport is restricted to the pure diffusion of a single species between the surface and a gas source (or sink) located at a large enough distance above the surface; at this place, it is considered that convection ensures the constancy of gas concentration. The relation between source concentration and source-to-surface distance may be obtained through a boundary-layer analysis. The criterion for the choice of the source-to-surface distance is based on a perturbation analysis argument, which shows that it can be as small as a few times the transverse characteristic length of the roughness pattern.

- Surface gas transfer is first order. It can be shown that oxidation and sublimation follow the same formalism in these conditions[14], as far as convection is not concerned.

Even though this model looks extremely restrictive, it has the merit of being easily tractable and of capturing the essentials of the bulk transport/interfacial transfer competition.

It has been possible to perform 3D simulations based on this model on rather large meshes. Also, some interesting analytical results have been produced.

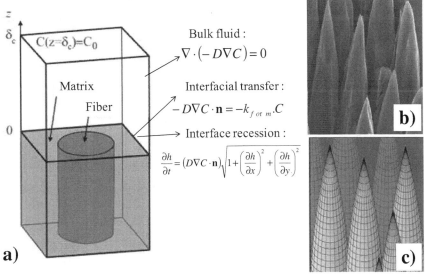

Fig. 1. Simple model used as a reference case: a periodic cell consisting of a vertical fiber embedded in matrix. a) Model geometry and equations; b) actual morphologies observed by SEM (one fiber diameter is ~ 8 μm); c) reproduction of fig. b) by the model after identification of the reactivities.

A very characteristic situation, which will serve as a key example throughout this document, is the "needle forest" situation: unidirectional bundles of fibers bonded by a weak matrix are oriented perpendicularly to the average surface; in an ablation experiment, the fibers are progressively denuded from the matrix and they acquired an ogive-like tip shape.

The peak-to-valley roughness is the fiber denudation height h_f^s and is given by the following formula[13]:

$$h_f^{\ s} = \frac{D}{k_m}\left(\sqrt{1 + 2Sh_f\sqrt{A^2 - 1} - Sh_f^{\ 2}} - 1\right) \tag{1}$$

where : D is the gas diffusion coefficient (m^2.s^{-1}), k_m and k_f are the matrix (or weak-phase) and fiber first-order reaction constants (m.s^{-1}). The number $A = \dfrac{k_m \upsilon_m}{k_f \upsilon_f}$ is the matrix-to-fiber contrast ratio ($\upsilon_i = M_i/\rho_i$ are molar volumes) and $Sh_f = k_m R_f/D$ is a Sherwood number, characteristic of matrix reaction/diffusion competition, based on the fiber radius R_f. These two dimensionless numbers play a central role in the approach.

Numerical simulations of this problem have been performed independently with two solvers. The first is based on a finite difference scheme for diffusion, a VOF (Volume-Of-Fluid) model for surface recession, and a PLIC (PLanar Interface Construction) discretization of the interface boundary[15,16]. The second one[12,17] is based on a Monte-Carlo-Random Walk (MC/RW) algorithm for gas diffusion, with a suited sticking probability law for surface reaction, and a Simplified Marching-Cube (SMC) discretization of the interface[18]. Both solvers yield the same result: a steady morphology in very good agreement with the analytical formula[13] is obtained.

In order to complete this first approach, another model[9,19], featuring momentum transfer equations and advection, has been treated numerically in 2D, with non-evolving surface geometries, using commercial finite element software.

One of the outputs of the presented models is, for a given recession rate J (in mol.m^{-2}.s^{-1}), the value of the local gas concentration C (mol.m^{-3}). By comparison, one obtains an effective reactivity k^{eff} (m.s^{-1}):

$$k^{eff}(z) = |J|/C(z) \qquad (2)$$

We will discuss the results, especially in the simple fiber-in-matrix case presented before, in three cases: first, the flat, initial surface; then the steady morphology, and finally in all the transient period between these two extremes.

FLAT (INITIAL) SURFACE

Results have been obtained for the initial reactivity of the composite, when its surface still is flat: in that case, the values range between the arithmetic average when reaction becomes limiting and the harmonic average when diffusion is limiting, as shown in figure 2. Those two limits are easy to understand. First, in the case of reaction limitation, there is no impact of the bulk transport on the effective transfer; so, all constituents act in a purely parallel fashion. The equivalent conductivity is the sum of the constituents' ones. On the other hand, in the case of diffusion limitation, the constituents are strongly in competition between themselves, since bulk transfer is slow: they are now acting as if they were in series, which explains the harmonic average behavior. From figure 2 it is seen that the role of gas flow is small with respect to bulk diffusion[9]: an increase by 4 orders of magnitude of the Reynolds number $Re = R_f U/\nu$ (U = fluid velocity, ν = kinematic viscosity) has the same effect as an increase by less than 1 order of magnitude of the Sherwood number Sh_f. However, the inspected values of Re did not enter the domain of turbulence, in which many other effects should appear, like the enhancement of the effective diffusion coefficient, etc ....

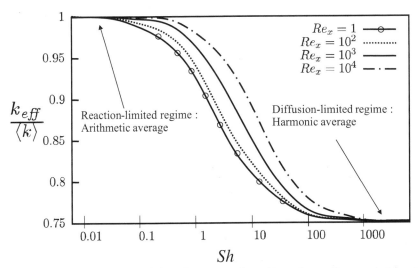

Figure 2. Effective reactivity of a composite with a flat surface, scaled by the arithmetic average. $A = 3$; Fiber volume fraction $\phi_f = 0.5$

STEADY, ROUGH MORPHOLOGY

One of the difficulties of defining an effective reactivity for the steady, rough surfaces is that one needs to choose a surface position for such a definition. When the surface is flat, such a problem does not exist; but when it is rough, then several choices are possible, since z may range between z_m (the lowest point) and $z_f = h_f + z_m$ (the highest one), as illustrated in figure 3. In any case one has:

$$1/k^{eff}(z) = 1/k_m + (z-z_m)/D \qquad (3)$$

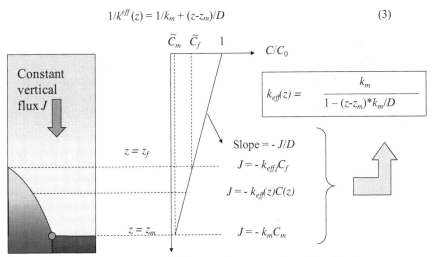

Figure 3 : The choice of an altitude influences the computation of the effective reactivity

Three positions are of particular interest:
- Choosing the equivalent surface at the lowest point of the rough surface yields an extremely simple result: indeed, the equivalent reactivity at that level is exactly equal to the reactivity of the weakest phase:

$$k^{eff}(z = z_m) = k_m \tag{4}$$

The practical difficulty is that such a level is not always easy to determine experimentally.
- Choosing the equivalent surface at the highest point of the surface yields an effective reactivity which can be estimated in a rather simple way:

$$k^{eff}(z = z_f) = k_m/(1 - Sh_f.h_f^s/R_f) = (1/k_m + h_f^s/D)^{-1} \tag{5}$$

It does not depend on the volume fractions of the constituents, under the assumptions of steady morphology and vertical flux approximation. When the contrast is low or diffusion is limiting, then the equivalent reactivity tends to the harmonic average; on the other hand, when reaction is limiting and the contrast is strong, then weakest-link rule works. This is illustrated in figure 4.
- Choosing the surface defined as the arithmetic average of all heights (*i.e.* an "average surface") yields a more complicated expression involving also the fiber volume fraction[9]. This is because of the computation of the position of the average surface. Nonetheless, the asymptotic results for steady morphologies are quite similar to the highest position reactivity: in the diffusion-controlled regime, the effective reactivity tends to the harmonic average (and the steady surface is flat), while in the reaction-controlled regime, the effective reactivity tends to the maximal value, *i.e.* to the weakest-link rule.

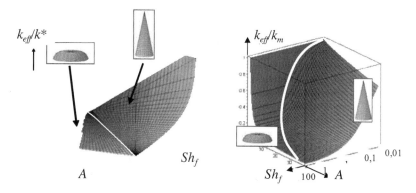

Figure 4: Evolution of the effective reactivity, as compared to the harmonic average (left) and weakest-phase reactivity (right) in the (Sh_f, A) parameter plane.

TRANSIENT BEHAVIOR

After having found out what the mixture rules are in initial and final states of surface, there remains to give some precisions on the transient behavior. The analytical resolution of the problem sketched at figure 1 may be extended to the transient case, by remarking that the transient surface features such as the ogive-shaped fiber tips are similar to the final ones, except that they are truncated, and by retaining the vertical flux hypothesis. Numerical simulation has shown an excellent agreement with the analytical results[10], which are briefly summarized below.

The geometry and notations of the transient problem are displayed in figure 5. Under the 1D vertical flux hypothesis, diffusion is accounted for by a vertical concentration gradient, starting from a fixed concentration C_0 some distance δ_c away from the highest point of the material surface. The top of the fiber is a disc with radius R. By comparing it to its initial value, a dimensionless indicator $\widetilde{R} = R/R_f$ of the morphology development is built: it decreases continuously from 1 to 0. It is related to the fiber height h_f above the matrix trough the following expression:

$$\widetilde{R}(h_f) = Sh_f^{-1}\left(\sqrt{A^2 - (1 + Sh_f \widetilde{h}_f)} - \sqrt{A^2 - 1}\right) + 1 \qquad (6)$$

Here, the dimensionless fiber tip height $\widetilde{h}_f = h_f / R_f$ has been introduced for convenience.

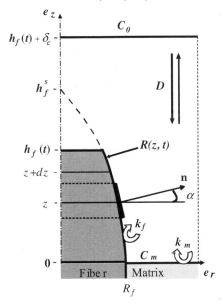

Fig. 5. Notations and geometry for the transient recession problem.

Solving the transport-reaction problem yields an analytical expression for the effective reactivity:

$$\widetilde{k}^{eff}(h_f) = k^{eff}(h_f)/k_m = \phi_f \widetilde{R}^2(h_f)\frac{1}{A} + (1 - \phi_f \widetilde{R}^2(h_f))\frac{1}{1 + Sh_f \widetilde{h}_f} \qquad (7)$$

where ϕ_f is the fiber volume fraction.

Figure 6 is a plot of the scaled effective reactivity as a function of the scaled fiber tip height. Moreover, the time evolution of the fiber tip height is expressed as the difference between the velocities of the highest and lowest point. This leads to the following differential equation:

$$\frac{d\tilde{h}_f}{d\tilde{t}} = \frac{1}{1+\dfrac{k^{eff}\left(h_f\right)\delta_c}{D}}\left(\frac{A}{1+Sh_f\tilde{h}_f}-1\right) \tag{8}$$

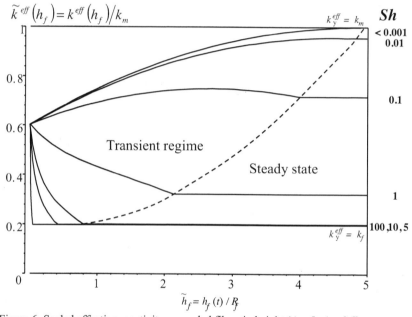

Figure 6. Scaled effective reactivity *vs.* scaled fiber tip height ($A = 5$, $\phi_f = 0.5$).

In this relation, the time has been scaled with respect to the reference value $\tau = R_f/(C_0 v_s k_f)$. Integration of eqns. (4-8) gives the time evolution of the fiber tip height, and of the effective reactivity. As a result, the total time needed to reach steady state is obtained. In the reaction-limited regime, it has been proved that its value is:

$$\tilde{t}_{tot} = \tau\frac{\sqrt{A^2-1}}{A-1} \tag{9}$$

Note that $\tilde{t}_{tot}/\tau$ is close to unity when the contrast is important, but increases without bounds when the contrast number A goes to 1. Figure 9 shows that, as the Sherwood number increases, this transient time first increases, and then decreases when diffusion becomes the limiting factor. The extreme value does not exceed 10 times the value given in eq. (9).

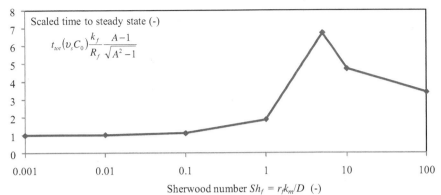

Sherwood number $Sh_f = r_f k_m/D$ (-)

Figure 9: Scaled time to steady state as a function of the fiber Sherwood number.

SUMMARY, CONCLUSION AND PERSPECTIVES

This paper has presented a study, principally based on an analytical model, which represents the morphological evolution of an ablative composite unit cell. Though many simplifying hypotheses have been used, some useful conclusions can be drawn on the effective reactivities (or resistances). For instance, the weakest-link rule[20] has been confirmed, but only when the heterogeneous transfer is limiting; moreover, it is shown that it does not hold either in the initial and transient period. Table 1 collects the results for the initial and final (steady) morphologies.

Table 1 – Average reactivity for a composite (at the highest point of surface) according to conditions

	Initial (flat)	Steady state
Reaction-limited ($Sh_f \ll 1$)	Arithmetic average	Weakest-link
Diffusion-limited ($Sh_f \gg 1$)	Harmonic average	Harmonic average

Other points of interest are:

(i) When reaction is limiting, the strongest (less reactive) phase brings a protection to the weakest phase at the beginning of ablation; this protective effect diminishes until the equilibrium morphology is attained.

(ii) In the converse case, one always has a harmonic average; however, no roughness is developed and the true limiting factor is not the material reactivity but the gas transfer to the wall.

(iii) The influence of the surrounding fluid velocity (Reynolds number) seems to be much less important than that of the diffusion/reaction ratio.

(iv) The transient time is a complex function of the contrast, regime, and fiber volume fraction; however it can be noted that its order of magnitude τ is mostly a function of the fiber reactivity and of the contrast number A, which diverges when A diminishes to 1 –that is, when contrast fades away.

Even if we have only presented exact results for a very restricted type of ideal material, they can be exploited as guidelines for a better ablative material conception in other structure. Validated numerical programs are also available for a more detailed computation.

Further work is oriented towards the incorporation of more physico-chemistry, such as: non-linear reaction rates, multicomponent diffusion, thermal gradients, and more effects of the fluid velocity, like turbulence.

REFERENCES

[1] L. M. Manocha and E. Fitzer, *Carbon reinforcement and C/C composites*, Springer, Berlin (1998).

[2] G. Duffa, *Ablation,* CEA, Le Barp, France (1998) ISBN 2-7272-0207-5.

[3] G. Savage, *Carbon/Carbon composites,* Chapman & Hall, London (1993).

[4] J. Couzi, J. de Winne and B. Leroy, Improvements in ablation predictions for reentry vehicle nosetip, *ESA Confs. Procs.* **SP-426**, 493-9 (1998).

[5] M. D. Jackson, Roughness Induced Transition on Blunt Axisymmetric Bodies, Interim Report SAMSO-TR-74-86 of Passive Nosetip Technology Program n°15 (1974).

[6] D. C. Reda, Sandia Report SAND 79-0649 (1979).

[7] R. G. Batt and H. H. Legner, A review of roughness-induced nosetip transition, *AIAA Journal* **21**, 7-22 (1983).

[8] V. Borie, Y. Maisonneuve, D. Lambert and G. Lengellé, Ablation des matériaux de tuyère de propulseurs à propergol solide, Technical Report N°13, ONERA, France (1990) (in French).

[9] Y. Aspa, PhD dissertation, INPT, Toulouse n°423 (2006) (in French).

[10] J. Lachaud, PhD dissertation, University Bordeaux 1, n°3291 (2006).

[11] G. L. Vignoles, J. Lachaud, Y. Aspa and J.-M. Goyhénèche, Ablation of carbon-based materials: multiscale roughness modelling, *Composites Science and Technology* **69**, 1470-1477 (2009).

[12] J. Lachaud and G.L. Vignoles, A Brownian motion technique to simulate gasification and its application to C/C composite ablation, *Computational Materials Science* **44**, 1034-1041 (2009).

[13] J. Lachaud, Y. Aspa and G. L. Vignoles, Analytical modeling of the steady state ablation of a 3D C/C composite, *Int. J. of Heat and Mass Transfer* **51**, 2618-2627 (2008).

[14] G. Duffa, G. L. Vignoles, J.-M. Goyhénèche and Y. Aspa, Ablation of carbon-based materials: Investigation of roughness set-up from heterogeneous reactions, *Int. J. Heat and Mass Transfer* **48**, 3387-401 (2005).

[15] Y. Aspa, M. Quintard, F. Plazanet, C. Descamps and G. L. Vignoles, *Ceram. Eng. Sci. Procs.* **26**, 99-106 (2005).

[16] Y. Aspa, J. Lachaud, G. L. Vignoles and M. Quintard, Simulation of C/C composites ablation using a VOF method with moving reactive interface, in *Procs 12th Eur. Conf. On Composite Materials*, J. Lamon and A. Torres-Marques, eds. (2006).

[17] J. Lachaud, G. L. Vignoles, J.-M. Goyhénèche and J.-F. Epherre, Ablation in C/C composites: microscopic observations and 3D numerical simulation of surface roughness evolution, *Ceram. Trans.* **191**, 149-60 (2005).

[18] G. L. Vignoles, Modelling binary, Knudsen and transition regime diffusion inside complex porous media, *J. Phys IV France* **C5**, 159-66 (1995).

[19] Y. Aspa, M. Quintard, J. Lachaud and G. L. Vignoles, Effective surface approach for the design of ablative composites, *Proc. 1st International ARA days*, Arcachon, 3-5 July (2006).

[20] W. H. Glime and J. D. Cawley. Oxidation of carbon fibers and films in ceramic matrix composites : a weak link process, *Carbon* **33**, 1053–1060 (1995).

MODEL OF THE INFLUENCE OF DAMAGE ON THE THERMAL PROPERTIES OF CERAMIC MATRIX COMPOSITES

Jalal El Yagoubi [a], Jacques Lamon [a], Jean-Christophe Batsale [b]
[a] University of Bordeaux/CNRS, Laboratoire des Composites Thermostructuraux.
[b] ENSAM, Laboratoire interdisciplinaire TREFLE.
Bordeaux, France

ABSTRACT

Ceramic matrix composites (CMC) are very attractive materials for structural applications at high temperatures. Not only must the CMC be damage tolerant, but also they must allow thermal management. For this purpose, heat transfers must be controlled in the presence of damage. Damage consists in multiple cracks that form in the matrix and ultimately in the fibers when the stresses exceed the proportional limit. Therefore thermal conductivity dependence on applied load is a factor of primary importance for the design of CMC components. The approach proposed in the present paper combines a model of matrix cracking with a model of heat transfer through an elementary cracked volume element containing a matrix crack and an interface crack. It is applied to 1D composites subject to tensile and thermal loading parallel to fiber direction.

INTRODUCTION

With a view to the development of the GEN IV next generation of nuclear reactors, additional requirements for structure design have to be met. The candidate structural materials must be able to withstand very severe environmental service conditions including high temperatures and irradiation. Furthermore, they must enable efficient heat transfers, particularly for fuel casing. Selection of materials which retain high thermal conductivity at high temperature in the presence of damage becomes a prerequisite to material or structure design.

Ceramic matrix composites reinforced by ceramic fibers, and more particularly the 2D woven SiC/SiC composites made of SiC fibers and matrix, are potential candidates. Therefore, predicting their thermal conductivity under load becomes of primary importance for designing SiC/SiC structures.

Figure 1 illustrates the significance of thermal management for composite material components in the presence of high heat flows. It shows the expected influence of material property degradation on component temperature by during service. In situation (a), temperature was controlled and kept below the bearable limit. In situation (b), temperature increases will lead to system failure. They result from thermal property degradation by stress-induced damage due to a too high damage sensitivity of thermal conductivity. Thermal property degradation can also result from environment induced material degradation, caused by microstucture changes.

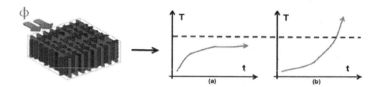

Figure 1. Thermal management for structural components: (a) in-service temperature remains below the tolerable limit ; (b) the limit was exceeded leading to system failure.

As damage results from cracks, it can be understood easily that since free surfaces appear, heat transfer is disturbed. Thus, one can draw a parallel between load and heat in the presence of a transverse crack. Figure 2 shows concentration of stress level lines and constriction of heat flux lines in the vicinity of the crack. At crack tip, the stress-state is characterized by the stress intensity factor, whereas heat flux is described by constriction resistance. At upper scale, material characteristics are affected by damage: rigidity (mechanical effect) and thermal conductivity (thermal effect) decrease.

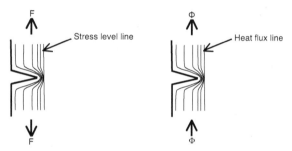

Figure 2: Local disturbances caused by a transverse crack.

In ceramic matrix composites, craks are created in the matrix when the load exceeds the limit of elasticity. Heat transfers involve conduction, radiation or convection through air or gas present within the cracks. The intent of the present paper is thus to propose a model of thermal behavior for a unidirectional composite subject to both thermal and mechanical loading parallel to fiber axis. This loading scheme was selected, in a first step, since the kinetics of matrix cracking is a factor of primary importance, which dictates the occurrence of singularities. Heat transferts through an elementary cracked volume are then described. Longitudinal thermal loading is considered first. Transverse thermal loading, which is less complex, will be introduced in a second step, using equations derived from those established in the present paper for the longitunal one.

Thermal conductivity – strain relations were established on minicomposite test specimens. Minicomposites are reinforced by single fiber bundles. Such specimens have great advantages for establishing microstructure-property relations, as demonstrated in several papers [1, 3]. They have simple linear shape when comparing to multidirectional composites, so that constitutive stress-strain relations are not polluted by structure parameters. The number of parameters necessary to modelling can be reduced, since the contribution of constituent properties to composite mechanical behavior is well identified. Then, minicomposite specimens represent the intermediate scale in 2D woven composites. The corresponding microstructure-property relations and micromechanics-based relations for local phenomena such as matrix cracking and fiber debonding can be introduced into computational models of multidirectional composites.

Experimental data on thermal conductivity versus damage have been produced [4]. Papers have examined the influence of environment and temperature on thermal conductivity of unidirectional reinforced composites [5, 6]. Thermal conductivity was not measured under load. Furthermore, the influence of transverse matrix cracks in longitudinal or transverse tows has been approached using computer codes in a few papers [4, 7].

The present paper combines micromechanics-based equations of matrix multiple cracking and equations of heat transfer in the vicinity of cracks to end up with thermal conductivity-deformation relations. The stress-strain relations during damage have been established in previous papers [2, 3] for unidirectional composites such as micro- (a single fiber reinforcement) or mini-composites. The equations of local heat transfers are established in the present paper.

MODEL OF DAMAGE BY MATRIX CRACKING

SiC/SiC minicomposite is an assembly of SiC matrix and a bundle of SiC fibers. A bundle contains 500, 600 or 800 fibers, depending on fiber manufacturer. Fibers and matrix are ceramic materials. Thus, they display typical features which control composite damage. Fracture of ceramics is a random phenomenon caused by flaws which are randomly distributed; as a consequence, tensile strength data exhibit a statistical distribution, which can be described using the Weibull model [2, 3, 8]:

$$P = 1 - \exp\left\{ -\frac{V}{V_0} \int_V \left[\frac{\sigma}{\sigma_0} \right]^m dV \right\} \tag{2}$$

where P is failure probability, σ_0 is a scale factor, m is a shape parameter, V is the stressed volume. V_0 is the reference volume ($V_0 = 1$ m^3 when International Units are used).

When the matrix experiences multiple cracking, equation (2) applies to the individual fragments (figure 3). It is assumed that fracture is dictated by a single population of flaws [3]. In the following the model is outlined. Details can be found in [2, 3, 8]. The matrix fragments have simple shapes so that $V_i = S_i L_i$, where S_i and L_i are fragment sectional area and length. The length of a fragment depends on the location of the critical flaw in the parent fragment. Since flaws have a random distribution in location and size, the length of successive fragments is a statistical variable. It can be derived from the spatial distribution of flaw strengths within the fragment using the following equation [3, 8]:

$$P(Li) = \int_{L_{i-1}} h(x)dx \tag{3}$$

where $P(L_i)$ is the probability that fragment length is L_i, $h(x)$ describes the spatial density of flaw strengths in fragments, and x is the abscissa along fragment axis. L_i is the length of the fragment generated from the failure of a parent fragment with length L_{i-1}.

When a crack is created in the matrix, it is perpendicular to the load direction and to fiber axis, and it propagates through the matrix, leading to two new fragments with cross section S_m. The stress state is affected by fiber/matrix debonding (figure 3). Stresses are given by:

$$
x \le \frac{u}{2} : \quad
\begin{bmatrix}
\sigma_m = 0 \\[6pt]
\sigma_f = \sigma_f^{max} = \dfrac{F}{S_f}
\end{bmatrix}
\qquad
x \ge \frac{u}{2} + l_d : \quad
\begin{bmatrix}
\sigma_m^{\infty} = \dfrac{F}{S_m}\dfrac{a}{1+a} \\[6pt]
\sigma_f^{\infty} = \dfrac{F}{S_f}\dfrac{1}{1+a}
\end{bmatrix}
\tag{4}
$$

where F is the applied load, E is Young's modulus, S is cross section et V_m and V_f are matrix and fiber volume fractions respectively, u is crack opening displacement, l_d is debond length, a is load sharing parameter, $a = E_m V_m / E_f V_f$.

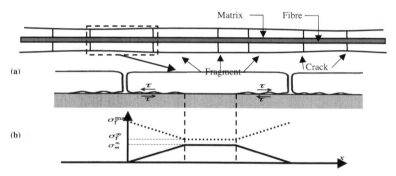

Figure 3: Schematic diagram showing multiple matrix cracking in a single fiber composite specimen: (a) fragments, and interface cracks, (b) stress lag in the fibers, and in the matrix induced by the cracks.

Debond length (l_d) is derived from transfer of load on the matrix through the debonded interface. Assuming a constant interfacial shear stress τ:

$$l_d = \frac{F.a}{2\pi\tau\, r_f\,(1+a)} \tag{5}$$

Crack opening displacement results from sliding of the matrix relative to the fiber in the debonded area:

$$u = 2 \int_0^{l_d+u} (\frac{\sigma_f(x)}{E_f} - \frac{\sigma_m(x)}{E_m})\, dx = \frac{F.l_d}{E_f S_f - F} \tag{6}$$

Occurence of a crack in a fragment is defined by equation (2), for the stress-state described by equations (4). P=0.5 leads to good approximation of fragment strength [3,8].
A crack is introduced at distance x from fragment end, using the following equation of probability of location for fracture inducing flaw in a fragment having length $2l_i$. Equation (7) is derived from (3) [3,8].

$$P(x) = \frac{x - l_d}{2(l_i - l_d)} \tag{7}$$

Fiber failure also obeys Weibull statistics. The stress-state induced by matrix cracks and debonding is introduced in equation (2). Ultimate failure of minicomposite happens when a critical amount of fibers (12 - 17% depending on fiber brand) has broken [3].

TENSILE STRESS-STRAIN BEHAVIOUR OF MINICOMPOSITE
    The tensile stress-strain behaviour of minicomposite is dictated by fiber deformation and failure. The number of fragments determines the stress state in fibers. Deformations are derived from the stress state in fibres (equations (4)):

$$\varepsilon = \frac{1}{LE_f} \int_L \sigma_f(x)\,dx = \sum_{i=1}^{n+1} \frac{1}{L_i E_f} \int_{L_i} \sigma_f^i(x)\,dx \tag{8}$$

with L is fiber length, n the number of fragments, L_i fragment length, and $\sigma_f^i(x)$ the stress-state in the ith fiber portion adjacent to fragment i.
When there is no overlapping of interface cracks, the minicomposite Young's modulus is given by:

$$\widetilde{E}(n,l_d) = \frac{F}{S\varepsilon} = \frac{E_f V_f (1+a)}{1 + \dfrac{n\, l_d\, a}{l_{comp}}(1 + \dfrac{F}{E_f S_f - F})} \tag{9}$$

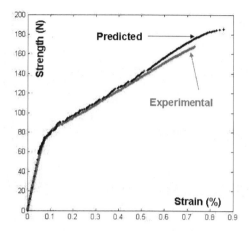

Figure 4 : Comparison of experimental and predicted tensile behavior for Hi-Nicalon/SiC minicomposite.

Table 1: Characteristics of SiC matrix and Hi-Nicalon fiber for predictions of mechanical behavior using the model.

E_f	r_f	V_f	m_f	σ_{of}	E_m	m_m	σ_{om}	τ
240GPa	7μm	0.22	7	60MPa	400GPa [3]	5.14 [3]	5.26MPa [3]	10MPa

Figure 4 shows prediction of the mechanical behavior of Hi-Nicalon/SiC minicomposite for the constituents properties given in table 1. As shown in previous work, a good agreement with experiment is obtained.

INFLUENCE OF MATRIX DAMAGE ON THERMAL CONDUCTIVITY.

Approach to thermal behavior during tension is parallel to the previous one. Determination on heat flux through the cracks (φ_{fiss}, Figure 5), is parallel to analysis of stress state (Figure 3). Heat transfers occur through interface cracks, but also through the intact portion of interface. This latter contribution must be considered for thermal equilibrium reasons, since the matrix flux φ_m^∞ coming out of matrix cracks does not correspond to the coming in one only. It is worth reminding that debond length is determined by mechanical equilibrium, but not by thermal equilibrium.

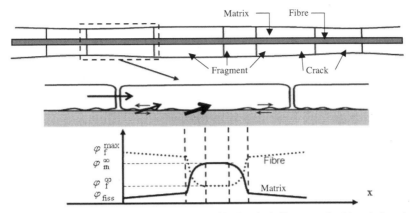

Figure 5 : Schematic diagram showing multiple matrix cracking in a single fiber composite: (a) cracked specimen with fragments and interface cracks, (b) Density of heat flux in the fibers, and in the matrix induced by the matrix and interface cracks.

For the elementary cell and the boundary conditions shown by figure 5, resolution of the heat equation led to thermal conductivity-damage relation (14) [9]. Equation (14) involves:

- Thermal conductivity of matrix (λ_m) and fibers (λ_f);
- Fiber characteristics: density (ρ) and radius (r_f).
- Composite damage characteristics: matrix crack spacing distance (u) and debond length (l_d).
- Coefficients of heat transfer through matrix cracks H_c and debond cracks h_i.

Coefficients of heat transfer:

$h_i = h_r + h_g + h_c$:

- heat transfer by radiation: $h_r = [4\varepsilon_e/(2 - \varepsilon_e)]\sigma_B n_r^2 T^3$ (10)

with ε_e emissivity, σ_s the Stefan-Boltzmann constant, n_r the refraction index, T temperature.

- heat transfer by conduction by the gas filling cracks: $h_g = \lambda_g / e$ (11)

with λ_g gas thermal conductivity, e thickness. e = 2 Ra was used as approximation. Ra is the mean interface crack surface roughness.

- heat transfer by contact: h_c. It is quite difficult to estimate this coefficient. It depends on both fiber surface roughness and residual stresses. Available experimental data [6] indicate that $h_c \approx h_g$. To a first approximation :

$$h_i = h_r + 2h_g \tag{12}$$

- $H_c = H_r + H_g$: there is no transfer by contact in the cracks,
  - heat transfer by radiation H_r (defined above).
  - heat transfer by conduction by the gas filling cracks : $H_g = \lambda_g / u$ $\tag{13}$

$$\frac{\lambda}{\lambda_o} = \left\{ 1 + \frac{(1-\rho)\lambda_m}{\rho\lambda_f} \frac{\tanh(\frac{\xi d}{2r_f})/(\frac{\xi d}{2r_f}) + F(\xi,\zeta,l_d/r_f,d/r_f)}{1 + (2\lambda_o B_c / \xi\rho\lambda_m)\tanh(\frac{\xi d}{2r_f})} \right\}^{-1}$$

with

$$F(\xi,\zeta,l_d/r_f,d/r_f) = \frac{\cosh(\zeta(0.5d - l_d)/r_f)}{\cosh(\zeta d/2r_f)}\left[\frac{\tanh(\xi(0.5d - l_d)/r_f)}{\xi d/2r_f} - \frac{\tanh(\zeta(0.5d - l_d)/r_f)}{\zeta d/2r_f} \right]$$

$$\tag{14}$$

$$\xi = \sqrt{\frac{8}{(1-\rho)\lambda_m}} \qquad and \qquad \zeta = \xi\sqrt{\frac{1}{1+4/B_i}}$$

$$B_c = \frac{(H_g + H_r)r_f}{\lambda_f}$$

$$B_i = \frac{(h_g + h_r + h_c)r_f}{\lambda_f}$$

B_c is the Biot number for heat flow through matrix cracks, and B_i is the Biot number for heat flow through interface cracks.

APPLICATION

The model was used to anticipate trends in the influence of constituent properties on thermal conductivity, including fiber properties, interfacial shear stress and environment.

Influence of fiber properties

Characteristics of both SiC-based fibers considered for this analysis, are summarized in table 2.

Both SiC/SiC minicomposites were assumed to possess identical interface characteristics ($\tau = 10$ MPa). Figure 6 describes the predicted tensile and thermal responses of Hi-Nicalon and Hi-Nicalon S fiber reinforced minicomposites. Damage kinetics are characterized by the number of cracks, and elastic and thermal properties versus applied strain. Note that Young's modulus and thermal conductivity (figures 6c and d) mirror cracking (figure 6b) and they display similar trend: a steep decrease to a plateau at saturation of matrix cracking. The Hi-Nicalon S minicomposites retain the highest properties during damage, i.e. load carrying capacity (indicated by high stresses), Young's modulus and thermal conductivity.

Table 2: Properties of Hi-Nicalon (HN) and Hi-Nicalon S (HNS) fibres

	E_f [10]	r_f	V_f	m_f [10]	σ_{of} [10]	k_{zf} [11]
HN	240GPa	7μm	0.47	7	60MPa	8W/m.K
HNS	420GPa	7μm	0.47	7.2	100MPa	18 W/m.K

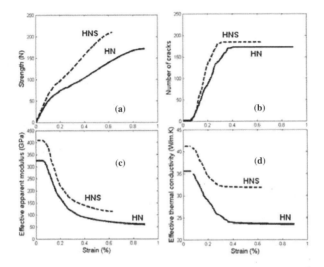

Figure 6 : Tensile stress-strain behavior (a), crack density (b), and mechanical (c) and thermal properties (d) of Hi-Nicalon/SiC and Hi-Nicalon S minicomposites (respectively HN and HNS).

Effect of interface characteristics

The influence of fiber/matrix interfaces on the mechanical behavior of composites has been demonstrated and quantified in previous papers [2, 3]. Practically, interface engineering uses fiber surface treatments aimed at tailoring fiber/matrix bonding. Figure 7 shows predictions of mechanical damage and properties degradation for various values of interfacial shear stresses: $\tau = 10 MPa$; $\tau = 30 MPa$; $\tau = 100 MPa$. Figure 7b shows that the density of cracks is commensurate with τ. This trend has been discussed in previous papers. Low crack density must be attributed to long debonds associated to small τ (equation (5)). When τ is high ($\tau = 100 MPa$), interface cracks are shorter, and cracking is denser (Figure 7b). Note that strong interface (high τ) is recommended for high mechanical performance, but the corresponding thermal conductivity is low, because of the presence of a large amount of matrix cracks. Thus, strong interface is detrimental to thermal conductivity (Figure 7d). Balance between both structural and thermal requirements needs to be found, which demonstrates the importance of modelling thermal property dependence on mechanical loading.

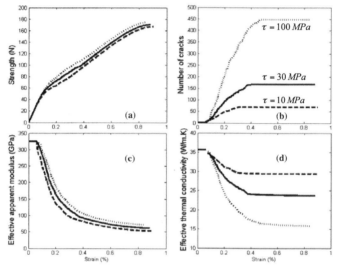

Figure 7 : Influence of interface shear stress on tensile stress-strain behavior (a), number of matrix cracks (b), mechanical (c) and thermal properties (d) for Hi-Nicalon/SiC minicomposites:

τ and h_c are related to contact topography in the interface crack. But, due to the lack of experimental data on heat transfer through contact points and on τ – interface roughness relations, it was quite difficult at this stage to establish quantitative correlation between τ and h_c.

Effects of heat conduction through the cracks

Heat transfers through the cracks occur preponderantly by conduction. The radiation heat transfer coefficient is negligible compared to the conduction one. Figure 8 shows predictions of Hi-Nicalon minicomposite thermal conductivity when the following gases are present within cracks: Argon $\lambda_g = 0.017\,W/mK$, Air $\lambda_g = 0.025\,W/mK$, Helium $\lambda_g = 0.14\,W/mK$. As logically expected, thermal conductivity is commensurate with λ.

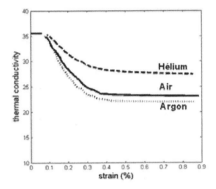

Figure 8 : Predictions of the influence of the nature of gas filling matrix cracks on the thermal conductivity of a Hi-Nicalon minicomposite.

CONCLUSION

Modelling the effects of damage on thermal conductivity involves statistical-probabilistic approach to matrix multiple cracking. Heat transfers through matrix and interface cracks were then introduced to determine thermal conductivity of cracked minicomposite. It was shown that the model allows the influence of mechanical loading on thermal conductivity to be calculated. Young's modulus and thermal conductivity were found to mirror cracking. Results were in agreement with logical expectation. The model was used to anticipate trends in the influence of fiber and interface on thermal behavior during tensile loading. It appeared that interface strength may have opposite effects on load carrying capacity and thermal behavior. The model is thus useful to identify appropriate interface characteristics. This approach is a first step to composite or structure design based on thermal behavior.

REFERENCES

[1] R. Naslain, J. Lamon, R. Pailler, X. Bourrat, A. Guette, F. Langlais, Micro/minicomposites: a useful approach to the design and development of non-oxide CMCs, *Composites: Part A*, **30**, 537-547 (1999).

[2] L. Guillaumat, J. Lamon. Fracture statistics applied to modelling the non-linear stress-strain behaviour in microcomposites: Influence of interfacial parameters, *Int. Journal of Fracture*, **82**, 297-316 (1996).

[3] N. Lissart, J. Lamon. Damage and failure in ceramic matrix minicomposites: exprimental study and model.
Acta Mater. **15**, 3, 1025-1044 (1997).

[4] J.K. Faroqui, M.A. Sheikh, Finite element modelling of thermal transport in ceramic matrix composites *Computational Materials Science*, **37**, 361-373 (2006).

[5] D.P.H. Hasselman, A. Venkateswaran, M. Yu, H. Tawil, Role of interfacial debonding and matrix cracking in the effective thermal diffusivity of SiC-Fiber-Reinforced chemical vapour deposited SiC matrix composites *Journal of Materials Science Letters*, **10**, 1037-1042 (1991).

[6] H. Bhatt, K.Y. Donaldson, D.P.H. Hasselman, R.T. Bhatt, *Role of interfacial debonding and matrix cracking in the effective thermal diffusivity of SiC-Fiber-Reinforced chemical vapour deposited SiC matrix composites, J. Am. Ceram. Soc.*, **73**, 312-316 (1990).

[7] B. Tomkova, M. Sejnoha, J. Novak, J. Zeman, Evaluation of Effective Thermal Conductivities of Porous Textile Composites, *Int. Journal for Multiscale Computational Eng*. (2008).

[8] J. Lamon, Stochastic approach to multiple cracking in composite systems based on the extreme-values theory *Composites Science and Technology*, **69**, 1607-1614 (2009).

[9] T.J. Lu, J.W. Hutchinson, Effect of matrix cracking on the overall thermal conductivity of fibre-reinforced composites, *Journ. Phil. Trans. of the Royal Society*, *351*, **1697**, 595-610 (1995).

[10] W. Gauthier, J. Lamon, Delayed failure of Hi-Nicalon and Hi-Nicalon S multifilament tows and single filaments at intermediate temperatures (500 - 800°C), *J. Am. Ceram. Soc*. 92, 3, 702-709 (2009).

[11] R. Yamada, N. Igawa, T. Taguchi, S. Jitsukawa. Highly thermal conductive, sintered SiC fiber-reinforced 3D-SiC/SiC composites: experiments and finite-element analysis of the thermal diffusivity/conductivity, *Journal of Nuclear Materials*, 307-311 (2002).

Author Index

Author Index